Isaac Todhunter

A treatise on the differential calculus with numerous examples

Isaac Todhunter

A treatise on the differential calculus with numerous examples

ISBN/EAN: 9783742892362

Manufactured in Europe, USA, Canada, Australia, Japa

Cover: Foto ©berggeist007 / pixelio.de

Manufactured and distributed by brebook publishing software
(www.brebook.com)

Isaac Todhunter

A treatise on the differential calculus with numerous examples

A TREATISE ON THE

DIFFERENTIAL CALCULUS

WITH NUMEROUS EXAMPLES.

BY

I. TODHUNTER, Sc.D., F.R.S.

London:

MACMILLAN AND CO.

AND NEW YORK

1890

PREFACE.

I HAVE endeavoured in the present work to exhibit a comprehensive view of the Differential Calculus on the method of Limits. In the more elementary portions I have entered into considerable detail in the explanations with the hope that a reader who is without the assistance of a tutor may be enabled to acquire a competent acquaintance with the subject. To the different Chapters will be found appended Examples sufficiently numerous to render another book unnecessary. These examples have been selected almost exclusively from the College and University Examination Papers; the greater part of them will be found to present no very serious difficulty to the student, although a few may require peculiar analytical skill.

I have frequently given more than one investigation of a theorem, because I believe that the student derives advantage from viewing the same proposition under different aspects, and that, in order to succeed in the examinations which he may have to undergo, he should be prepared for a considerable variety in the order of arranging the several branches of the subject, and for a corresponding variety in the mode of demonstration.

In the composition of the first edition of this work, while

trusting mainly to independent knowledge and judgment, I derived assistance from the labours of well known authors on the subject, especially Cournot, De Morgan, Moigno, Navier, and Schlömilch.   In the subsequent editions a considerable amount of fresh matter has been introduced, and this rests almost exclusively on my own authority; increased experience as a teacher naturally gave stronger confidence to the writer. Thus the work now contains on the whole much that is original in substance, and much that is new in form.

The present edition has been carefully revised and some-, what enlarged.   I have examined with attention and interest treatises on the Differential Calculus recently published by eminent mathematicians, in order to discover if the methods of explaining and developing the principles of the subject had gained any real improvement during the last twenty years.   I have not however found reason for concluding that I could with advantage make any essential change in this elementary work.

I have much reason to be grateful for the approbation bestowed by teachers and students on this volume, the first of a long series relating to various branches of mathematics.   My thanks are especially due to Professor Battaglini of Naples for the honour which he has conferred on me by translating my treatises on the Differential and the Integral Calculus into Italian.

I. TODHUNTER.

St John's College,
    April, 1871.

Since the foregoing Preface to the fifth edition was printed the work has obtained increased favour both at home and abroad, and translations of it have appeared in Russia and in India.   An elementary treatise on Laplace's Functions, Lamé's Functions, and Bessel's Functions, designed as a sequel to the volumes on the Differential and Integral Calculus, has since been published.

January, 1878.

CONTENTS.

Students reading this work for the first time may omit the following Articles:

DIFFERENTIAL CALCULUS.

CHAPTER I.

DEFINITIONS. LIMIT. INFINITE.

1. SUPPOSE two quantities which are susceptible of change so connected that if we alter one of them there is a consequent alteration in the other, this second quantity is called a *function* of the first. Thus if x be a symbol to which we can assign different numerical values, such expressions as x^2, 3^x, log x, and sin x, are all *functions* of x. If, a function of x is supposed equal to another quantity, as for example $\sin x = y$, then both quantities are called *variables*, one of them being the *independent variable* and the other the *dependent variable*. An *independent variable* is a quantity to which we may suppose any value arbitrarily assigned; a *dependent* variable is a quantity the value of which is determined as soon as that of some independent variable is known. Frequently when we are considering two or more variables it is in our power to fix upon whichever we please as the *independent* variable, but having once made our choice we must admit no change in this respect throughout our operations; at least such a change would require certain precautions and transformations.

2. We generally denote functions by such symbols as $F(x)$, $f(x)$, $\phi(x)$, $\psi(x)$, and the like, the variable being denoted by x. Such an equation as $y = \phi(x)$ implies that the dependent variable y is so connected with the independent variable x, that the value of y becomes known as soon as that of x is given, and that if any change be made in the numerical value assigned to x, the consequent change in y can be found.

3. The student has probably already had occasion to consider the meaning of the terms "variable quantity" and "function" which we have here introduced. In treatises on the conic sections, for example, the equation $y = 2\sqrt{ax}$ occurs, where x is a general symbol to which different numerical values may be assigned, and a is a symbol to which we suppose some invariable numerical value assigned, and which is therefore called a "constant." For every value given to x we can deduce the corresponding numerical value of y. In the equation $y = 2\sqrt{ax}$, since the value of y depends upon that of a as well as that of x, we may say that y is a function of a and x. Hence such symbols may be used as $F(a, x)$ to denote a function of both a and x; and such an equation as $y = \phi(x, z, t)$ indicates that y is a function of the three quantities denoted by the symbols x, z, and t.

4. In the equation $y = 2\sqrt{ax}$, if we know that a is to be a *constant* quantity throughout any investigation on which we may be engaged, we shall frequently not require to be reminded of this constant, and shall continue to speak of y as a function of x. So the equation $y = \dfrac{b}{a}\sqrt{(a^2 - x^2)}$ may be represented by $y = \phi(x)$, where we express only that symbol x which throughout our investigations will be considered variable.

5. If the equation connecting the variables x and y be such that y *alone* occurs on one side, and on the other side some expression involving x and not y, we say that y is an *explicit* function of x. When an equation connecting x and y is not of this form, we say that y is an *implicit* function of x. Thus if $y = ax^2 + bx + c$, we have y an explicit function of x. If $ay^2 - 2bxy + cx^2 + g = 0$ we have y an *implicit* function of x. The words *implicit function* assume that y really is a *function* of x in the sense in which we have used the word function. This assumption may be seen to be true in the example given, for we can by the solution of a quadratic equation exhibit y as a function of x; or rather we can infer that y must be one of two explicit functions of x, namely either $\dfrac{bx + \sqrt{\{(b^2 - ac)\,x^2 - ag\}}}{a}$ or $\dfrac{bx - \sqrt{\{(b^2 - ac)\,x^2 - ag\}}}{a}$. We shall return to this point hereafter, in Art. 58.

6. Explicit functions may be divided into *algebraical* and *transcendental*. The former are those in which the only operations indicated are addition, subtraction, multiplication, division, and the raising of a quantity to a known power or the extraction of a known root; the latter are those which involve other operations, as exponential functions, logarithmic functions, and trigonometrical functions. We suppose here that the number of the operations indicated is *finite;* for as we shall see hereafter a transcendental function may be equivalent to an infinite series of algebraical functions.

To the independent variable in an equation we may suppose any value assigned, either positive or negative, as great as we please or as small as we please. If we suppose a series of different values assigned to x, beginning with some negative value numerically very large and gradually increasing algebraically up to some large positive value, the series of values we obtain for y may present to us very different results. For example, if $y = x^3$, then the values of y will form a series beginning with a negative value numerically large, and increasing algebraically up to a large positive value. If $y = x^2$, the values of y are always positive, and form a series first decreasing and then again increasing. If $y = \sqrt{(a^2 - x^2)}$, then the values of y are unreal for every value of x not contained between $-a$ and $+a$.

7. We proceed to another example more important for our purpose. Suppose $y = \dfrac{x}{1+x}$, and consider the series of values which y assumes when to x are assigned different *positive* values. When $x = 0$, $y = 0$, and when x has any positive value, y is a positive proper fraction. If we put y in the form $1 - \dfrac{1}{1+x}$, we see that as x increases so does y, but y being a proper fraction can never be so great as unity. The difference of y from unity is $\dfrac{1}{1+x}$; this fraction diminishes as x increases, and *can be made smaller than any assigned fraction, however small, by giving a sufficiently great value to x.* Thus if we wish y to differ from unity by a quantity less than $\dfrac{1}{100,000}$,

make $x = 100{,}000$, and the required result is obtained. If we wish y to differ from unity by a quantity less than $\dfrac{1}{1{,}000{,}000}$, make $x = 1{,}000{,}000$, and the required result is obtained. Under these circumstances we say "the *limit* of y when x increases indefinitely is unity."

8. The importance of the notion of a *limit* cannot be over-estimated; in fact the whole of the Differential Calculus consists in tracing the consequences which follow from that notion. The student has probably already fallen upon cases in which the word *limit* has been used, to which it will be useful to recur. For example, the sum of the geometrical progression $1 + \frac{1}{2} + \frac{1}{4} + \frac{1}{8} + \ldots$ continued to n terms is $2 - \dfrac{1}{2^{n-1}}$, and hence he has deduced the result that the *limit* of the series when the number of terms is indefinitely increased is 2.

9. A very important example of a *limit* occurs in works on Trigonometry. It is there shewn that if θ denote the circular measure of an angle, the fraction $\dfrac{\sin \theta}{\theta}$ will, if θ be diminished indefinitely, approach as near as we please to unity. In other words the *limit* of $\dfrac{\sin \theta}{\theta}$, as θ continually diminishes, is unity. We shall express this by saying "the limit of $\dfrac{\sin \theta}{\theta}$, *when* $\theta = 0$, is unity;" that is, we use the words "*when* $\theta = 0$" as an abbreviation for "*when θ is continually diminished towards zero*," or for "*when θ is diminished without limit*."

10. The proposition "the limit of $\dfrac{\sin \theta}{\theta}$, when $\theta = 0$, is unity" is sometimes expressed thus, "$\dfrac{\sin \theta}{\theta} = 1$, when $\theta = 0$," or "$\sin \theta = \theta$, when $\theta = 0$." *It must however be most carefully remembered that such expressions are only abbreviations and cannot be understood absolutely.* In like manner the result

obtained in Art. 7, namely that the limit of $\frac{x}{1+x}$ when x increases indefinitely is unity, would be sometimes expressed thus, "when x is infinite $\frac{x}{1+x}$ equals unity." Here both parts of the sentence are abbreviations: "when x is infinite" can only be considered as meaning "when x is increased without limit," and "$\frac{x}{1+x}$ equals unity" means strictly "$\frac{x}{1+x}$ can be made to differ from unity by as small a quantity as we please."

11. In the example $y = \frac{x}{x+1}$ let us now ascribe to x negative values. Put $-z$ for x; thus $y = \frac{z}{z-1}$. Now suppose z to change gradually from 0 to 1; the numerator of y is positive and continually increasing, while the denominator is negative and numerically continually diminishing. The value of y then is negative and numerically continually increases, and by taking z sufficiently near to unity we may make y *as great as we please;* that is, as z approaches unity y has no finite limit. For the sake of shortness, this is sometimes expressed thus, "y is infinite, when $z = 1$;" but it must not be forgotten that this last phrase *is an abbreviation*, and must be considered to mean: "by taking z sufficiently near to unity y can be made to exceed any assigned magnitude, however great." We shall not proceed further with the example; the reader will see that when z is greater than unity y is positive, that y continually diminishes as z increases, and approaches the limit unity when z increases indefinitely.

12. The student has already seen an example of the same kind as that brought forward in the last Article, for he has probably been accustomed to say, "the tangent of an angle of 90° is infinity." On reflexion he will see that the only way in which a meaning can be given to this statement is to consider it an abbreviation of the following: "as we increase an angle gradually up to 90°, the tangent of the angle increases, and by taking the angle near enough to 90° we may make the tangent as great as we please." We can

form no distinct conception of an *infinite* magnitude, and the word can only be used in Mathematics as an abbreviation in the manner of the examples here given.

If to x the independent variable be ascribed values beginning with zero and increasing without limit, this is sometimes expressed for abbreviation by saying that x increases from zero to infinity.

13.   The meaning of the word "limit," or its equivalent "limiting value," will be understood from its use in the preceding Articles. The following may be given as a definition: "The limit of a function for an assigned value of the independent variable, is that value from which the function can be made to differ as little as we please, by making the independent variable approach its assigned value."

14.   In the example "the limit of $\dfrac{\sin \theta}{\theta} = 1$ when $\theta = 0$," it is obvious that $\dfrac{\sin \theta}{\theta}$ is never equal to 1 so long as θ has any value different from zero, and if we actually make $\theta = 0$, we render the expression $\dfrac{\sin \theta}{\theta}$ unmeaning. In other words, although $\dfrac{\sin \theta}{\theta}$ approaches as nearly as we please to the limit unity *it never actually attains that limit.* Sometimes in the definition of a limit the words "that value which the function never actually attains" have been introduced. But it is more convenient to omit them; for if we take any function of x, say $\dfrac{x}{x+1}$, and ascribe to x any value, say 1, we can determine the *actual* value of the function, which in this case would be $\frac{1}{2}$. According to the definition we have given in the preceding Article we may if we please call $\frac{1}{2}$ the *limit* of $\dfrac{x}{x+1}$ when x approaches unity. The same holds for any finite value of any function, and generally according to the definition of a limit laid down in Art. 13, *any actual value of a function may be considered as a limiting value.*

15. *Limit of* $\left(1 + \dfrac{1}{x}\right)^{x}$. The following theorem, which we proceed to demonstrate, is very important. *When x increases indefinitely the expression* $\left(1 + \dfrac{1}{x}\right)^{x}$ *approaches a certain limit which lies between 2 and 3.*

In the first place suppose x a positive whole number, $= m$ say; we shall prove that the above expression continually increases with m, but can never reach the value 3. Assuming the Binomial Theorem for positive integral exponents, we have

$$\left(1 + \frac{1}{m}\right)^{m} = 1 + m\frac{1}{m} + \frac{m(m-1)}{1.2}\left(\frac{1}{m}\right)^{2} + \frac{m(m-1)(m-2)}{1.2.3}\left(\frac{1}{m}\right)^{3} +$$

$$\dots + \frac{m(m-1)(m-2)\dots\{m-(m-1)\}}{1.2\dots m}\left(\frac{1}{m}\right)^{m}$$

which may be written

$$\left(1 + \frac{1}{m}\right)^{m} = 1 + \frac{1}{1} + \frac{1 - \dfrac{1}{m}}{1.2} + \frac{\left(1 - \dfrac{1}{m}\right)\left(1 - \dfrac{2}{m}\right)}{1.2.3} + \dots$$

$$+ \frac{\left(1 - \dfrac{1}{m}\right)\left(1 - \dfrac{2}{m}\right)\dots\left(1 - \dfrac{m-1}{m}\right)}{1.2\dots m}\dots(1)$$

Similarly

$$\left(1 + \frac{1}{1+m}\right)^{m+1} = 1 + \frac{1}{1} + \frac{1 - \dfrac{1}{m+1}}{1.2} + \frac{\left(1 - \dfrac{1}{m+1}\right)\left(1 - \dfrac{2}{m+1}\right)}{1.2.3} +$$

$$\dots + \frac{\left(1 - \dfrac{1}{m+1}\right)\left(1 - \dfrac{2}{m+1}\right)\dots\left(1 - \dfrac{m}{m+1}\right)}{1.2\dots(m+1)}\dots(2).$$

Now in the last two series we see that their first and second terms are equal, but the third term in (2) is greater than the third term in (1); also the fourth term in (2) is greater than the fourth term in (1), and so on; moreover in (2) there is one term more than in (1). Hence

$$\left(1 + \frac{1}{1+m}\right)^{m+1} \text{ is greater than } \left(1 + \frac{1}{m}\right)^{m}.$$

Therefore if we put m successively equal to 2, 3, 4, &c. the expression $\left(1 + \dfrac{1}{m}\right)^m$ continually increases.

But since $1 - \dfrac{1}{m}, 1 - \dfrac{2}{m}, 1 - \dfrac{3}{m}, \ldots$ are all positive and all less than unity it follows that the series in (1) cannot be greater than

$$1 + \frac{1}{1} + \frac{1}{1.2} + \frac{1}{1.2.3} + \frac{1}{1.2.3.4} + \ldots + \frac{1}{1.2\ldots m} \ldots\ldots (3),$$

however great m may be.

But the series in (3) is less than the following series, which forms a geometrical progression, beginning at the second term,

$$1 + \frac{1}{1} + \frac{1}{2} + \frac{1}{2^2} + \frac{1}{2^3} + \ldots + \frac{1}{2^{m-1}},$$

that is, the series in (3) is less than

$$1 + \frac{1 - \dfrac{1}{2^m}}{1 - \dfrac{1}{2}} \text{ or } 3 - \frac{1}{2^{m-1}}.$$

Hence $\left(1 + \dfrac{1}{m}\right)^m$ is less than 3, however great m may be.

Since then the expression $\left(1 + \dfrac{1}{m}\right)^m$ continually increases with m, but at the same time cannot exceed 3, there must be some "limit" towards which it approaches as m is increased indefinitely. We shall use the symbol e to denote this limit, and shall hereafter shew how to calculate its approximate value: we say *approximate*, for it will prove to be an incommensurable number. See Art. 115.

16. We might perhaps leave it to the student to convince himself that the limiting value of $\left(1 + \dfrac{1}{x}\right)^x$ must be the same whether we attribute to x a succession of *integral* or of *fractional* values increasing without limit. But it may be formally shewn thus. Whatever fractional value be ascribed to x there must be two consecutive integers, say m and $m + 1$, between which such fractional value lies. Suppose then

$1 + \frac{1}{x}$ greater than $1 + \frac{1}{n}$ and less than $1 + \frac{1}{m}$, where n is put for $m + 1$.

Then $\left(1 + \frac{1}{x}\right)^x$ lies between $\left(1 + \frac{1}{n}\right)^x$ and $\left(1 + \frac{1}{m}\right)^x$.

Suppose $x = m + a = n - \beta$, so that a and β are proper fractions, then

$$\left(1 + \frac{1}{x}\right)^x \text{ lies between } \left(1 + \frac{1}{n}\right)^{n-\beta} \text{ and } \left(1 + \frac{1}{m}\right)^{m+a},$$

that is, between

$$\left\{\left(1 + \frac{1}{n}\right)^n\right\}^{1-\frac{\beta}{n}} \text{ and } \left\{\left(1 + \frac{1}{m}\right)^m\right\}^{1+\frac{a}{m}}.$$

If x be now supposed to increase without limit, so also do m and n. The limit of $\left(1 + \frac{1}{n}\right)^n$ and of $\left(1 + \frac{1}{m}\right)^m$ is e, and as $1 - \frac{\beta}{n}$ and $1 + \frac{a}{m}$ have unity for their limit it follows that the limit of $\left(1 + \frac{1}{x}\right)^x$ is e.

17. We may shew that the limit of $\left(1 + \frac{1}{x}\right)^x$ is also e when x is *negative* and increases without limit. For put $x = -z$, then we have to find the limit of $\left(1 - \frac{1}{z}\right)^{-z}$ when z increases without limit.

Now $\left(1 - \frac{1}{z}\right)^{-z} = \left(\frac{z-1}{z}\right)^{-z} = \left(\frac{z}{z-1}\right)^z$,

$$= \left(\frac{1+y}{y}\right)^{y+1}, \text{ where } y = z - 1,$$

$$= \left\{\left(1 + \frac{1}{y}\right)^y\right\}^{1+\frac{1}{y}}.$$

Let now x increase numerically without limit, then z, and consequently y, do the same. The limit of $\left(1 + \frac{1}{y}\right)^y$ is e, and that of $1 + \frac{1}{y}$ is unity, and therefore the limit of $\left(1 - \frac{1}{z}\right)^{-z}$ is e.

18. Since the limit of $\left(1 + \dfrac{1}{x}\right)^{x}$ when x increases indefi-

nitely is e, we see, by putting $\dfrac{1}{x} = z$, that the limit of $(1 + z)^{\frac{1}{z}}$

when z is *diminished* indefinitely is also e. Hence we can

deduce the limit when $z = 0$ of $(1 + az)^{\frac{1}{z}}$, where a is any

constant quantity. For

$$(1 + az)^{\frac{1}{z}} = \left\{(1 + az)^{\frac{1}{az}}\right\}^{a}.$$

Now as z diminishes without limit, so also does az, therefore

the limit of $(1 + az)^{\frac{1}{az}}$ is e,

and   the limit of $(1 + az)^{\frac{1}{z}}$ is e^{a}.

19. Since $\log_a (1 + z)^{\frac{1}{z}} = \dfrac{1}{z} \log_a (1 + z),$

a being any base, we have, by diminishing z indefinitely,

  the limit of $\dfrac{\log_a (1 + z)}{z}$ = the limit of $\log_a (1 + z)^{\frac{1}{z}}$,

$$= \log_a e \,;$$

and, putting e for a,

  the limit of $\dfrac{\log_e (1 + z)}{z} = 1.$

20. From the equation

$$\log_a (1 + z)^{\frac{1}{z}} = \dfrac{\log_a (1 + z)}{z},$$

we deduce, by assuming $1 + z = a^{v}$,

$$\log_a (1 + z)^{\frac{1}{z}} = \dfrac{v}{a^{v} - 1}.$$

Now suppose z to diminish without limit, and therefore also v.
We have then

  the limit of $\dfrac{v}{a^{v} - 1}$ when $v = 0$

$= $ limit of $\log_a (1 + z)^{\frac{1}{z}}$ when $z = 0$

$=$ $\log_a e.$

Therefore the limit of $\dfrac{a^v - 1}{v}$ when $v = 0$

$$= \dfrac{1}{\log_a e}$$

$$= \log_e a.$$

Suppose $\qquad\qquad a = e^\mu,$

therefore $\qquad\qquad \mu = \log_e a,$

and $\qquad$ the limit of $\dfrac{e^{\mu v} - 1}{v}$ when $v = 0$ is $\mu.$

21. The following results will be found in works on Trigonometry. If the variable x diminish indefinitely

$$\text{the limit of } \dfrac{\tan x}{x} = 1,$$

$$\text{the limit of } \dfrac{\sin^{-1} x}{x} = 1,$$

$$\text{the limit of } \dfrac{\tan^{-1} x}{x} = 1.$$

22. A few general remarks may be made at the close of this Introductory Chapter. It frequently happens that a person commencing this subject is discouraged at the outset because he cannot discover or imagine any *practical application* of the somewhat abstruse points to which his attention is directed. From what he remembers of the early portions of those branches of mathematics with which he is already acquainted, he is led to expect that almost as soon as he begins the Differential Calculus, he will be able to comprehend its general scope, and to make use of it in solving algebraical and geometrical examples; and being disappointed in this expectation, he is apt to imagine as a reason for it, that he has not correctly understood the elementary principles of the subject. It may, therefore, be of some service to assure him, that the difficulty of which he complains is probably owing much more to the nature of the subject than to his own want of comprehension. The student must, of course, leave to his teacher the task of arranging

the different portions of the subject he is studying, and of selecting the definitions necessary to be understood; and in reading a work on the Differential Calculus, he must be satisfied at first with reflecting upon the meaning of the definitions, and examining whether the deductions drawn by the writer from those definitions are correct. There are innumerable applications of the elementary principles of the Differential Calculus, as will be seen in the Chapter on Expansions and those following it, but we shall at first confine ourselves merely to the logical exercise of tracing the consequences of certain definitions.

A difficulty of a more serious kind which is connected with the notion of a limit, appears to embarrass many students of this subject, namely, a suspicion that the methods employed are only approximative, and therefore a doubt as to whether the results are absolutely true. This objection is certainly very natural, but at the same time by no means easy to meet, on account of the inability of the reader to point out any definite place at which his uncertainty commences. In such a case all he can do is, to fix his attention very carefully on some part of the subject, as the theory of expansions for example, where specific important formulæ are obtained. He must examine the demonstrations, and if he can find no flaw in them, he must allow that results *absolutely true and free from all approximation* can be legitimately derived by the doctrine of Limits.

23. The demonstration in Arts. 15, 16 of the proposition that $\left(1 + \dfrac{1}{x}\right)^x$ tends to some fixed limit as x increases indefinitely, has been given in several elementary works on the Differential Calculus, and it is accordingly retained here. But the following method, in which the Binomial Theorem is not assumed, is worthy of notice.

We shall first establish two inequalities.

If β and λ are positive quantities, and λ greater than unity,

$$(1 + \beta)^\lambda \text{ is greater than } 1 + \lambda\beta \dots\dots\dots\dots (1).$$

If β and μ are positive quantities, and $\mu\beta$ less than unity,

$$(1+\beta)^\mu \text{ is less than } \frac{1}{1-\mu\beta} \quad \ldots\ldots\ldots\ldots (2).$$

To establish these inequalities we shall use the known theorem that the arithmetical mean of any number of positive quantities is greater than the geometrical mean; see *Algebra*, Chapter LI.

Let $\lambda = \dfrac{p}{q}$, where p and q are positive integers. Take p quantities, q of which are equal to $1 + \dfrac{p}{q}\beta$, and $p - q$ equal to unity. Then their arithmetical mean is $\dfrac{p - q + q\left(1 + \dfrac{p}{q}\beta\right)}{p}$,

that is $1 + \beta$; their geometrical mean is $\left(1 + \dfrac{p}{q}\beta\right)^{\frac{q}{p}}$. The former is the greater; and therefore $(1+\beta)^{\frac{p}{q}}$ is greater than $1 + \dfrac{p}{q}\beta$. Thus (1) is established.

Let $\mu = \dfrac{s}{t}$, and $\mu\beta = \dfrac{r}{t}$, where r, s and t are positive integers; thus $\beta = \dfrac{r}{s}$. Take $s + t$ quantities, s of which are equal to $1 + \dfrac{r}{s}$, and t equal to $1 - \dfrac{r}{t}$. Then their arithmetical mean is $\dfrac{s\left(1 + \dfrac{r}{s}\right) + t\left(1 - \dfrac{r}{t}\right)}{s + t}$, that is unity; their geometrical mean is $\left\{\left(1 + \dfrac{r}{s}\right)^s\left(1 - \dfrac{r}{t}\right)^t\right\}^{\frac{1}{s+t}}$. The former is the greater; therefore $\left(1 + \dfrac{r}{s}\right)^s\left(1 - \dfrac{r}{t}\right)^t$ is less than unity; and therefore $\left(1 + \dfrac{r}{s}\right)^{\frac{s}{t}}$ is less than $\dfrac{1}{1 - \dfrac{r}{t}}$. Thus (2) is established.

In (1) put $\beta = \dfrac{1}{\lambda\gamma}$, and raise both sides to the power γ; then

$$\left(1 + \frac{1}{\lambda\gamma}\right)^{\lambda\gamma} \text{ is greater than } \left(1 + \frac{1}{\gamma}\right)^{\gamma};$$

that is, if δ be greater than γ,

$$\left(1 + \frac{1}{\delta}\right)^{\delta} \text{ is greater than } \left(1 + \frac{1}{\gamma}\right)^{\gamma} \quad\ldots\ldots\ldots\ldots (3).$$

From (3) we see that $\left(1 + \dfrac{1}{x}\right)^{x}$ continually increases as x increases. It does not, however, pass beyond a certain finite limit; for in (2) write $\dfrac{1}{\mu\gamma}$ for β, and raise both sides to the power γ; then

$$\left(1 + \frac{1}{\mu\gamma}\right)^{\mu\gamma} \text{ is less than } \frac{1}{\left(1 - \dfrac{1}{\gamma}\right)^{\gamma}} \text{ if } \gamma \text{ be greater than 1.}$$

Hence, if we put $\gamma = 2$, we find that $\left(1 + \dfrac{1}{x}\right)^{x}$ can never exceed 4. By ascribing to γ greater values we shall obtain a closer limit for $\left(1 + \dfrac{1}{x}\right)^{x}$. If we put $\gamma = 6$ we see that $\left(1 + \dfrac{1}{x}\right)^{x}$ must be less than $\left(\dfrac{6}{5}\right)^{6}$, and therefore less than 3. Since then the limit of $\left(1 + \dfrac{1}{x}\right)^{x}$, as x becomes indefinitely great, must lie between $\left(1 + \dfrac{1}{n}\right)^{n}$ and $\left(\dfrac{n}{n-1}\right)^{n}$, where n has any positive value, we may, by ascribing successive integral values to n, easily approximate to the numerical value of the limit.

CHAPTER II.

24. WE shall now lay down the fundamental definition of the Differential Calculus, and deduce from it various inferences.

DEFINITION. Let $\phi(x)$ denote any function of x, and $\phi(x+h)$ the same function of $x+h$; then the limiting value of $\dfrac{\phi(x+h)-\phi(x)}{h}$, when h is made indefinitely small, is called the differential coefficient of $\phi(x)$ with respect to x.

This definition assumes that the above fraction really *has a limit*. Strictly speaking, we should use an enunciation of this form—" If $\dfrac{\phi(x+h)-\phi(x)}{h}$ have a limit when h is made indefinitely small, that limit is called the differential coefficient of $\phi(x)$ with respect to x." We shall shew, however, that the limit does exist in functions of every kind, by examining them in detail in this and the following two Chapters. We give two examples for the purpose of illustrating the definition.

Suppose $\qquad \phi(x) = x^2$;

therefore $\qquad \phi(x+h) = (x+h)^2$;

therefore $\qquad \dfrac{\phi(x+h)-\phi(x)}{h} = \dfrac{(x+h)^2 - x^2}{h}$

$$= \frac{2xh + h^2}{h} = 2x + h,$$

and the limit of $2x+h$ when $h=0$, is $2x$; therefore $2x$ is the differential coefficient of x^2 with respect to x.

Again, suppose $\phi(x) = \dfrac{a}{b+x}$;

therefore $\phi(x+h) = \dfrac{a}{b+x+h}$,

therefore $\dfrac{\phi(x+h) - \phi(x)}{h} = -\dfrac{a}{(b+x)(b+x+h)}$.

The limit of this when $h = 0$ is

$$-\frac{a}{(b+x)^2},$$

which is therefore the differential coefficient of $\dfrac{a}{b+x}$ with respect to x.

25.   We now give the notation which usually accompanies the definition in Art. 24.

Let $\phi(x) = y$, then $\phi(x+h) - \phi(x)$ is the *difference* of the two values of the dependent variable y corresponding to the two values, x and $x+h$, of the independent variable. This difference may be conveniently denoted by the symbol Δy, where Δ may be taken as an abbreviation of the word *difference*. We have thus

$$\Delta y = \phi(x+h) - \phi(x).$$

Agreeably with this notation, h may be denoted by Δx, so that

$$\frac{\Delta y}{\Delta x} = \frac{\phi(x+h) - \phi(x)}{h}.$$

It may appear a superfluity of notation to use both h and Δx to denote the same thing, but in finding the limit of the right-hand side we shall sometimes have to perform several analytical transformations, and thus a single letter is more convenient. On the left-hand side Δx is recommended by considerations of symmetry.

We say then, according to the definition in Art. 24, that the limit of $\dfrac{\Delta y}{\Delta x}$, when Δx is diminished indefinitely, is the differential coefficient of y or $\phi(x)$ with respect to x. *This limit is denoted by the symbol* $\dfrac{dy}{dx}$.

26. The symbol $\dfrac{dy}{dx}$ we consider as a *whole*, and we do not assign a separate meaning to dy and dx. As, however, $\dfrac{\Delta y}{\Delta x}$ is a real fraction in which Δy and Δx have definite meanings, the student will very possibly suspect that some meanings may be given to dy and dx which will enable him to regard $\dfrac{dy}{dx}$ as a fraction. This suspicion will probably be strengthened as he proceeds in the subject and finds that in many cases $\dfrac{dy}{dx}$ possesses the properties of an algebraical fraction. We remark that there are indeed methods of treating the Differential Calculus in which meanings are given to dy and dx, and we shall recur to them hereafter (see Chap. XXVII.), but at present we define the symbol $\dfrac{dy}{dx}$ as above, and only leave to the reader the task of examining whether we are consistent with ourselves in the inferences we proceed to draw and express by means of our definitions and symbols.

The following notation is also frequently used. If $\phi(x)$ denote any function of x, then $\phi'(x)$ denotes the differential coefficient of $\phi(x)$ with respect to x.

The operation of finding the differential coefficient of a function is called "differentiating" that function.

27. *Differential coefficient of a sum of Functions.*

Let y and z denote two functions of x, and u their sum. Suppose that y', z', u', denote the values these functions assume when x is changed into $x + h$. Then

$$u = y + z,$$
$$u' = y' + z',$$

therefore $\qquad u' - u = y' - y + z' - z \cdot$

that is $\qquad \Delta u = \Delta y + \Delta z.$

Divide by h or Δx, then

$$\frac{\Delta u}{\Delta x} = \frac{\Delta y}{\Delta x} + \frac{\Delta z}{\Delta x}.$$

C

Now let h diminish without limit, and we have

$$\frac{du}{dx} = \frac{dy}{dx} + \frac{dz}{dx}.$$

Hence *the differential coefficient of the sum of two functions is the sum of the differential coefficients of the functions.*

Similarly, if $u = y - z$

$$\frac{du}{dx} = \frac{dy}{dx} - \frac{dz}{dx}.$$

28.   The results of Art. 27 may be extended to the case of any number of functions connected by the signs of addition or subtraction.   For example, let

$$u = w + y + z,$$

then, as before, $\Delta u = \Delta w + \Delta y + \Delta z$;

therefore $\dfrac{\Delta u}{\Delta x} = \dfrac{\Delta w}{\Delta x} + \dfrac{\Delta y}{\Delta x} + \dfrac{\Delta z}{\Delta x}$;

therefore, proceeding to the limit,

$$\frac{du}{dx} = \frac{dw}{dx} + \frac{dy}{dx} + \frac{dz}{dx}.$$

29.   *Differential coefficient of the product of two Functions.*

Let $\phi(x)$ and $\psi(x)$ denote two functions of x, and let

$$u = \phi(x)\ \psi(x).$$

Change x into $x + h$, and let $u + \Delta u$ denote the new product,

then $u + \Delta u = \phi(x + h)\ \psi(x + h)$,

therefore $\Delta u = \phi(x + h)\ \psi(x + h) - \phi(x)\ \psi(x)$

$$= \{\phi(x + h) - \phi(x)\}\ \psi(x + h) + \phi(x)\ \{\psi(x + h) - \psi(x)\};$$

therefore $\dfrac{\Delta u}{\Delta x} = \dfrac{\phi(x+h) - \phi(x)}{h}\ \psi(x+h) + \dfrac{\psi(x+h) - \psi(x)}{h}\ \phi(x).$

Suppose now h diminished indefinitely; then the limit of $\dfrac{\phi(x + h) - \phi(x)}{h}$ is the differential coefficient of $\phi(x)$ with respect to x, or $\phi'(x)$; the limit of $\dfrac{\psi(x+h)-\psi(x)}{h}$ is the differential coefficient of $\psi(x)$ with respect to x, or $\psi'(x)$; the limit of $\psi(x+h)$ is $\psi(x)$;

therefore $\qquad \dfrac{du}{dx}=\phi'(x)\psi(x)+\psi'(x)\phi(x).$

Hence *the differential coefficient of the product of two functions is found by multiplying each factor by the differential coefficient of the other factor and adding the resulting products.*

Divide each side of the last result by u or $\phi(x)\psi(x)$; thus

$$\frac{1}{u}\frac{du}{dx}=\frac{\phi'(x)}{\phi(x)}+\frac{\psi'(x)}{\psi(x)}.$$

30. An equation similar to that just obtained holds for the product of any number of functions. For example, let

$$u=wyz,$$

w, y, z being all functions of x.

Assume $\qquad\qquad v=wy,$

therefore $\qquad\qquad u=vz$;

then, by Art. 29,

$$\frac{1}{u}\frac{du}{dx}=\frac{1}{v}\frac{dv}{dx}+\frac{1}{z}\frac{dz}{dx};$$

also $\qquad\qquad \dfrac{1}{v}\dfrac{dv}{dx}=\dfrac{1}{w}\dfrac{dw}{dx}+\dfrac{1}{y}\dfrac{dy}{dx};$

therefore $\qquad \dfrac{1}{u}\dfrac{du}{dx}=\dfrac{1}{w}\dfrac{dw}{dx}+\dfrac{1}{y}\dfrac{dy}{dx}+\dfrac{1}{z}\dfrac{dz}{dx};$

therefore $\qquad \dfrac{du}{dx}=yz\dfrac{dw}{dx}+wz\dfrac{dy}{dx}+wy\dfrac{dz}{dx}.$

Proceeding in this manner we have as a rule: *The differential coefficient of the product of any number of functions is found by multiplying the differential coefficient of each factor by all the other factors and adding the products thus formed.*

31. *Differential coefficient of a quotient.*

Let $\phi(x)$ and $\psi(x)$ denote two functions of x, and let

$$u = \frac{\phi(x)}{\psi(x)}.$$

Suppose x changed into $x + h$, and let $u + \Delta u$ denote the new value of the quotient. Then

$$u + \Delta u = \frac{\phi(x+h)}{\psi(x+h)},$$

therefore $\Delta u = \dfrac{\phi(x+h)\,\psi(x) - \phi(x)\,\psi(x+h)}{\psi(x+h)\,\psi(x)}$

$$= \frac{\{\phi(x+h) - \phi(x)\}\,\psi(x) - \{\psi(x+h) - \psi(x)\}\,\phi(x)}{\psi(x+h)\,\psi(x)};$$

therefore $\dfrac{\Delta u}{\Delta x} = \dfrac{\dfrac{\phi(x+h) - \phi(x)}{h}\,\psi(x) - \dfrac{\psi(x+h) - \psi(x)}{h}\,\phi(x)}{\psi(x+h)\,\psi(x)}.$

Let h diminish without limit, then

$$\frac{du}{dx} = \frac{\phi'(x)\,\psi(x) - \psi'(x)\,\phi(x)}{\{\psi(x)\}^2}.$$

Hence we have this rule : *To find the differential coefficient of a quotient ; multiply the denominator by the differential coefficient of the numerator and the numerator by the differential coefficient of the denominator ; subtract the second product from the first and divide the result by the square of the denominator.*

32.   The result of Art. 31 may also be obtained thus :

Since $u = \dfrac{\phi(x)}{\psi(x)},$

therefore $\phi(x) = u\psi(x);$

therefore, by Art. 29,

$$\phi'(x) = \frac{du}{dx}\,\psi(x) + u\psi'(x);$$

therefore $\psi(x)\dfrac{du}{dx} = \phi'(x) - \dfrac{\phi(x)}{\psi(x)}\,\psi'(x),$

therefore $\dfrac{du}{dx} = \dfrac{\phi'(x)\,\psi(x) - \psi'(x)\,\phi(x)}{\{\psi(x)\}^2}.$

33.   *Differentiation of a constant.*

If $y = c$ where c is a constant, then $\dfrac{dy}{dx} = 0$.   For to say that y is equal to a constant is the same thing as saying that y cannot vary ; hence $\Delta y = 0$, therefore

$$\frac{\Delta y}{\Delta x} = 0$$

whatever be the value of Δx ; therefore

$$\frac{dy}{dx} = 0.$$

Hence, making $\phi(x) = $ a constant c in Art. 29, we have

$$\frac{dc\psi(x)}{dx} = c\psi'(x).$$

This may of course be obtained directly thus :

Let $\qquad\qquad u = c\psi(x),$

then $\qquad\qquad u + \Delta u = c\psi(x + h) ;$

therefore $\qquad \dfrac{\Delta u}{\Delta x} = c\dfrac{\psi(x + h) - \psi(x)}{h},$

therefore $\qquad \dfrac{du}{dx} = c\psi'(x).$

So by putting $\phi(x) = c$ in Art. 31, we obtain

$$d\frac{\dfrac{c}{\psi(x)}}{dx} = -\frac{c\psi'(x)}{\{\psi(x)\}^2},$$

which likewise may be found independently.

34.   We have now defined a differential coefficient and have shewn how the differential coefficient of a compound function can be found as soon as we know the differential coefficients of the component functions.   Before we proceed to the rules for determining the differential coefficient of any known algebraical expression, we shall give some geometrical illustrations which will assist in forming a conception of the meaning of a differential coefficient and afford some hints as to the applications which can be made of the doctrine of limits.

35. Suppose we have given the equation $y = \phi(x)$, and that we attribute to the independent variable x all possible values between $-\infty$ and $+\infty$ and notice the corresponding values of y. Geometry gives us the means of representing distinctly this succession of values. We can take x for an abscissa measured from a fixed origin along a certain axis, and y for the corresponding ordinate measured along an axis at right angles to the first. The values of y corresponding to those of x in the equation $y = \phi(x)$ will belong to a curve AMN, the form of which will indicate the series of values we are considering. It is necessary to have always present in our mind not merely any particular value of x and the corresponding value of y, but the whole series of corresponding values of these two variables.

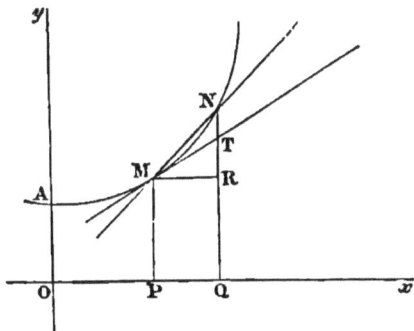

36. Among the properties which the function $\phi(x)$, or the line which represents it, possesses, the most remarkable, that in fact which is the object of the differential calculus and the consideration of which is perpetually occurring in all applications of this calculus, is *the degree of rapidity with which the function varies when the variable begins to vary from any assigned value.* The degree of rapidity of increase of the function when the variable is made to increase may differ not only in different functions but also in the same function according to the value attributed to the variable from which the increase is supposed to commence. Suppose we give to x a particular value denoted by OP, to which corresponds a determinate value of y or $\phi(x)$ represented by MP. Let x, starting from the value assigned, increase by a quantity which we denote by Δx, and which is represented by PQ. The function y will vary in consequence by a certain quantity which we denote by Δy, so that

$$y + \Delta y = \phi(x + \Delta x),$$

therefore $\qquad \Delta y = \phi(x + \Delta x) - \phi(x).$

The new value of the ordinate is represented in the figure by NQ, and NR represents Δy. The fraction $\dfrac{\Delta y}{\Delta x}$ represents the ratio of the increase of the function to the increase of the variable, and is equal to the trigonometrical tangent of the angle NMR formed by the secant MN with the axis of x.

37. It is evident that this fraction is a natural measure of the degree of rapidity with which the function y increases when the independent variable x increases; for the greater this fraction is, the greater will be the increase of the function y corresponding to the given increase Δx of the variable. But it is important to observe that the value of $\dfrac{\Delta y}{\Delta x}$ will depend not only on the value given to x, *but also on the magnitude of the increment* Δx, except in the case in which the curve becomes a straight line.

If then we left this increment arbitrary, it would be impossible to assign to the fraction $\dfrac{\Delta y}{\Delta x}$ any definite value, and it is thus necessary to adopt some convention which will remove this uncertainty.

38. Suppose that after giving to Δx a certain value, to which will correspond a certain value for Δy and a certain direction for the secant MN, we make the value of Δx gradually diminish and become ultimately zero. The value of Δy will also gradually diminish and become ultimately zero. The point N will move along the curve towards M, and we shall find *in every example we consider, that the straight line MN will approach towards some limiting position MT*. This is in fact equivalent to the assertion made in Art. 24, that by examining every case in detail we could shew that every function has a differential coefficient. The limiting position which the secant assumes when N coincides with M is called *the tangent to the curve at the point M*, and thus $\dfrac{dy}{dx}$ is the trigonometrical tangent of the inclination to the axis of x of the tangent line to the curve.

39. The limit of the fraction $\dfrac{\Delta y}{\Delta x}$, when Δx is diminished indefinitely, may be considered as affording a precise measure of the rapidity with which the function increases when the independent variable increases, for there remains no longer anything arbitrary in the expression. The limit $\dfrac{dy}{dx}$ does not depend on the value assigned to Δx nor upon the form of the curve at any finite distance from the point whose co-ordinates are x and y; it depends only on the *direction* of the curve at this point, that is to say, on the inclination of the tangent line to the axis of x.

40. As an example of the preceding, we will determine the differential coefficient of $\sqrt{(a^2 - x^2)}$, and point out its geometrical application.

Let $\qquad y = \sqrt{(a^2 - x^2)}$,

then $\qquad y + \Delta y = \sqrt{\{a^2 - (x + h)^2\}}$;

therefore $\qquad \Delta y = \sqrt{\{a^2 - (x + h)^2\}} - \sqrt{(a^2 - x^2)}$,

$$= \frac{x^2 - (x + h)^2}{\sqrt{\{a^2 - (x + h)^2\}} + \sqrt{(a^2 - x^2)}},$$

$$= \frac{-(2xh + h^2)}{\sqrt{\{a^2 - (x + h)^2\}} + \sqrt{(a^2 - x^2)}};$$

therefore $\qquad \dfrac{\Delta y}{\Delta x} = -\dfrac{2x + h}{\sqrt{\{a^2 - (x + h)^2\}} + \sqrt{(a^2 - x^2)}}.$

The limit of this when h is made indefinitely small is

$$-\frac{x}{\sqrt{(a^2 - x^2)}};$$

therefore $\qquad \dfrac{dy}{dx} = -\dfrac{x}{\sqrt{(a^2 - x^2)}}.$

It will be seen that we have in the above example used an algebraical artifice, namely, that of multiplying both numerator and denominator of a fraction by $\sqrt{\{a^2 - (x+h)^2\}} + \sqrt{(a^2 - x^2)}$, in order to obtain $\dfrac{\Delta y}{\Delta x}$ in a form the limit of which can be easily seen. In treating any example without the aid of general rules, we should frequently find our success dependent upon our readiness in effecting such transformations; but the next two Chapters will explain methods of making the problem of ascertaining any differential coefficient depend upon the knowledge of those of a few standard functions.

41. From analytical geometry we know that the equation $y = \sqrt{(a^2 - x^2)}$ represents a circle, and it is also known from the principles of that subject that the tangent at the point (x, y) of a circle is inclined to the axis of x at an angle whose trigonometrical tangent is $-\dfrac{x}{\sqrt{(a^2 - x^2)}}$. Also in the case of a circle the straight line which we have defined as the tangent is the same straight line as that which fulfils the condition of "touching the circle," given in *Euclid*, Book III.

42. In the Chapters on the geometrical application of the Differential Calculus we shall recur to the subject of tangents. We have given the above example here that the student may at this early period acquire the conviction that important uses may be made of a differential coefficient.

43. The following is another geometrical application. The area $OAMP$, see the diagram to Art. 35, must be *some* function of x, since it is a definite quantity when we assign a definite value to x, and varies when x varies. Denote this function by u, and PQ by Δx; then

$$u + \Delta u = \text{area } OANQ,$$

therefore $\qquad \Delta u = \text{area } MNQP;$

therefore Δu lies between $MP.PQ$ and $NQ.PQ,$

that is, between $y\Delta x$ and $(y + \Delta y)\Delta x$;

therefore $\qquad \dfrac{\Delta u}{\Delta x}$ lies between y and $y + \Delta y.$

Hence, diminishing Δx, and therefore Δy, without limit, we have

$$\frac{du}{dx} = y.$$

CHAPTER III.

DIFFERENTIAL COEFFICIENTS OF SIMPLE FUNCTIONS.

44. *Differential coefficient of x^n where n is a positive integer.*

Let $y = x^n$, therefore

$$y + \Delta y = (x + h)^n,$$

therefore $\quad \Delta y = (x + h)^n - x^n$

$$= nx^{n-1}h + \frac{n(n-1)}{1 \cdot 2} x^{n-2}h^2 + .. + h^n;$$

therefore $\quad \dfrac{\Delta y}{\Delta x} = nx^{n-1} + \dfrac{n(n-1)}{1 \cdot 2} x^{n-2}h + ... + h^{n-1}.$

Diminish h without limit, and we have

$$\frac{dy}{dx} = nx^{n-1}.$$

45. The same result may also be obtained by means of Art. 30. For let

$$u = y_1 y_2 \cdots y_n,$$

where the n quantities $y_1, y_2, \cdots y_n$, are all functions of x; we have then

$$\frac{1}{u}\frac{du}{dx} = \frac{1}{y_1}\frac{dy_1}{dx} + \frac{1}{y_2}\frac{dy_2}{dx} + ... + \frac{1}{y_n}\frac{dy_n}{dx}.$$

If now $y_1 = x$, we have

$$\Delta y_1 = \Delta x,$$

therefore $\quad \dfrac{\Delta y_1}{\Delta x} = 1,$

therefore $\quad \dfrac{dy_1}{dx} = 1.$

Put then $y_1,\ y_2,\ \dots\ y_n$, all equal to x; thus u becomes x^n, and we obtain

$$\frac{1}{u}\frac{du}{dx} = \frac{n}{x},$$

therefore

$$\frac{du}{dx} = nx^{n-1}.$$

46. If n be not a positive integer, we may by assuming the truth of the Binomial Theorem for fractional exponents proceed as in Art. 44 to determine $\dfrac{dx^n}{dx}$. But in that case we shall require to assume that "if we have a series containing an infinite number of terms and each term becomes ultimately indefinitely small, the sum of the terms becomes so too." To avoid this assumption we adopt another mode.

47. *Differential coefficient of x^n the exponent n being unrestricted.*

Let $y = x^n$, therefore

$$y + \Delta y = (x + h)^n,$$

therefore

$$\frac{\Delta y}{\Delta x} = \frac{(x + h)^n - x^n}{h}$$

$$= x^n \frac{\left(\dfrac{x + h}{x}\right)^n - 1}{h}.$$

Now whatever be the value of n, positive or negative, whole or fractional, it may be supposed $= \dfrac{p - q}{r}$, where p, q, r, are positive integers.

Let

$$\frac{x + h}{x} = z,$$

therefore

$$h = x(z - 1),$$

and

$$\frac{\Delta y}{\Delta x} = x^{n-1}\frac{z^n - 1}{z - 1}.$$

As h diminishes indefinitely z approaches the limit 1, and we have to find in that case the limit of $\dfrac{z^n - 1}{z - 1}$.

Suppose $v = z^{\frac{1}{r}}$, then

$$\frac{z^n - 1}{z - 1} \text{ or } \frac{z^{\frac{p-q}{r}} - 1}{z - 1} = \frac{v^{p-q} - 1}{v^r - 1} = \frac{v^p - v^q}{v^q (v^r - 1)}$$

$$= \frac{v^p - 1 - (v^q - 1)}{v^q (v^r - 1)}$$

$$= \frac{v^{p-1} + v^{p-2} + \ldots + 1 - (v^{q-1} + v^{q-2} + \ldots + 1)}{v^q (v^{r-1} + v^{r-2} + \ldots + 1)} .$$

This last result is obtained by dividing both numerator and denominator of the preceding fraction by $v - 1$. Let now v approach the limit 1, then the limit of the last fraction is

$$\frac{p - q}{r} ,$$

therefore

$$\frac{dy}{dx} = \frac{p - q}{r} x^{n-1} = n x^{n-1}.$$

48. *Differential coefficient of x^n. Second method.*

Let $y = x^n$, therefore

$$y + \Delta y = (x + h)^n,$$

therefore

$$\frac{\Delta y}{\Delta x} = \frac{(x + h)^n - x^n}{h}$$

$$= \frac{x^n}{h} \left\{ \left(1 + \frac{h}{x} \right)^n - 1 \right\}.$$

Assume $\frac{h}{x} = z$ and $(1 + z)^n - 1 = v$, then z and v are quantities which diminish indefinitely with h. Thus

$$\frac{\Delta y}{\Delta x} = x^{n-1} \frac{v}{z} .$$

From the above assumptions

$$(1 + z)^n = 1 + v,$$

therefore

$$\log_e (1 + v) = n \log_e (1 + z).$$

From Art. 19 the expressions

$$\frac{\log_e(1+z)}{z} \text{ and } \frac{\log_e(1+v)}{v}$$

both tend to the limit unity.   Hence we may assume

$$\frac{\log_e(1+v)}{v} = 1 + \gamma,$$

$$\frac{\log_e(1+z)}{z} = 1 + \delta,$$

where each of the quantities γ and δ has zero for its limit. Hence

$$\frac{v}{z} = \frac{1+\delta}{1+\gamma} \cdot \frac{\log_e(1+v)}{\log_e(1+z)}$$

$$= n\frac{1+\delta}{1+\gamma} \text{ from above;}.$$

therefore the limit of $\dfrac{v}{z}$ is n, and

$$\frac{dy}{dx} = nx^{n-1}.$$

49.   *Differential coefficient of a^x.*

Let $y = a^x$, therefore

$$y + \Delta y = a^{x+h} = a^x a^h,$$

therefore

$$\frac{\Delta y}{\Delta x} = a^x \frac{a^h - 1}{h}.$$

Now, by Art. 20, the limit of $\dfrac{a^h - 1}{h}$, when h is indefinitely diminished is $\log_e a$;   therefore

$$\frac{dy}{dx} = a^x \log_e a.$$

Next let $y = a^{ex}$;   then

$$y = (a^e)^x;$$

hence by the rule just proved

$$\frac{dy}{dx} = (a^c)^x \log_e a^c$$

$$= a^{cx} c \log_e a.$$

Hence if $y = e^{cx}$,

$$\frac{dy}{dx} = ce^{cx};$$

and if $\qquad\qquad y = e^x,$

$$\frac{dy}{dx} = e^x.$$

50. *Differential coefficient of* $\log_a x$.

Let $y = \log_a x$, therefore

$$y + \Delta y = \log_a (x + h),$$

therefore $\qquad\qquad \Delta y = \log_a (x + h) - \log_a x$

$$= \log_a \frac{x + h}{x};$$

therefore $\qquad\qquad \dfrac{\Delta y}{\Delta x} = \dfrac{\log_a \dfrac{x + h}{x}}{h}.$

Assume $h = xz$, therefore

$$\frac{\Delta y}{\Delta x} = \frac{1}{x} \frac{\log_a (1 + z)}{z}.$$

By Art. 19 the limit of $\dfrac{\log_a (1 + z)}{z}$ when z diminishes indefinitely is $\log_a e$, therefore

$$\frac{dy}{dx} = \frac{1}{x} \log_a e$$

$$= \frac{1}{x} \cdot \frac{1}{\log_e a}.$$

Hence if $y = \log_e x$

$$\frac{dy}{dx} = \frac{1}{x}.$$

51. *Differential coefficient of* sin x.

Let $y = \sin x$, therefore

$$y + \Delta y = \sin (x + h),$$

therefore $\quad\quad \Delta y = \sin (x + h) - \sin x$

$$= 2 \cos \left(x + \frac{h}{2} \right) \sin \frac{h}{2}, \text{ by Trigonometry,}$$

therefore $\quad\quad \dfrac{\Delta y}{\Delta x} = \cos \left(x + \dfrac{h}{2} \right) \dfrac{\sin \dfrac{h}{2}}{\dfrac{h}{2}}.$

Now when h is indefinitely diminished, the limit of $\dfrac{\sin \dfrac{h}{2}}{\dfrac{h}{2}}$ is unity by Art. 9, therefore

$$\frac{dy}{dx} = \cos x.$$

52. *Differential coefficient of* cos x.

Let $y = \cos x$, therefore

$$y + \Delta y = \cos (x + h),$$

therefore $\quad\quad \Delta y = \cos (x + h) - \cos x$

$$= - 2 \sin \left(x + \frac{h}{2} \right) \sin \frac{h}{2},$$

therefore $\quad\quad \dfrac{\Delta y}{\Delta x} = - \sin \left(x + \dfrac{h}{2} \right) \dfrac{\sin \dfrac{h}{2}}{\dfrac{h}{2}},$

therefore $\quad\quad \dfrac{dy}{dx} = - \sin x.$

53. *Differential coefficient of* tan x.

Let $y = \tan x$, therefore

$$y + \Delta y = \tan (x + h),$$

therefore $\quad \Delta y = \tan (x+h) - \tan x$

$$= \frac{\sin (x+h)}{\cos (x+h)} - \frac{\sin x}{\cos x}$$

$$= \frac{\sin (x+h-x)}{\cos (x+h) \cos x} = \frac{\sin h}{\cos (x+h) \cos x},$$

therefore $\quad \dfrac{\Delta y}{\Delta x} = \dfrac{\sin h}{h} \dfrac{1}{\cos (x+h) \cos x},$

therefore $\quad \dfrac{dy}{dx} = \dfrac{1}{\cos^2 x}.$

54. *Differential coefficient of* cot x.

By proceeding as in the last Example, we find that if

$$y = \cot x,$$

$$\frac{dy}{dx} = -\frac{1}{\sin^2 x}.$$

55. *Differential coefficient of* sec x.

Let $y = \sec x$, therefore

$$y + \Delta y = \sec (x+h),$$

therefore $\quad \Delta y = \sec (x+h) - \sec x$

$$= \frac{1}{\cos (x+h)} - \frac{1}{\cos x} = \frac{\cos x - \cos (x+h)}{\cos x \cos (x+h)}$$

$$= \frac{2 \sin \left(x + \dfrac{h}{2}\right) \sin \dfrac{h}{2}}{\cos x \cdot \cos (x+h)};$$

therefore $\quad \dfrac{\Delta y}{\Delta x} = \dfrac{\sin \left(x + \dfrac{h}{2}\right)}{\cos x \cos (x+h)} \dfrac{\sin \dfrac{h}{2}}{\dfrac{h}{2}},$

therefore $\quad \dfrac{dy}{dx} = \dfrac{\sin x}{\cos^2 x}.$

56. *Differential coefficient of* cosec x.

Let $y = \operatorname{cosec} x$; proceed as in the last example, and we find

$$\frac{dy}{dx} = -\frac{\cos x}{\sin^2 x}.$$

57. Since $\tan x$, $\cot x$, $\sec x$, and $\operatorname{cosec} x$ are all fractional forms, we may deduce the differential coefficient of each of these functions by Art. 31 from those of $\sin x$ and $\cos x$. Thus, let

$$y = \tan x = \frac{\sin x}{\cos x},$$

therefore

$$\frac{dy}{dx} = \frac{\cos x \dfrac{d \sin x}{dx} - \sin x \dfrac{d \cos x}{dx}}{\cos^2 x}, \text{ Art. 31,}$$

$$= \frac{\cos^2 x + \sin^2 x}{\cos^4 x}, \text{ Arts. 51 and 52,}$$

$$= \frac{1}{\cos^2 x}.$$

Similarly we may proceed with $\cot x$, $\sec x$, and $\operatorname{cosec} x$.

Since vers $x = 1 - \cos x$, the differential coefficient of vers x by Arts. 27 and 33

$$= - \text{ differential coefficient of } \cos x$$

$$= \sin x.$$

CHAPTER IV.

DIFFERENTIAL COEFFICIENTS OF THE INVERSE TRIGONOME-TRICAL FUNCTIONS AND OF COMPLEX FUNCTIONS.

58. LET $y = \phi(x)$, so that y is a known function of x; it follows from this that x must be *some* function of y, although we may not be able to express that function in any simple form. The best mode for the reader to convince himself of this will be to recur to algebraical geometry and suppose x and y to be the co-ordinates of a point in a curve the equation to which is $y = \phi(x)$. For every value of x there will be generally one or more values of y, positive or negative, as the case may be. So for any value of y there will be generally one or more definite values of x, which, as they really exist, may be made the subjects of our investigations, even although our present powers of mathematical expression may not always furnish us with simple modes of representing them.

59. A simple example will be given in the equation

$$y = x^2 - 2x + 1 \quad \dots\dots\dots\dots\dots\dots (1).$$

Solve this equation with respect to x, and we have

$$x = 1 \pm y^{\frac{1}{2}} \quad \dots\dots\dots\dots\dots\dots (2).$$

Here (2) shews that if any value be assigned to y we must have for x one of two definite values.

Now in (1), x being the independent variable and y the dependent variable, we have by Arts. 28, 33, and 44,

$$\frac{dy}{dx} = 2x - 2 \quad \dots\dots\dots\dots\dots\dots (3).$$

In equation (2) we may treat y as the independent variable and x as the dependent variable, and we find, by Art. 47,

$$\frac{dx}{dy} = \pm \tfrac{1}{2} y^{-\frac{1}{2}} \dots\dots\dots\dots\dots(4).$$

From (2) $\qquad x - 1 = \pm y^{\frac{1}{2}},$

therefore $\qquad \dfrac{1}{x-1} = \pm y^{-\frac{1}{2}}.$

Hence, from (4), $\qquad \dfrac{dx}{dy} = \dfrac{1}{2(x-1)} \dots\dots\dots\dots\dots(5).$

Comparing (5) with (3), we see that

$$\frac{dy}{dx} \times \frac{dx}{dy} = 1.$$

The theorem which holds in this simple case we shall now prove to be universally true.

60. *To prove* $\dfrac{dy}{dx} \times \dfrac{dx}{dy} = 1.$

Let $\qquad\qquad y = \phi(x) \ \dots\dots\dots\dots\dots (1),$

since from this it follows that x must be *some* function of y,

suppose $\qquad\qquad x = \psi(y) \ \dots\dots\dots\dots\dots (2).$

Let x in (1) be changed into $x + \Delta x$, in consequence of which y becomes $y + \Delta y$, then

$$y + \Delta y = \phi(x + \Delta x) \ \dots\dots\dots\dots\dots (3).$$

Now in (2) it may happen that x *has more than one* value for any assigned value of y, but if the value of y in (2) be the same as that in (1), then among the values which x can have, *one must be the value we supposed assigned to x in* (1). Hence we may suppose x and y in (2) to have the same values as the same symbols respectively had in (1). In equation (2) change y into $y + \Delta y$, where y has the same value as in (1) and (3), and Δy the same value as in (3). Then among the values which the dependent variable is susceptible of in (2), *one must be* $x + \Delta x$, the symbols having the same values as in (3).

Hence $\qquad x + \Delta x = \psi\,(y + \Delta y)$ .... (4).

From (1) and (3)

$$\frac{\Delta y}{\Delta x} = \frac{\phi\,(x + \Delta x) - \phi\,(x)}{\Delta x} \qquad .... (5).$$

From (2) and (4)

$$\frac{\Delta x}{\Delta y} = \frac{\psi\,(y + \Delta y) - \psi\,(y)}{\Delta y} \qquad .... (6).$$

In (5) and (6) the same symbols have the same values, and since in that case $\dfrac{\Delta y}{\Delta x} \times \dfrac{\Delta x}{\Delta y} = 1$, we have

$$\frac{\phi\,(x + \Delta x) - \phi\,(x)}{\Delta x} \times \frac{\psi\,(y + \Delta y) - \psi\,(y)}{\Delta y} = 1.$$

Now diminish Δx and Δy without limit, and we have

$$\phi'\,(x) \times \psi'\,(y) = 1\,;$$

or, as it may be written,

$$\frac{dy}{dx} \times \frac{dx}{dy} = 1.$$

61. The demonstration given in the last Article may appear laborious. In reviewing it, the student will perceive that this arises from the necessity of proving that the x, y, Δx, and Δy, which occur in (5), *have the same numerical values as the quantities denoted by the same symbols respectively in* (6). This point is sometimes assumed, and it is considered sufficient to say " since $\dfrac{\Delta y}{\Delta x} \times \dfrac{\Delta x}{\Delta y} = 1$ always, we have, by proceeding to the limit, $\dfrac{dy}{dx} \times \dfrac{dx}{dy} = 1$," but it would appear necessary at least that the assumption should be noticed.

62. Suppose $\qquad z = \phi\,(x),$

$$y = \psi\,(z),$$

so that y is a function of z, and z a function of x. It follows that if we substitute for z its value in $\psi\,(z)$, we make y an explicit function of x, and consequently y must have a differential coefficient with respect to x. For example, if $z = x^2$ and $y = z^3$, we have by substitution $y = x^6$. Now this is a function of x of which we know the differential coefficient, by Art. 44. Hence $\dfrac{dy}{dx} = 6x^5$. But if $z = \cos x$ and $y = a^z$, we find $y = a^{\cos x}$, a function of x which we have not yet seen how to differentiate. Hence the necessity and use of the rule demonstrated in the next Article.

63. *Differential coefficient of a function of a function.*

Let $$z = \phi(x) \ \dots\dots\dots\dots\dots\dots\dots (1),$$

and $$y = \psi(z) \ \dots\dots\dots\dots\dots\dots\dots (2),$$

so that y is a function of x; required the differential coefficient of y with respect to x.

Let x be changed into $x + \Delta x$, in consequence of which z becomes $z + \Delta z$, and suppose in consequence of this change in z, that y becomes $y + \Delta y$; thus

$$z + \Delta z = \phi(x + \Delta x) \ \dots\dots\dots\dots (3),$$

$$y + \Delta y = \psi(z + \Delta z) \ \dots\dots\dots\dots (4).$$

Now suppose that by putting for z its value in (2), we obtain

$$y = F(x) \ \dots\dots\dots\dots\dots\dots (5),$$

where $F(x)$ denotes some function of x. From the mode in which equation (5) is obtained it follows that we may *suppose x and y to have respectively the same values in (5) as in (1) and (2), and also that*

$$y + \Delta y = F(x + \Delta x) \ \dots\dots\dots\dots (6),$$

where Δx and Δy are the same quantities as have already occurred in (3) and (4).

From these equations we deduce

$$\frac{\Delta y}{\Delta x} = \frac{F(x + \Delta x) - F(x)}{\Delta x} \text{ from (5) and (6)},$$

$$\frac{\Delta y}{\Delta z} = \frac{\psi(z + \Delta z) - \psi(z)}{\Delta z} \text{ from (2) and (4)},$$

$$\frac{\Delta z}{\Delta x} = \frac{\phi(x + \Delta x) - \phi(x)}{\Delta x} \text{ from (1) and (3)},$$

where the same symbols denote throughout the same quantities. Hence, since

$$\frac{\Delta y}{\Delta x} = \frac{\Delta y}{\Delta z} \times \frac{\Delta z}{\Delta x},$$

we have

$$\frac{F(x + \Delta x) - F(x)}{\Delta x} = \frac{\psi(z + \Delta z) - \psi(z)}{\Delta z} \times \frac{\phi(x + \Delta x) - \phi(x)}{\Delta x}.$$

Now let Δx, Δz, and Δy, diminish without limit, and we obtain $\qquad F'(x) = \psi'(z)\,\phi'(x)$;

or, as it may be written,

$$\frac{dy}{dx} = \frac{dy}{dz}\frac{dz}{dx}.$$

Hence the differential coefficient of y with respect to x is equal to the product of the differential coefficient of y with respect to z, and of the differential coefficient of z with respect to x.

64. We may make a remark on the demonstration of the last Article similar to that in Art. 61. It is often considered sufficient to say that " $\dfrac{\Delta y}{\Delta x} = \dfrac{\Delta y}{\Delta z} \times \dfrac{\Delta z}{\Delta x}$ by the properties of fractions, and therefore, by taking the limit, $\dfrac{dy}{dx} = \dfrac{dy}{dz}\dfrac{dz}{dx}$. "

65. *Differential coefficient of* $\sin^{-1}x$.

Let $y = \sin^{-1}x$, therefore

$$\sin y = x,$$

therefore $\qquad\qquad \dfrac{dx}{dy} = \cos y$,   Art. 51,

therefore $\qquad\qquad \dfrac{dy}{dx} = \dfrac{1}{\cos y}$,   Art. 60

Since $\sin y = x$, $\cos y = \pm\sqrt{(1 - x^2)}$; the proper sign to be taken will of course depend on the value of y; we may therefore put

$$\frac{dy}{dx} = \frac{1}{\sqrt{(1 - x^2)}},$$

remembering that the radical must have a negative sign if $\cos y$ be negative.

66. *Differential coefficient of* $\cos^{-1}x$.

Let $y = \cos^{-1}x$, therefore

$$\cos y = x,$$

therefore $\qquad \dfrac{dx}{dy} = -\sin y,$ Art. 52,

therefore $\qquad \dfrac{dy}{dx} = -\dfrac{1}{\sin y},$ Art. 60,

$$= -\dfrac{1}{\sqrt{(1 - x^2)}}.$$

See the preceding Article.

67. *Differential coefficient of* $\tan^{-1}x$ *and* $\cot^{-1}x$.

Let $y = \tan^{-1}x$, therefore

$$x = \tan y,$$

therefore $\qquad \dfrac{dx}{dy} = \dfrac{1}{\cos^2 y},$ Art. 53,

therefore $\qquad \dfrac{dy}{dx} = \cos^2 y,$ Art. 60,

$$= \dfrac{1}{1 + \tan^2 y}$$

$$= \dfrac{1}{1 + x^2}.$$

Similarly, if $\qquad y = \cot^{-1}x,$

$$\dfrac{dy}{dx} = -\dfrac{1}{1 + x^2}.$$

68. *Differential coefficient of* $\sec^{-1}x$ *and* $\operatorname{cosec}^{-1}x$.

Let $y = \sec^{-1}x$, therefore

$$x = \sec y,$$

therefore $\qquad \dfrac{dx}{dy} = \dfrac{\sin y}{\cos^2 y},$ Art. 55,

therefore $\qquad \dfrac{dy}{dx} = \dfrac{\cos^2 y}{\sin y},$ Art. 60.

But sec $y = x$, therefore $\cos y = \dfrac{1}{x}$, and $\sin y = \dfrac{\sqrt{(x^2-1)}}{x}$, see Art. 65, thus

$$\frac{dy}{dx} = \frac{1}{x\sqrt{(x^2-1)}}.$$

Similarly, if $\qquad y = \operatorname{cosec}^{-1}x,$

$$\frac{dy}{dx} = -\frac{1}{x\sqrt{(x^2-1)}}.$$

69. In the manner given in the preceding Articles the differential coefficients of the inverse trigonometrical functions are usually determined. They may however be found without using Art. 60.

For example, suppose

$$y = \tan^{-1}x,$$

therefore $\qquad y + \Delta y = \tan^{-1}(x+h),$

therefore $\qquad \Delta y = \tan^{-1}(x+h) - \tan^{-1}x$

$$= \tan^{-1}\frac{h}{1+x(x+h)}, \text{ by Trigonometry,}$$

therefore $\qquad \dfrac{\Delta y}{\Delta x} = \dfrac{1}{h}\tan^{-1}\dfrac{h}{1+x(x+h)}$

$$= \frac{1}{1+x^2+xh}\cdot\frac{\tan^{-1}\dfrac{h}{1+x(x+h)}}{\dfrac{h}{1+x(x+h)}}.$$

Now let h diminish without limit, then

the limit of $\dfrac{\tan^{-1}\dfrac{h}{1+x(x+h)}}{\dfrac{h}{1+x(x+h)}} = 1$, Art. 21,

therefore $\qquad \dfrac{dy}{dx} = \dfrac{1}{1+x^2}.$

70.   Again, suppose $y = \sin^{-1}x$,

therefore $y + \Delta y = \sin^{-1}(x + h)$,

therefore $\Delta y = \sin^{-1}(x + h) - \sin^{-1}x$

$$= \sin^{-1}[(x + h)\sqrt{(1 - x^2)} - x\sqrt{\{1 - (x + h)^2\}}],$$

by Trigonometry,

therefore $\dfrac{\Delta y}{\Delta x} = \dfrac{\sin^{-1}[(x + h)\sqrt{(1 - x^2)} - x\sqrt{\{1 - (x + h)^2\}}]}{h}$;

put $(x + h)\sqrt{(1 - x^2)} - x\sqrt{\{1 - (x + h)^2\}} = z$ for abbreviation,

then $\dfrac{\Delta y}{\Delta x} = \dfrac{\sin^{-1}z}{h} = \dfrac{\sin^{-1}z}{z} \cdot \dfrac{z}{h}$.

Now $\dfrac{z}{h} = \dfrac{(x + h)\sqrt{(1 - x^2)} - x\sqrt{\{1 - (x + h)^2\}}}{h}$

$$= \dfrac{(x + h)^2(1 - x^2) - x^2\{1 - (x + h)^2\}}{h[(x + h)\sqrt{(1 - x^2)} + x\sqrt{\{1 - (x + h)^2\}}]}$$

$$= \dfrac{2x + h}{(x + h)\sqrt{(1 - x^2)} + x\sqrt{\{1 - (x + h)^2\}}};$$

thus the limit of $\dfrac{z}{h}$, when $h = 0$, is $\dfrac{x}{x\sqrt{(1 - x^2)}}$ or $\dfrac{1}{\sqrt{(1 - x^2)}}$;

and the limit of $\dfrac{\sin^{-1}z}{z}$ is 1, Art. 21;   therefore

$$\frac{dy}{dx} = \frac{1}{\sqrt{(1 - x^2)}}.$$

71.   *Differential coefficient of* $\mathrm{vers}^{-1}x$.

Let $y = \mathrm{vers}^{-1}x$, therefore

$$\mathrm{vers}\, y = x,$$

therefore $1 - \cos y = x$,

therefore $\dfrac{dx}{dy} = \sin y$,

therefore $\dfrac{dy}{dx} = \dfrac{1}{\sin y} = \dfrac{1}{\sqrt{(1 - \cos^2 y)}} = \dfrac{1}{\sqrt{\{1 - (1 - x)^2\}}}$

$$= \frac{1}{\sqrt{(2x - x^2)}}.$$

72. *Differential coefficient of z^v.*

Let $y = z^v$, where v and z are both functions of x.

Take the logarithms of both members of the equation, hence

$$\log_e y = v \log_e z.$$

Now since these two functions of x are always equal, their differential coefficients with respect to x must be so.

And $\qquad \dfrac{d \log_e y}{dx} = \dfrac{d \log_e y}{dy} \dfrac{dy}{dx}$, Art. 63,

$$= \frac{1}{y} \frac{dy}{dx}, \text{ Art. 50.}$$

Also the differential coefficient of $v \log_e z$

$$= \frac{dv}{dx} \log_e z + v \frac{d \log_e z}{dx}, \text{ Art. 29,}$$

$$= \frac{dv}{dx} \log_e z + \frac{v}{z} \frac{dz}{dx}, \text{ Art. 63 ;}$$

therefore . $\qquad \dfrac{1}{y} \dfrac{dy}{dx} = \dfrac{dv}{dx} \log_e z + \dfrac{v}{z} \dfrac{dz}{dx}$,

and $\qquad \dfrac{dy}{dx} = z^v \left(\dfrac{dv}{dx} \log_e z + \dfrac{v}{z} \dfrac{dz}{dx} \right)$.

73. If we compare Arts. 29...31 with Art. 72 we may deduce a general rule for the differential coefficient of a composite function. Differentiate in order each component function, treating all the others as if they were constant; then add the results thus obtained.

It is advisable to call the attention of the student explicitly to three different cases which beginners are apt to confound.

(1) If $y = z^a$ where z is a function of x and a *is a constant*, then by Arts. 47 and 63

$$\frac{dy}{dx} = az^{a-1} \frac{dz}{dx}.$$

(2) If $y = a^z$ where z is a function of x and a *is a constant*, then by Arts. 49 and 63

$$\frac{dy}{dx} = a^z \log_e a \frac{dz}{dx}.$$

(3) If $y = z^v$ where both z and v are functions of x, then by Art. 72

$$\frac{dy}{dx} = z^v \left(\frac{dv}{dx} \log_e z + \frac{v}{z} \frac{dz}{dx} \right).$$

74. *Differential coefficient of* x^n. *Third method.* For the other methods see Arts. 47 and 48.

The differential coefficient of x^n is sometimes found thus:

First prove as in Art. 44 or 45 that if n be a positive integer, the differential coefficient of x^n is nx^{n-1}.

If then n be fractional and positive, suppose it $= \frac{p}{q}$ where p and q are positive integers.

Let $$y = x^n = x^{\frac{p}{q}},$$
therefore $$y^q = x^p.$$

Hence taking the differential coefficients of both sides

$$qy^{q-1} \frac{dy}{dx} = px^{p-1},$$

therefore $$\frac{dy}{dx} = \frac{px^{p-1}}{qy^{q-1}} = \frac{p}{q} \cdot \frac{x^{p-1}}{x^{\frac{p}{q}(q-1)}}$$

$$= \frac{p}{q} x^{\frac{p}{q}-1} = nx^{n-1}.$$

The rule is thus established so long as n is positive.

If n be negative suppose it $= -m$, so that m is positive. Let $y = x^{-m}$, therefore

$$\frac{1}{y} = x^m,$$

therefore $$1 = yx^m.$$

Differentiate both sides, and we have

$$0 = x^m \frac{dy}{dx} + ymx^{m-1}, \text{ Arts. 29 and 33,}$$

therefore $$\frac{dy}{dx} = -\frac{my}{x} = -mx^{-m-1}$$

$$= nx^{n-1}.$$

Hence the rule for differentiating x^n is universally established.

75. We shall now give some examples of the preceding rules for finding differential coefficients.

(1) Let $y = \sin ax$.

Put $ax = z$; therefore $y = \sin z$,

and $\qquad\qquad \dfrac{dy}{dx} = \dfrac{dy}{dz}\dfrac{dz}{dx}$, Art. 63.

But $\qquad\qquad \dfrac{dy}{dz} = \cos z$, Art. 51,

and $\qquad\qquad \dfrac{dz}{dx} = a$, Art. 33,

therefore $\qquad\qquad \dfrac{dy}{dx} = a\cos z = a\cos ax$.

(2) Let $y = \sin(\log x)$.

By $\log x$ without any base specified, we mean $\log_e x$.

Put $\qquad\qquad \log x = z$,

therefore $\qquad\qquad y = \sin z$,

and $\qquad\qquad \dfrac{dy}{dx} = \dfrac{dy}{dz}\dfrac{dz}{dx}$, Art. 63.

But $\qquad\qquad \dfrac{dy}{dz} = \cos z$, Art. 51,

$\qquad\qquad \dfrac{dz}{dx} = \dfrac{1}{x}$, Art. 50,

therefore $\qquad\qquad \dfrac{dy}{dx} = \dfrac{\cos z}{x} = \dfrac{\cos(\log x)}{x}$.

(3) $y = \log(\sin x)$.

Put $\qquad\qquad \sin x = z$,

therefore $\qquad\qquad y = \log z$,

and $\qquad\qquad \dfrac{dy}{dx} = \dfrac{dy}{dz}\dfrac{dz}{dx}$, Art. 63,

$\qquad\qquad = \dfrac{1}{z}\cos x = \dfrac{\cos x}{\sin x} = \cot x$.

(4) $y = \log \dfrac{a + bx}{a - bx}$.

Put $\qquad\qquad \dfrac{a + bx}{a - bx} = z,$

therefore $\quad \dfrac{dz}{dx} = \dfrac{b\,(a - bx) + b\,(a + bx)}{(a - bx)^2}$, Art. 31,

$$= \dfrac{2ab}{(a - bx)^2},$$

therefore $\quad \dfrac{dy}{dx} = \dfrac{1}{z}\,\dfrac{2ab}{(a - bx)^2} = \dfrac{2ab}{a^2 - b^2 x^2}.$

This example may also be solved by putting

$$y = \log(a + bx) - \log(a - bx),$$

therefore $\quad \dfrac{dy}{dx} = \dfrac{b}{a + bx} + \dfrac{b}{a - bx} = \dfrac{2ab}{a^2 - b^2 x^2}.$

(5) $y = \cos^{-1} \dfrac{4 - 3x^2}{x^3}$.

Put $\qquad\qquad \dfrac{4 - 3x^2}{x^3} = z,$

therefore $\qquad\qquad y = \cos^{-1} z,$

and $\qquad\qquad \dfrac{dy}{dx} = \dfrac{dy}{dz}\,\dfrac{dz}{dx}.$

Now $\quad \dfrac{dy}{dz} = -\dfrac{1}{\sqrt{(1 - z^2)}}$, Art. 66,

$$= -\dfrac{1}{\sqrt{\left\{1 - \left(\dfrac{4 - 3x^2}{x^3}\right)^2\right\}}} = \dfrac{-\,x^3}{\sqrt{(x^6 - 9x^4 + 24x^2 - 16)}};$$

$$\dfrac{dz}{dx} = \dfrac{-6x^4 - 3x^2\,(4 - 3x^2)}{x^6}, \text{ Art. 31,}$$

$$= \dfrac{3\,(x^2 - 4)}{x^4};$$

therefore $\dfrac{dy}{dx} = -\dfrac{x^3}{\sqrt{(x^6 - 9x^4 + 24x^2 - 16)}} \cdot \dfrac{3(x^2 - 4)}{x^4}$

$\qquad = \dfrac{-3(x^2 - 4)}{x\sqrt{\{(x^2 - 1)(x^2 - 4)^2\}}} = -\dfrac{3}{x\sqrt{(x^2 - 1)}}$.

In differentiating $\dfrac{4 - 3x^2}{x^3}$ we made use of the rule for finding the differential coefficient of a fraction. By putting the expression in the form

$$\frac{4}{x^3} - \frac{3}{x},$$

that is, $\qquad 4x^{-3} - 3x^{-1}$,

we obtain for the differential coefficient

$$-12x^{-4} + 3x^{-2}, \text{ Art. 47,}$$

or $\qquad \dfrac{3(x^2 - 4)}{x^4}$, as above.

It may be observed that cases of this kind frequently occur in which we may adopt more than one method. The student will find it very useful in rendering him familiar with the rules, to obtain his results, if possible, by different methods.

(6) $\quad y = \dfrac{\sqrt{\{ax(x - 3a)\}}}{\sqrt{(x - 4a)}}$.

It is often convenient to take the logarithms of both sides of an equation before differentiating. Thus, from the above, we have

$$\log y = \tfrac{1}{2}\{\log a + \log x + \log(x - 3a) - \log(x - 4a)\}.$$

Take the differential coefficient of each member of the equation, therefore

$$\frac{1}{y}\frac{dy}{dx} = \frac{1}{2}\left\{\frac{1}{x} + \frac{1}{x - 3a} - \frac{1}{x - 4a}\right\}$$

$$= \frac{x^2 - 8ax + 12a^2}{2x(x - 3a)(x - 4a)},$$

therefore $\dfrac{dy}{dx} = \dfrac{\sqrt{a} \cdot (x^2 - 8ax + 12a^2)}{2 \{x(x - 3a)\}^{\frac{1}{2}} (x - 4a)^{\frac{1}{2}}}$

(7) $y = \tan^{-1} \dfrac{x}{a}$.

Put $\dfrac{x}{a} = z$, therefore $y = \tan^{-1} z$,

therefore $\dfrac{dy}{dx} = \dfrac{1}{1 + z^2} \dfrac{dz}{dx}$

$$= \dfrac{1}{1 + \dfrac{x^2}{a^2}} \dfrac{1}{a} = \dfrac{a}{a^2 + x^2}.$$

(8) Let $y = \tan^{-1} \dfrac{3a^2 x - x^3}{a(a^2 - 3x^2)}$.

Put $\dfrac{3xa^2 - x^3}{a(a^2 - 3x^2)} = z$; therefore $y = \tan^{-1} z$,

and $\dfrac{dy}{dx} = \dfrac{dy}{dz} \dfrac{dz}{dx} = \dfrac{1}{1 + z^2} \dfrac{dz}{dx}$.

Now $\dfrac{dz}{dx} = \dfrac{3(a^2 - x^2)(a^2 - 3x^2) + 6x(3xa^2 - x^3)}{a(a^2 - 3x^2)^2}$, Art. 31,

$$= \dfrac{3(a^4 + 2a^2 x^2 + x^4)}{a(a^2 - 3x^2)^2}.$$

And by reduction we find that

$$\dfrac{1}{1 + z^2} = \dfrac{a^2 (a^2 - 3x^2)^2}{(a^2 + x^2)^3}.$$

Therefore $\dfrac{dy}{dx} = \dfrac{3a}{a^2 + x^2}$.

In fact we have from Trigonometry

$$\tan^{-1} \dfrac{3a^2 x - x^3}{a(a^2 - 3x^2)} = 3 \tan^{-1} \dfrac{x}{a},$$

and therefore the value of $\dfrac{dy}{dx}$ ought to be $\dfrac{3a}{a^2 + x^2}$

It is obvious that other self-verifying examples may be constructed on the model of this example.

(9) $\quad y = \tan^{-1}\left(\dfrac{e^x \cos x}{1 + e^x \sin x}\right)$.

Put $\qquad \dfrac{e^x \cos x}{1 + e^x \sin x} = z$,

thus $\qquad y = \tan^{-1} z$,

therefore $\qquad \dfrac{dy}{dx} = \dfrac{1}{1 + z^2}\dfrac{dz}{dx}$.

Now $\dfrac{dz}{dx} = \dfrac{(e^x \cos x - e^x \sin x)(1 + e^x \sin x) - e^x \cos x (e^x \cos x + e^x \sin x)}{(1 + e^x \sin x)^2}$

$\qquad = \dfrac{e^x (\cos x - \sin x - e^x)}{(1 + e^x \sin x)^2}$;

and $\qquad \dfrac{1}{1 + z^2} = \dfrac{(1 + e^x \sin x)^2}{1 + 2e^x \sin x + e^{2x}}$;

therefore $\qquad \dfrac{dy}{dx} = \dfrac{e^x (\cos x - \sin x - e^x)}{1 + 2e^x \sin x + e^{2x}}$.

(10) $\quad y = \sin x \tan^{-1} x \, a^x \log x$.

$\dfrac{dy}{dx} = \cos x \tan^{-1} x \, a^x \log x + \dfrac{\sin x \, a^x \log x}{1 + x^2}$

$\quad + \sin x \tan^{-1} x \, a^x \log a \log x + \dfrac{\sin x \tan^{-1} x \, a^x}{x}$. Art. 30.

76. The differential coefficients of the simple functions are here collected for the sake of reference.

$y = x^n$. $\qquad\qquad \dfrac{dy}{dx} = nx^{n-1}$.

$y = \log_a x$. $\qquad\qquad \dfrac{dy}{dx} = \dfrac{1}{x \log_e a}$.

$y = a^x$. $\qquad\qquad \dfrac{dy}{dx} = a^x \log_e a$.

$y = \sin \dfrac{x}{a}.$ $\qquad \dfrac{dy}{dx} = \dfrac{1}{a} \cos \dfrac{x}{a}.$

$y = \cos \dfrac{x}{a}.$ $\qquad \dfrac{dy}{dx} = -\dfrac{1}{a} \sin \dfrac{x}{a}.$

$y = \tan \dfrac{x}{a}.$ $\qquad \dfrac{dy}{dx} = \dfrac{1}{a} \sec^2 \dfrac{x}{a}.$

$y = \cot \dfrac{x}{a}.$ $\qquad \dfrac{dy}{dx} = -\dfrac{1}{a} \operatorname{cosec}^2 \dfrac{x}{a}.$

$y = \sec \dfrac{x}{a}.$ $\qquad \dfrac{dy}{dx} = \dfrac{1}{a} \dfrac{\sin \dfrac{x}{a}}{\cos^2 \dfrac{x}{a}}.$

$y = \operatorname{cosec} \dfrac{x}{a}.$ $\qquad \dfrac{dy}{dx} = -\dfrac{1}{a} \dfrac{\cos \dfrac{x}{a}}{\sin^2 \dfrac{x}{a}}.$

$y = \sin^{-1} \dfrac{x}{a}.$ $\qquad \dfrac{dy}{dx} = \dfrac{1}{\sqrt{(a^2 - x^2)}}.$

$y = \cos^{-1} \dfrac{x}{a}.$ $\qquad \dfrac{dy}{dx} = -\dfrac{1}{\sqrt{(a^2 - x^2)}}.$

$y = \tan^{-1} \dfrac{x}{a}.$ $\qquad \dfrac{dy}{dx} = \dfrac{a}{a^2 + x^2}.$

$y = \cot^{-1} \dfrac{x}{a}.$ $\qquad \dfrac{dy}{dx} = -\dfrac{a}{a^2 + x^2}.$

$y = \sec^{-1} \dfrac{x}{a}.$ $\qquad \dfrac{dy}{dx} = \dfrac{a}{x \sqrt{(x^2 - a^2)}}.$

$y = \operatorname{cosec}^{-1} \dfrac{x}{a}.$ $\qquad \dfrac{dy}{dx} = -\dfrac{a}{x \sqrt{(x^2 - a^2)}}.$

$y = \operatorname{vers}^{-1} \dfrac{x}{a}.$ $\qquad \dfrac{dy}{dx} = \dfrac{1}{\sqrt{(2ax - x^2)}}.$

EXAMPLES.

1. $y = c \sqrt{x}$. $\qquad$ $\dfrac{dy}{dx} = \dfrac{c}{2\sqrt{x}}$.

2. $y = \dfrac{a - x}{x}$. $\qquad$ $\dfrac{dy}{dx} = -\dfrac{a}{x^2}$.

3. $y = \dfrac{1 + x}{1 + x^2}$. $\qquad$ $\dfrac{dy}{dx} = \dfrac{1 - 2x - x^2}{(1 + x^2)^2}$.

4. $y = x \log x$. $\qquad$ $\dfrac{dy}{dx} = 1 + \log x$.

5. $y = \log \cotan x$. $\qquad$ $\dfrac{dy}{dx} = -\dfrac{2}{\sin 2x}$.

6. $y = \dfrac{x}{\sqrt{(a^2 - x^2)}}$. $\qquad$ $\dfrac{dy}{dx} = \dfrac{a^2}{(a^2 - x^2)^{\frac{3}{2}}}$.

7. $y = \dfrac{x^3}{(1 - x^2)^{\frac{3}{2}}}$. $\qquad$ $\dfrac{dy}{dx} = \dfrac{3x^2}{(1 - x^2)^{\frac{5}{2}}}$.

8. $y = e^x (1 - x^3)$. $\qquad$ $\dfrac{dy}{dx} = e^x (1 - 3x^2 - x^3)$.

9. $y = (x - 3) e^{2x} + 4xe^x + x + 3$.

$$\dfrac{dy}{dx} = (2x - 5) e^{2x} + 4 (x + 1) e^x + 1.$$

10. $y = (2x - 5) e^{2x} + 4 (x + 1) e^x + 1$.

$$\dfrac{dy}{dx} = 4e^x \{(x - 2) e^x + x + 2\}.$$

11. $y = \left(\dfrac{x}{n}\right)^{nx}$. $\qquad$ $\dfrac{dy}{dx} = n \left(\dfrac{x}{n}\right)^{nx} \left\{1 + \log \dfrac{x}{n}\right\}$.

12. $y = \dfrac{x^n}{(1 + x)^n}$. $\qquad$ $\dfrac{dy}{dx} = \dfrac{nx^{n-1}}{(1 + x)^{n+1}}$.

13. $y = \dfrac{e^x - e^{-x}}{e^x + e^{-x}}.$ $\qquad \dfrac{dy}{dx} = \dfrac{4}{(e^x + e^{-x})^2}.$

14. $y = \log (e^x + e^{-x}).$ $\qquad \dfrac{dy}{dx} = \dfrac{e^x - e^{-x}}{e^x + e^{-x}}.$

15. $y = x^2 (a + x)^3 (b - x)^4.$

$\qquad \dfrac{dy}{dx} = \{2ab - (6a - 5b) x - 9x^2\} x (a + x)^2 (b - x)^3.$

16. $y = (a + x)^m (b + x)^n.$

$\qquad \dfrac{dy}{dx} = (a + x)^{m-1} (b + x)^{n-1} \{m (b + x) + n (a + x)\}.$

17. $y = \dfrac{1}{(a + x)^m} \dfrac{1}{(b + x)^n}.$ $\qquad \dfrac{dy}{dx} = -\dfrac{m (b + x) + n (a + x)}{(a + x)^{m+1} (b + x)^{n+1}}.$

18. $y = \dfrac{\tan^3 x}{3} - \tan x + x.$ $\qquad \dfrac{dy}{dx} = \tan^4 x.$

19. $y = \dfrac{1}{x + \sqrt{(1 - x^2)}}.$ $\qquad \dfrac{dy}{dx} = \dfrac{x - \sqrt{(1 - x^2)}}{\sqrt{(1 - x^2)}\{1 + 2x \sqrt{(1 - x^2)}\}}.$

20. $y = (a^2 + x^2) \tan^{-1} \dfrac{x}{a}.$ $\qquad \dfrac{dy}{dx} = 2x \tan^{-1} \dfrac{x}{a} + a.$

21. $y = \sqrt{\left(a + \dfrac{b}{x} + \dfrac{c}{x^2}\right)}.$ $\qquad \dfrac{dy}{dx} = -\dfrac{1}{2x^2} \dfrac{bx + 2c}{\sqrt{(ax^2 + bx + c)}}.$

22. $y = \log \{\log (a + bx^n)\}.$ $\qquad \dfrac{dy}{dx} = \dfrac{nbx^{n-1}}{(a + bx^n) \log (a + bx^n)}.$

23. $y = \log \tan \left(\dfrac{\pi}{4} + \dfrac{x}{2}\right).$ $\qquad \dfrac{dy}{dx} = \dfrac{1}{\cos x}.$

24. $y = e^{(a+x)^2} \sin x.$ $\qquad \dfrac{dy}{dx} = e^{(a+x)^2} \{2 (a + x) \sin x + \cos x\}.$

25. $y = \dfrac{\sqrt{(a + x)}}{\sqrt{a} + \sqrt{x}}.$ $\qquad \dfrac{dy}{dx} = \dfrac{\sqrt{a} (\sqrt{x} - \sqrt{a})}{2 \sqrt{x} \sqrt{(a + x)} (\sqrt{a} + \sqrt{x})^2}.$

26. $y = \sqrt{\left(\dfrac{1 + x}{1 - x}\right)}.$ $\qquad \dfrac{dy}{dx} = \dfrac{1}{\sqrt{(1 - x^2)} (1 - x)}.$

27. $y = \sqrt{\left\{\dfrac{1-x^2}{(1+x^2)^3}\right\}}$. $\dfrac{dy}{dx} = \dfrac{-2x(2-x^2)}{(1-x^2)^{\frac{1}{2}}(1+x^2)^{\frac{5}{2}}}$.

28. $y = \dfrac{x}{e^x - 1}$. $\dfrac{dy}{dx} = \dfrac{e^x(1-x)-1}{(e^x-1)^2}$.

29. $y = e^x\dfrac{(x-2)e^x + x + 2}{(e^x-1)^3}$.

$\dfrac{dy}{dx} = -\dfrac{e^x}{(e^x-1)^4}\{(x-3)e^{2x} + 4xe^x + x + 3\}$.

30. $y = \log\dfrac{\sqrt{(1+x)}+\sqrt{(1-x)}}{\sqrt{(1+x)}-\sqrt{(1-x)}}$. $\dfrac{dy}{dx} = -\dfrac{1}{x\sqrt{(1-x^2)}}$.

31. $y = \{x+\sqrt{(1-x^2)}\}^n$. $\dfrac{dy}{dx} = n\{x+\sqrt{(1-x^2)}\}^{n-1}\dfrac{\sqrt{(1-x^2)}-x}{\sqrt{(1-x^2)}}$

32. $y = \left\{\dfrac{x}{1+\sqrt{(1-x^2)}}\right\}^n$. $\dfrac{dy}{dx} = \dfrac{ny}{x\sqrt{(1-x^2)}}$.

33. $y = \dfrac{x}{\sqrt{(1-x^2)}}\left\{\dfrac{x}{1+\sqrt{(1-x^2)}}\right\}^n$.

$\dfrac{dy}{dx} = \left\{\dfrac{x}{1+\sqrt{(1-x^2)}}\right\}^n\dfrac{1+n\sqrt{(1-x^2)}}{(1-x^2)^{\frac{3}{2}}}$.

34. $y = a^{\sqrt{(a^2-x^2)}}$. $\dfrac{dy}{dx} = \dfrac{xy}{(a^2-x^2)^{\frac{1}{2}}}\log_e a$.

35. $y = \tan a^{\frac{1}{x}}$. $\dfrac{dy}{dx} = -\dfrac{\sec^2 a^{\frac{1}{x}}}{x^2}\log_e a\,.\,a^{\frac{1}{x}}$.

36. $y = \log\{\sqrt{(1+x^2)}+\sqrt{(1-x^2)}\}$. $\dfrac{dy}{dx} = \dfrac{1}{x}\left\{1-\dfrac{1}{\sqrt{(1-x^4)}}\right\}$.

37. $y = (2a^{\frac{1}{4}}+x^{\frac{1}{4}})\sqrt{(a^{\frac{1}{4}}+x^{\frac{1}{4}})}$. $\dfrac{dy}{dx} = \dfrac{4a^{\frac{1}{2}}+3x^{\frac{1}{2}}}{4\sqrt{x}\sqrt{(a^{\frac{1}{2}}+x^{\frac{1}{2}})}}$.

38. $y = x+\log\cos\left(\dfrac{\pi}{4}-x\right)$. $\dfrac{dy}{dx} = \dfrac{2}{1+\tan x}$.

39. $y = \dfrac{\sqrt{(1+x^2)} + \sqrt{(1-x^2)}}{\sqrt{(1+x^2)} - \sqrt{(1-x^2)}}$. $\dfrac{dy}{dx} = -\dfrac{2}{x^3}\left\{1 + \dfrac{1}{\sqrt{(1-x^4)}}\right\}$.

40. $y = x \sin^{-1}x$. $\dfrac{dy}{dx} = \sin^{-1}x + \dfrac{x}{\sqrt{(1-x^2)}}$.

41. $y = \tan x \tan^{-1}x$. $\dfrac{dy}{dx} = \sec^2 x \tan^{-1}x + \dfrac{\tan x}{1+x^2}$.

42. $y = \sin nx (\sin x)^n$. $\dfrac{dy}{dx} = n (\sin x)^{n-1}\sin (n+1)x$.

43. $y = \dfrac{(\sin nx)^m}{(\cos mx)^n}$. $\dfrac{dy}{dx} = \dfrac{mn (\sin nx)^{m-1} \cos (mx - nx)}{(\cos mx)^{n+1}}$.

44. $y = e^{-a^2x^2} \cos rx$. $\dfrac{dy}{dx} = -e^{-a^2x^2}(2a^2x\cos rx + r \sin rx)$.

45. $y = \dfrac{x - \sin^{-1}x}{(\sin x)^3}$.

$$\dfrac{dy}{dx} = \dfrac{\sin x\left\{1 - \dfrac{1}{\sqrt{(1-x^2)}}\right\} - 3 (x - \sin^{-1}x) \cos x}{(\sin x)^4}.$$

46. $y = \log\left\{\dfrac{a + b \tan \dfrac{x}{2}}{a - b \tan \dfrac{x}{2}}\right\}$. $\dfrac{dy}{dx} = \dfrac{ab}{a^2 \cos^2 \dfrac{x}{2} - b^2 \sin^2 \dfrac{x}{2}}$.

47. $y = x^x$. $\dfrac{dy}{dx} = x^x (1 + \log x)$.

48. $y = x^{\frac{1}{x}}$. $\dfrac{dy}{dx} = \dfrac{x^{\frac{1}{x}} (1 - \log x)}{x^2}$.

49. $y = x^{\sin^{-1}x}$. $\dfrac{dy}{dx} = x^{\sin^{-1}x}\left\{\dfrac{\sin^{-1}x}{x} + \dfrac{\log x}{\sqrt{(1-x^2)}}\right\}$.

50. $y = e^{e^x}$. $\dfrac{dy}{dx} = e^{e^x} e^x$.

51. $y = e^{x^x}$. $\dfrac{dy}{dx} = e^{x^x} x^x (1 + \log x)$.

52. $y = x^{x^x}$. $\dfrac{dy}{dx} = yx^x \left\{ \dfrac{1}{x} + \log x + (\log x)^2 \right\}$.

53. $y = x^{e^x}$. $\dfrac{dy}{dx} = x^{e^x} e^x \dfrac{1 + x \log x}{x}$.

54. $y = \tan^{-1} \dfrac{2x}{1 + x^2}$. $\dfrac{dy}{dx} = \dfrac{2(1 - x^2)}{1 + 6x^2 + x^4}$.

55. $y = \sin^{-1} \dfrac{x + 1}{\sqrt{2}}$. $\dfrac{dy}{dx} = \dfrac{1}{\sqrt{(1 - 2x - x^2)}}$.

56. $y = \tan \sqrt{(1 - x)}$. $\dfrac{dy}{dx} = \dfrac{-\{\sec \sqrt{(1 - x)}\}^2}{2\sqrt{(1 - x)}}$.

57. $y = \tan^{-1} \dfrac{x}{\sqrt{(1 - x^2)}}$. $\dfrac{dy}{dx} = \dfrac{1}{\sqrt{(1 - x^2)}}$.

58. $y = \tan^{-1}(n \tan x)$. $\dfrac{dy}{dx} = \dfrac{n}{\cos^2 x + n^2 \sin^2 x}$.

59. $y = \sec^{-1} \dfrac{a}{\sqrt{(a^2 - x^2)}}$. $\dfrac{dy}{dx} = \dfrac{1}{\sqrt{(a^2 - x^2)}}$.

60. $y = (x + a) \tan^{-1} \sqrt{\dfrac{x}{a}} - \sqrt{(ax)}$. $\dfrac{dy}{dx} = \tan^{-1} \sqrt{\dfrac{x}{a}}$.

61. $y = \tan^{-1} \dfrac{x}{a} + \log \left(\dfrac{x - a}{x + a} \right)^{\frac{1}{4}}$. $\dfrac{dy}{dx} = \dfrac{2ax^2}{x^4 - a^4}$.

62. $y = \sin^{-1} \sqrt{(\sin x)}$. $\dfrac{dy}{dx} = \frac{1}{2} \sqrt{(1 + \operatorname{cosec} x)}$.

63. $y = \tan^{-1} \dfrac{2x}{1 - x^2}$. $\dfrac{dy}{dx} = \dfrac{2}{1 + x^2}$.

64. $y = \sin^{-1} \dfrac{ax}{b + cx^2}$. $\dfrac{dy}{dx} = \dfrac{a(b - cx^2)}{\sqrt{\{b^2 + (2bc - a^2)x^2 + c^2 x^4\}}} \cdot \dfrac{1}{b + cx^2}$.

65. $y = \sqrt{(1 - x^2)} \sin^{-1} x - x$. $\dfrac{dy}{dx} = -\dfrac{x \sin^{-1} x}{\sqrt{(1 - x^2)}}$.

66. $y = \dfrac{x \sin^{-1} x}{\sqrt{(1 - x^2)}} + \log \sqrt{(1 - x^2)}$. $\dfrac{dy}{dx} = \dfrac{\sin^{-1} x}{(1 - x^2)^{\frac{3}{2}}}$.

67. $y = \tan^{-1}\{x + \sqrt{(1-x^2)}\}.$ $\quad \dfrac{dy}{dx} = \dfrac{\sqrt{(1-x^2)}-x}{2\sqrt{(1-x^2)}\{1+x\sqrt{(1-x^2)}\}}.$

68. $y = \sin^{-1}\dfrac{x\tan\alpha}{\sqrt{(a^2-x^2)}}.$ $\quad \dfrac{dy}{dx} = \dfrac{a^2\tan\alpha}{a^2-x^2}\cdot\dfrac{1}{\sqrt{(a^2-x^2\sec^2\alpha)}}.$

69. $y = \sin^{-1}\sqrt{\left(\dfrac{a^2-x^2}{b^2-x^2}\right)}.$ $\quad \dfrac{dy}{dx} = -\dfrac{x\sqrt{(b^2-a^2)}}{(b^2-x^2)\sqrt{(a^2-x^2)}}.$

70. $y = \tan^{-1}\sqrt{\left(\dfrac{1-\cos x}{1+\cos x}\right)}.$ $\quad \dfrac{dy}{dx} = \dfrac{1}{2}.$

71. $y = \sin^{-1}\dfrac{b+a\cos x}{a+b\cos x}.$ $\quad \dfrac{dy}{dx} = \dfrac{-\sqrt{(a^2-b^2)}}{a+b\cos x}.$

72. $y = \tan^{-1}\left\{\dfrac{\sqrt{(a^2-b^2)}\sin x}{b+a\cos x}\right\}.$ $\quad \dfrac{dy}{dx} = \dfrac{\sqrt{(a^2-b^2)}}{a+b\cos x}.$

73. $y = \cos^{-1}\dfrac{x^{2n}-1}{x^{2n}+1}.$ $\quad \dfrac{dy}{dx} = -\dfrac{2nx^{n-1}}{x^{2n}+1}.$

74. $y = \sec^{-1}\dfrac{1}{2x^2-1}.$ $\quad \dfrac{dy}{dx} = -\dfrac{2}{\sqrt{(1-x^2)}}.$

75. $y = \tan^{-1}\dfrac{\sqrt{(1+x^2)}-1}{x}.$ $\quad \dfrac{dy}{dx} = \dfrac{1}{2(1+x^2)}.$

76. $y = \log\dfrac{1+x\sqrt{2}+x^2}{1-x\sqrt{2}+x^2} + 2\tan^{-1}\dfrac{x\sqrt{2}}{1-x^2}.$ $\quad \dfrac{dy}{dx} = \dfrac{4\sqrt{2}}{1+x^4}.$

77. If $u = \tfrac{1}{6}\log\dfrac{(y+1)^2}{y^2-y+1} - \dfrac{1}{\sqrt{3}}\tan^{-1}\dfrac{2y-1}{\sqrt{3}},$

where $\qquad y = \dfrac{\sqrt[3]{(1+3x+3x^2)}}{x},$

shew that $\qquad \dfrac{du}{dx} = \dfrac{1}{xy(1+x)}.$

78.   Given $\sin x + \sin 2x + \ldots + \sin nx = \dfrac{\sin \dfrac{n+1}{2}\, x \sin \dfrac{nx}{2}}{\sin \dfrac{x}{2}}$,

deduce, by taking the differential coefficients of both sides, the sum of

$$\cos x + 2 \cos 2x + \ldots + n \cos nx.$$

Result. $\dfrac{\dfrac{n+1}{2} \sin \dfrac{x}{2} \sin \dfrac{2n+1}{2}\, x - \dfrac{1}{2} \left(\sin \dfrac{n+1}{2}\, x \right)^2}{\sin^2 \dfrac{x}{2}}.$

79.   Having given (see *Plane Trigonometry*, Chap. XXIII.)

$$\sin x \sin \left(\frac{\pi}{m} + x \right) \sin \left(\frac{2\pi}{m} + x \right) \ldots \sin \left(\frac{m-1}{m} \pi + x \right) = \frac{\sin mx}{2^{m-1}},$$

where m is a positive integer, shew that

$$\cot x + \cot \left(\frac{\pi}{m} + x \right) + \ldots + \cot \left(\frac{m-1}{m} \pi + x \right) = m \cot mx.$$

80.   From the preceding result deduce that

$$\mathrm{cosec}^2\, x + \mathrm{cosec}^2 \left(\frac{\pi}{m} + x \right) + \ldots + \mathrm{cosec}^2 \left(\frac{m-1}{m} \pi + x \right)$$

$$= m^2 \,\mathrm{cosec}^2\, mx.$$

CHAPTER V.

SUCCESSIVE DIFFERENTIATION.

77. In the preceding Chapters we have shewn how from any given function of a variable another function may be deduced, called the differential coefficient of the first. This second function, by the same rules, has *its* differential coefficient, which is called the *second differential coefficient* of the original function.

Thus, if $y = x^n$, we have $\dfrac{dy}{dx} = nx^{n-1}$. The differential coefficient of nx^{n-1} with respect to x is $n(n-1)x^{n-2}$, which is therefore the *second* differential coefficient of y or x^n with respect to x. The second differential coefficient of y with respect to x is denoted by $\dfrac{d^2y}{dx^2}$, which is to be considered as an abbreviation for $\dfrac{d\dfrac{dy}{dx}}{dx}$.

What we said of $\dfrac{dy}{dx}$ in Art. 26, we now say of $\dfrac{d^2y}{dx^2}$, that it is to be looked upon as a *whole symbol, not admitting of decomposition into a numerator d^2y and a denominator dx^2*.

As $\dfrac{d^2y}{dx^2}$ will be generally a function of x it will have *its* differential coefficient with respect to x. This is called the third differential coefficient of y with respect to x, and is denoted by $\dfrac{d^3y}{dx^3}$, as an abbreviation for $\dfrac{d\dfrac{d^2y}{dx^2}}{dx}$.

This process and notation may be carried on to any extent.

The successive differential coefficients of a function are often conveniently denoted by accents on the function. Thus, if $\phi(x)$ be any function of x, then $\phi'(x)$, $\phi''(x)$, $\phi'''(x)$, $\phi^{IV}(x)$, ... denote the first, second, third, fourth, ... differential coefficients of $\phi(x)$ with respect to x.

78. In some cases the nth differential coefficient of a function admits of a simple algebraical expression. For example, suppose

$$y = \sin x ;$$

therefore

$$\frac{dy}{dx} = \cos x = \sin\left(x + \frac{\pi}{2}\right),$$

$$\frac{d^2y}{dx^2} = \frac{d \sin\left(x + \frac{\pi}{2}\right)}{dx} = \cos\left(x + \frac{\pi}{2}\right)$$

$$= \sin\left(x + \frac{2\pi}{2}\right),$$

so

$$\frac{d^3y}{dx^3} = \sin\left(x + \frac{3\pi}{2}\right),$$

and generally $\dfrac{d^ny}{dx^n} = \sin\left(x + \dfrac{n\pi}{2}\right).$

So also, if $\qquad y = \sin ax,$

$$\frac{d^ny}{dx^n} = a^n \sin\left(ax + \frac{n\pi}{2}\right).$$

In like manner, if

$$y = \cos x,$$

$$\frac{d^ny}{dx^n} = \cos\left(x + \frac{n\pi}{2}\right);$$

and if $\qquad y = \cos ax, \qquad \dfrac{d^ny}{dx^n} = a^n \cos\left(ax + \dfrac{n\pi}{2}\right).$

79. Suppose $\qquad y = a^x$;

therefore $\qquad \dfrac{dy}{dx} = a^x \log a,$

$$\dfrac{d^2y}{dx^2} = a^x (\log a)^2,$$

and $\qquad \dfrac{d^ny}{dx^n} = a^x (\log a)^n.$

Similarly, if $\quad y = e^{ax}, \quad \dfrac{d^ny}{dx^n} = a^n e^{ax}.$

If $\qquad\qquad y = \log x,$

$$\dfrac{dy}{dx} = \dfrac{1}{x} = x^{-1},$$

$$\dfrac{d^2y}{dx^2} = - x^{-2},$$

$$\dfrac{d^3y}{dx^3} = 2x^{-3},$$

and $\qquad \dfrac{d^ny}{dx^n} = \dfrac{\lfloor n-1 \, (-1)^{n-1}}{x^n},$

where $\lfloor n-1$ stands for $1 . 2 . 3 \ldots (n - 1)$.

80. *Differential coefficient of the product of two functions.*

Suppose $\qquad\qquad u = yz,$

where y and z are functions of x; we have

$$\dfrac{du}{dx} = y \dfrac{dz}{dx} + \dfrac{dy}{dx} z.$$

Differentiating both sides of the equation with respect to x, we have

$$\dfrac{d^2u}{dx^2} = y \dfrac{d^2z}{dx^2} + \dfrac{dy}{dx} \dfrac{dz}{dx} + \dfrac{dy}{dx} \dfrac{dz}{dx} + \dfrac{d^2y}{dx^2} z$$

$$= y \dfrac{d^2z}{dx^2} + 2 \dfrac{dy}{dx} \dfrac{dz}{dx} + \dfrac{d^2y}{dx^2} z.$$

Similarly

$$\frac{d^3u}{dx^3} = y\frac{d^3z}{dx^3} + \frac{dy}{dx}\frac{d^2z}{dx^2} + 2\frac{dy}{dx}\frac{d^2z}{dx^2} + 2\frac{d^2y}{dx^2}\frac{dz}{dx} + \frac{d^2y}{dx^2}\frac{dz}{dx} + \frac{d^3y}{dx^3}z,$$

$$= y\frac{d^3z}{dx^3} + 3\frac{dy}{dx}\frac{d^2z}{dx^2} + 3\frac{d^2y}{dx^2}\frac{dz}{dx} + \frac{d^3y}{dx^3}z.$$

So far, then, as we have proceeded, the numerical coefficients follow the same law as those of the Binomial Theorem. We may prove by the method of Induction that such will always be the case. For assume

$$\frac{d^nu}{dx^n} = y\frac{d^nz}{dx^n} + n\frac{dy}{dx}\frac{d^{n-1}z}{dx^{n-1}} + \frac{n(n-1)}{1.2}\frac{d^2y}{dx^2}\frac{d^{n-2}z}{dx^{n-2}} + \ldots$$

$$+ \frac{n(n-1)\ldots(n-r+1)}{\underline{r}}\frac{d^ry}{dx^r}\frac{d^{n-r}z}{dx^{n-r}}$$

$$+ \frac{n(n-1)\ldots(n-r)}{\underline{r+1}}\frac{d^{r+1}y}{dx^{r+1}}\frac{d^{n-r-1}z}{dx^{n-r-1}} + \ldots + \frac{d^ny}{dx^n}z \ldots.(1).$$

Differentiate both sides with respect to x: then

$$\frac{d^{n+1}u}{dx^{n+1}} = y\frac{d^{n+1}z}{dx^{n+1}} + \frac{dy}{dx}\frac{d^nz}{dx^n} + n\frac{dy}{dx}\frac{d^nz}{dx^n} + n\frac{d^2y}{dx^2}\frac{d^{n-1}z}{dx^{n-1}} + \ldots$$

$$+ \frac{n(n-1)\ldots(n-r+1)}{\underline{r}}\left\{\frac{d^ry}{dx^r}\frac{d^{n-r+1}z}{dx^{n-r+1}} + \frac{d^{r+1}y}{dx^{r+1}}\frac{d^{n-r}z}{dx^{n-r}}\right\}$$

$$+ \frac{n(n-1)\ldots(n-r)}{\underline{r+1}}\left\{\frac{d^{r+1}y}{dx^{r+1}}\frac{d^{n-r}z}{dx^{n-r}} + \frac{d^{r+2}y}{dx^{r+2}}\frac{d^{n-r-1}z}{dx^{n-r-1}}\right\}$$

$$+ \ldots\ldots + \frac{d^ny}{dx^n}\frac{dz}{dx} + \frac{d^{n+1}y}{dx^{n+1}}z \ldots\ldots\ldots (2).$$

Rearranging the terms, we have

$$\frac{d^{n+1}u}{dx^{n+1}} = y\frac{d^{n+1}z}{dx^{n+1}} + (n+1)\frac{dy}{dx}\frac{d^nz}{dx^n} + \ldots$$

$$+ \frac{(n+1)n\ldots(n+1-r)}{\underline{r+1}}\frac{d^{r+1}y}{dx^{r+1}}\frac{d^{n-r}z}{dx^{n-r}}$$

$$+ \ldots\ldots + \frac{d^{n+1}y}{dx^{n+1}}z \ldots\ldots(3).$$

Now the series (3) follows the same law as (1). Hence if for any value of n the formula in (1) is true, it is true also for the next greater value of n. But we have proved that it holds when $n = 3$; therefore it holds when $n = 4$, therefore when $n = 5$, and so on; that is, it is universally true.

This theorem is called after the name of its discoverer, Leibnitz.

81. If $u = e^{ax} \cos bx$; we have by Arts. 78 and 80,

$$\frac{d^n u}{dx^n} = e^{ax} \left\{ a^n \cos bx + nba^{n-1} \cos\left(bx + \frac{\pi}{2}\right) + \frac{n(n-1)}{1 \cdot 2} a^{n-2} b^2 \cos\left(bx + \frac{2\pi}{2}\right) \right.$$
$$\left. + \ldots\ldots + b^n \cos\left(bx + \frac{n\pi}{2}\right) \right\}.$$

We may also find another form for this n^{th} differential coefficient as follows:

$$\frac{du}{dx} = e^{ax}(a \cos bx - b \sin bx);$$

assume
$$a = r \cos \phi,$$
$$b = r \sin \phi,$$

so that
$$r = (a^2 + b^2)^{\frac{1}{2}},$$

thus
$$\frac{du}{dx} = re^{ax} \cos(bx + \phi),$$

where r and ϕ are constant quantities.

Similarly $\dfrac{d^2 u}{dx^2} = re^{ax} \{a \cos(bx + \phi) - b \sin(bx + \phi)\}$

$$= r^2 e^{ax} \cos(bx + 2\phi),$$

and generally

$$\frac{d^n e^{ax} \cos bx}{dx^n} = r^n e^{ax} \cos(bx + n\phi).$$

82. The following is an important example of Art. 80.

Let
$$u = e^{ax} y;$$

then, remembering that $\dfrac{d^n e^{ax}}{dx^n} = a^n e^{ax}$, we have

$$\frac{d^n u}{dx^n} = e^{ax} \left\{ a^n y + na^{n-1} \frac{dy}{dx} + \frac{n(n-1)}{1 \cdot 2} a^{n-2} \frac{d^2 y}{dx^2} + \ldots + \frac{d^n y}{dx^n} \right\} \ldots (1).$$

If now the expression

$$\left(a + \frac{d}{dx}\right)^{n} y$$

be expanded by the Binomial Theorem, and the symbols

$$\left(\frac{d}{dx}\right) y, \quad \left(\frac{d}{dx}\right)^{2} y, \quad \left(\frac{d}{dx}\right)^{3} y, \ldots$$

replaced by

$$\frac{dy}{dx}, \quad \frac{d^{2}y}{dx^{2}}, \quad \frac{d^{3}y}{dx^{3}}, \ldots \text{ respectively,}$$

the result will be the same as the series in parentheses in (1).

Hence, we may write

$$\frac{d^{n}(e^{ax}y)}{dx^{n}} = e^{ax}\left(a + \frac{d}{dx}\right)^{n} y \quad\ldots\ldots\ldots\ldots\ldots\ldots (2),$$

as a convenient abbreviated method of stating the equation (1).

83. The following theorem is sometimes of use in the higher branches of mathematics.

If n be any positive integer

$$v\frac{d^{n}u}{dx^{n}} = \frac{d^{n}uv}{dx^{n}} - n\frac{d^{n-1}}{dx^{n-1}}\left(u\frac{dv}{dx}\right) + \frac{n(n-1)}{1 \cdot 2}\frac{d^{n-2}}{dx^{n-2}}\left(u\frac{d^{2}v}{dx^{2}}\right)$$

$$- \ldots\ldots\ldots\ldots + (-1)^{n} u\frac{d^{n}v}{dx^{n}} \ldots\ldots\ldots\ldots (1).$$

This theorem may be readily established by Induction. For it is obviously true when $n = 1$, and if we assume it to be true for a specific value of n we can shew that it will be true when n is changed into $n + 1$. Assume that (1) is true and differentiate both sides ; thus

$$v\frac{d^{n+1}u}{dx^{n+1}} + \frac{dv}{dx}\frac{d^{n}u}{dx^{n}} = \frac{d^{n+1}uv}{dx^{n+1}} - n\frac{d^{n}}{dx^{n}}\left(u\frac{dv}{dx}\right) + \frac{n(n-1)}{1 \cdot 2}\frac{d^{n-1}}{dx^{n-1}}\left(u\frac{d^{2}v}{dx^{2}}\right)$$

$$- \ldots\ldots\ldots\ldots + (-1)^{n}\frac{d}{dx}\left(u\frac{d^{n}v}{dx^{n}}\right) \ldots\ldots\ldots\ldots (2).$$

Also since the theorem is supposed to hold for the value n we have from (1), by changing v into $\dfrac{dv}{dx}$,

$$\frac{dv}{dx}\frac{d^n u}{dx^n} = \frac{d^n}{dx^n}\left(u\frac{dv}{dx}\right) - n\frac{d^{n-1}}{dx^{n-1}}\left(u\frac{d^2 v}{dx^2}\right) + \frac{n(n-1)}{1.2}\frac{d^{n-2}}{dx^{n-2}}\left(u\frac{d^3 v}{dx^3}\right)$$
$$- \ldots\ldots\ldots\ldots + (-1)^n u\frac{d^{n+1}v}{dx^{n+1}}\ldots\ldots\ldots (3).$$

Now suppose the right-hand members of (2) and (3) written so that the first term of (3) is immediately under the second term of (2), the second term of (3) under the third term of (2), and so on. Then by subtracting we have

$$v\frac{d^{n+1}u}{dx^{n+1}} = \frac{d^{n+1}uv}{dx^{n+1}} - (n+1)\frac{d^n}{dx^n}\left(u\frac{dv}{dx}\right) + \frac{(n+1)n}{1.2}\frac{d^{n-1}}{dx^{n-1}}\left(u\frac{d^2 v}{dx^2}\right)$$
$$- \ldots\ldots\ldots\ldots + (-1)^{n+1}u\frac{d^{n+1}v}{dx^{n+1}}.$$

This shews that if the theorem is true for a specific value of n it is also true when n is changed into $n+1$. Therefore since it is true when $n=1$ it is universally true.

EXAMPLES.

1. If $y = \tan x + \sec x,\qquad \dfrac{d^2 y}{dx^2} = \dfrac{\cos x}{(1-\sin x)^2}.$

2. Let $y = \sin^3 x = \dfrac{3\sin x - \sin 3x}{4},$

   then $\dfrac{d^n y}{dx^n} = \dfrac{3}{4}\sin\left(x + \dfrac{n\pi}{2}\right) - \dfrac{3^n}{4}\sin\left(3x + \dfrac{n\pi}{2}\right).$

3. If $y = x^2 \log x,\qquad \dfrac{d^3 y}{dx^3} = \dfrac{\lfloor 2}{x}.$

4. If $y = x^3 \log x,\qquad \dfrac{d^4 y}{dx^4} = \dfrac{\lfloor 3}{x}.$

5. If $y = (x^2 + a^2)\tan^{-1}\dfrac{x}{a},\qquad \dfrac{d^3 y}{dx^3} = \dfrac{4a^3}{(a^2 + x^2)^2}.$

6.   If $y = e^{-x} \cos x$, $\dfrac{d^4y}{dx^4} + 4y = 0$.

7.   If $y = \sqrt{\left(\dfrac{x^3}{x-a}\right)}$, $\dfrac{d^2y}{dx^2} = \dfrac{3a^2}{4\sqrt{x}\,(x-a)^{\frac{5}{2}}}$.

8.   If $y = \{x + \sqrt{(x^2-1)}\}^n$, $(x^2-1)\dfrac{d^2y}{dx^2} + x\dfrac{dy}{dx} - n^2y = 0$.

9.   If $y = x^{n-1} \log x$, $\dfrac{d^ny}{dx^n} = \dfrac{\lfloor n-1}{x}$.

10.   If $y = \dfrac{1-x}{1+x}$, $\dfrac{d^ny}{dx^n} = \dfrac{2(-1)^n\lfloor n}{(1+x)^{n+1}}$.

11.   If $u_n = (e^x + e^{-x})^n$, $\dfrac{d^2u_n}{dx^2} = n^2u_n - 4n(n-1)u_{n-2}$.

12.   If $y = e^{2\sqrt{x}}$, $\dfrac{d^2y}{dx^2} = \dfrac{2\sqrt{x}-1}{2x\sqrt{x}}e^{2\sqrt{x}}$.

13.   If $y = \dfrac{x^3}{1-x}$, $\dfrac{d^4y}{dx^4} = \dfrac{24}{(1-x)^5}$.

14.   If $y^2 = \sec 2x$, $y + \dfrac{d^2y}{dx^2} = 3y^5$.

15.   If $y^2(1+x^2) = (1-x+x^2)^3$, $\dfrac{d^2y}{dx^2} = \dfrac{1+3x+x^2}{(1+x^2)^{\frac{3}{2}}}$.

16.   If $y = \dfrac{ax+b}{x^2-c^2}$, $\dfrac{d^ny}{dx^n} = \dfrac{(-1)^n\lfloor n}{2c}\left\{\dfrac{b+ac}{(x-c)^{n+1}} - \dfrac{b-ac}{(x+c)^{n+1}}\right\}$.

17.   If $y = x^n \sin x$,

$$\dfrac{d^ny}{dx^n} = \lfloor n\left\{\sin x + \dfrac{n}{1}x\sin\left(x+\dfrac{\pi}{2}\right) + \dfrac{n(n-1)}{\lfloor 2\,\lfloor 2}x^2\sin\left(x+\dfrac{2\pi}{2}\right)\right.$$

$$\left. + \dfrac{n(n-1)(n-2)}{\lfloor 3\,\lfloor 3}x^3\sin\left(x+\dfrac{3\pi}{2}\right) + \ldots\right\}.$$

18.   If $\dfrac{y}{a} = \tan^{-1}\dfrac{x}{a}$,

then $\dfrac{dy}{dx} = \dfrac{a^2}{a^2+x^2} = \cos^2\dfrac{y}{a}$,

hence
$$\frac{d^2y}{dx^2} = -\frac{2}{a}\cos\frac{y}{a}\sin\frac{y}{a}\frac{dy}{dx}$$

$$= -\frac{1}{a}\sin\frac{2y}{a}\frac{dy}{dx}$$

$$= \frac{1}{a}\cos\left(\frac{2y}{a}+\frac{\pi}{2}\right)\cos^2\frac{y}{a}.$$

Shew that $\dfrac{d^3y}{dx^3} = \dfrac{2}{a^2}\cos\left(\dfrac{3y}{a}+2\cdot\dfrac{\pi}{2}\right)\cos^3\dfrac{y}{a}$,

and generally that $\dfrac{d^ny}{dx^n} = \dfrac{\lfloor n-1}{a^{n-1}}\cos\left\{\dfrac{ny}{a}+(n-1)\dfrac{\pi}{2}\right\}\cos^n\dfrac{y}{a}$.

Now $\tan^{-1}\dfrac{x}{a} = \dfrac{\pi}{2} - \tan^{-1}\dfrac{a}{x} = \dfrac{\pi}{2} - \theta$ suppose;

thus $\cos\left\{\dfrac{ny}{a}+(n-1)\dfrac{\pi}{2}\right\} = \sin\left(\dfrac{ny}{a}+\dfrac{n\pi}{2}\right) = \sin(n\pi - n\theta)$

$$= (-1)^{n-1}\sin n\theta; \text{ and } \cos^n\frac{y}{a} = \frac{a^n}{(a^2+x^2)^{\frac{n}{2}}};$$

therefore $\dfrac{d^ny}{dx^n} = a(-1)^{n-1}\dfrac{\lfloor n-1}{(a^2+x^2)^{\frac{n}{2}}}\sin n\theta.$

19. Since

$$\frac{d\tan^{-1}\dfrac{x}{a}}{dx} = \frac{a}{a^2+x^2}, \quad \frac{d^n}{dx^n}\left(\frac{1}{a^2+x^2}\right) = \frac{1}{a}\frac{d^{n+1}\tan^{-1}\left(\dfrac{x}{a}\right)}{dx^{n+1}}.$$

Hence, shew that

$$\frac{d^n}{dx^n}\left(\frac{1}{a^2+x^2}\right) = \frac{(-1)^n\lfloor n\sin(n+1)\theta}{a(a^2+x^2)^{\frac{n+1}{2}}},$$

where $\tan\theta = \dfrac{a}{x}.$

The nth differential coefficient of $\dfrac{1}{a^2 + x^2}$ with respect to x is sometimes obtained thus:

$$\frac{1}{a^2 + x^2} = \frac{1}{2a\sqrt{(-1)}}\left\{\frac{1}{x - a\sqrt{(-1)}} - \frac{1}{x + a\sqrt{(-1)}}\right\};$$

therefore

$$\frac{d^n}{dx^n}\left(\frac{1}{a^2 + x^2}\right) = \frac{(-1)^n\lfloor n}{2a\sqrt{(-1)}}\left[\frac{1}{\{x - a\sqrt{(-1)}\}^{n+1}} - \frac{1}{\{x + a\sqrt{(-1)}\}^{n+1}}\right].$$

Now assume $x = r\cos\theta$, $a = r\sin\theta$, so that

$$r^2 = a^2 + x^2 \text{ and } \tan\theta = \frac{a}{x}.$$

Then $\{x + a\sqrt{(-1)}\}^{n+1} = r^{n+1}\{\cos\theta + \sqrt{(-1)}\sin\theta\}^{n+1}$

$$= r^{n+1}\{\cos(n+1)\theta + \sqrt{(-1)}\sin(n+1)\theta\}$$

by De Moivre's Theorem.

Hence

$$\frac{1}{\{x - a\sqrt{(-1)}\}^{n+1}} - \frac{1}{\{x + a\sqrt{(-1)}\}^{n+1}} = \frac{2\sqrt{(-1)}\sin(n+1)\theta}{r^{n+1}};$$

and we obtain the same result as before for the proposed n^{th} differential coefficient.

20. $$\frac{d^n\dfrac{x}{a^2 + x^2}}{dx^n} = x\frac{d^n\left(\dfrac{1}{a^2 + x^2}\right)}{dx^n} + n\frac{d^{n-1}\left(\dfrac{1}{a^2 + x^2}\right)}{dx^{n-1}}. \quad \text{Art. 80.}$$

Hence, by means of the preceding Example, shew that

$$\frac{d^n}{dx^n}\left(\frac{x}{a^2 + x^2}\right) = \frac{(-1)^n\lfloor n\cos(n+1)\theta}{(a^2 + x^2)^{\frac{n+1}{2}}}.$$

We may also proceed in the second manner indicated for the preceding Example, starting with

$$\frac{x}{a^2 + x^2} = \frac{1}{2}\left\{\frac{1}{x + a\sqrt{(-1)}} + \frac{1}{x - a\sqrt{(-1)}}\right\}.$$

21. Find the 4th differential coefficient of $\dfrac{1}{e^x - 1}$ and of $e^{-\frac{1}{x^2}}$.

Results

$$\frac{e^x + 11e^{2x} + 11e^{3x} + e^{4x}}{(e^x - 1)^5} \text{ and } e^{-\frac{1}{x^2}}\{16x^{-12} - 144x^{-10} + 300x^{-8} - 120x^{-6}\}.$$

22. $\dfrac{d^n (x^2 a^x)}{dx^n} = \{x^2 c^n + 2nxc^{n-1} + n(n-1) c^{n-2}\} a^x,$

where $c = \log a$. Art. 80.

23. If $y = \sin(m \sin^{-1} x)$, shew that

$$(1 - x^2) \frac{d^2y}{dx^2} = x \frac{dy}{dx} - m^2 y.$$

Apply Leibnitz's theorem, Art. 80, and deduce

$$(1 - x^2) \frac{d^{n+2}y}{dx^{n+2}} = (2n+1) x \frac{d^{n+1}y}{dx^{n+1}} + (n^2 - m^2) \frac{d^n y}{dx^n}.$$

24. If $y = a \cos(\log x) + b \sin(\log x)$, shew that

$$x^2 \frac{d^2y}{dx^2} + x \frac{dy}{dx} + y = 0,$$

and that $x^2 \dfrac{d^{n+2}y}{dx^{n+2}} + (2n+1) x \dfrac{d^{n+1}y}{dx^{n+1}} + (n^2 + 1) \dfrac{d^n y}{dx^n} = 0.$

CHAPTER VI.

EXPANSION OF FUNCTIONS IN SERIES.

84. In the Binomial Theorem, we are furnished with a series proceeding according to powers of h, which is equivalent to the expression $(x + h)^n$. Other series have also presented themselves in Algebra and Trigonometry, such as the expansion of e^x in powers of x and of $\log (1 + x)$ in powers of x. In the previous Articles of this book, we have, however, not assumed the knowledge of any expansions, *except the Binomial Theorem in the case of a positive integral exponent;* but we are now about to investigate the expansion of $f(x + h)$ in powers of h, where $f(x)$ denotes any function of x, and it will appear that all the isolated examples which the student may have seen hitherto, are but cases of this general theorem.

85. Before we offer a strict demonstration of the theorem in question, we shall notice the method which it was usual to adopt in treatises on the Differential Calculus not based on the doctrine of limits. Such treatises commenced with an unsatisfactory demonstration of the proposition that $f(x + h)$ could generally be expanded in a series proceeding according to ascending integral positive powers of h; it remained then to determine the coefficients of the different powers of h, and that was accomplished in the manner given in the next two Articles.

86. We have first to establish the following theorem. If $f(x + h)$ be any function of $x + h$, we obtain the same result whether we differentiate it with respect to x, considering h constant, or differentiate it with respect to h, considering x constant.

For put $\qquad x + h = z.$

In the first case

$$\frac{df(x+h)}{dx} = \frac{df(z)}{dz} \cdot \frac{dz}{dx}$$

$$= f'(z), \text{ since } \frac{dz}{dx} = 1.$$

In the second case,

$$\frac{df(x+h)}{dh} = \frac{df(z)}{dz} \cdot \frac{dz}{dh}$$

$$= f'(z), \text{ since } \frac{dz}{dh} = 1.$$

87. *To expand $f(x+h)$ in a series of ascending powers of h.*

Assume (Art. 85) that

$$f(x+h) = A_0 + A_1 h + A_2 h^2 + A_3 h^3 + \ldots\ldots (1),$$

where $\qquad A_0, A_1, A_2, \ldots,$ do not contain h.

Then

$$\frac{df(x+h)}{dx} = \frac{dA_0}{dx} + h\frac{dA_1}{dx} + h^2\frac{dA_2}{dx} + h^3\frac{dA_3}{dx} + \ldots\ldots (2),$$

and $\quad \dfrac{df(x+h)}{dh} = A_1 + 2A_2 h + 3A_3 h^2 + \ldots\ldots\ldots (3).$

By Art. 86, the series (2) and (3) must be equal. Hence, equating the coefficients of like powers of h, we have

$$A_1 = \frac{dA_0}{dx},$$

$$A_2 = \frac{1}{2}\frac{dA_1}{dx} = \frac{1}{1.2}\frac{d^2A_0}{dx^2},$$

$$A_3 = \frac{1}{3}\frac{dA_2}{dx} = \frac{1}{1.2.3}\frac{d^3A_0}{dx^3},$$

$$\ldots\ldots\ldots\ldots\ldots\ldots\ldots\ldots$$

And by putting $h = 0$ in (1), we find

$$A_0 = f(x).$$

Hence, substituting the values of $A_0, A_1, \dots$ in (1), we have

$$f(x+h) = f(x) + hf'(x) + \frac{h^2}{1.2} f''(x) + \frac{h^3}{1.2.3} f'''(x) + \dots (4),$$

the general term being

$$\frac{h^n}{\underline{|n}} \frac{d^n f(x)}{dx^n}.$$

This result is called Taylor's Theorem.

88.   There are numerous objections to the method of the preceding Articles, and especially the use of an infinite series, without ascertaining that it is convergent, is inadmissible; we proceed then to a rigorous investigation.

89.   Let $y = F(x)$, and suppose Δx and Δy to represent the simultaneous increments of x and y; then the fraction $\frac{\Delta y}{\Delta x}$, *since it has for its limit the differential coefficient* $F'(x)$, *will ultimately when* Δx *is taken small enough have the same sign as this limit*, and therefore will be positive if the differential coefficient be positive, and negative if the differential coefficient be negative. In the former case, the quantities Δy and Δx being of the *same* sign, the function y will increase or diminish according as x increases or diminishes. In the latter case, Δy and Δx being of *contrary* signs, y will increase if x diminishes and will diminish if x increases.

The above supposes that there really is a *finite* limit to which $\frac{\Delta y}{\Delta x}$ tends; in other words we assume that $F'(x)$ is *not infinite*. The limitation that the functions with which we are concerned are not to become *infinite* is one which ought to be understood in most theorems in mathematics, even if it is not formally enunciated. In the present subject however it is usual to state this limitation expressly at the more important stages of the investigations.

It may be observed that we may sometimes obtain useful information respecting the sign of a function by examining the differential coefficient of the function. For example, suppose $y = (x-1) e^x + 1$, then $\frac{dy}{dx} = xe^x$; as $\frac{dy}{dx}$ is positive for all positive values of x, it follows by the present Article that y is always increasing so long as x is positive; but $y = 0$ when $x = 0$; therefore y *is positive for all positive values of x.*

Similarly we can shew that $x - \log(1 + x)$ is positive for all positive values of x.

90. A function of a variable is said to be *continuous* between certain values of the variable when it fulfils the following conditions: the function must have a single finite value for every value of the variable, and the function must change *gradually* as the variable passes from one value to the other, so that corresponding to an indefinitely small change in the variable there must be an indefinitely small change in the function.

91. Suppose $\phi(x)$ a function which vanishes when $x = a$ and when $x = b$, and is continuous between those values. Suppose also that $\phi'(x)$ is continuous between those values. Then $\phi'(x)$ will vanish for some value of x between a and b.

For $\phi'(x)$ cannot be always positive between those values, for then $\phi(x)$ would be constantly increasing as the variable increased from the lower value to the higher (Art. 89), which is inconsistent with the supposition that $\phi(x)$ vanishes at the two specified values. Similarly $\phi'(x)$ cannot be always negative. Hence $\phi'(x)$ must change from positive to negative or from negative to positive between the assigned values; and since it is continuous it cannot become infinite and must therefore pass through the value zero.

If a denote some constant quantity, such expressions as $f'(a), f''(a), \ldots$ may occur in our investigations, the meaning to be attached to them being that $f(x)$ is to be differentiated once, twice, $\ldots$ with respect to x, and in the result x changed into a.

We can now demonstrate Taylor's Theorem. The proof which we give in the next Article is due to Mr Homersham Cox; it was published by him in the 6th volume of the *Cambridge and Dublin Mathematical Journal,* and subsequently in his *Manual of the Differential Calculus.*

92. Suppose $f(a+x)$ and its differential coefficients up to the $(n+1)^{\text{th}}$ to be continuous between the values 0 and h of the variable x. The expression

$$f(a+x)-f(a)-xf'(a)-\frac{x^2}{\lfloor 2}f''(a)\ldots-\frac{x^n}{\lfloor n}f^n(a)-\frac{x^{n+1}R}{\lfloor n+1}\ldots(1),$$

vanishes when $x=h$ if $R=$

$$\frac{\lfloor n+1}{h^{n+1}}\left\{f(a+h)-f(a)-hf'(a)-\frac{h^2}{\lfloor 2}f''(a)\ldots-\frac{h^n}{\lfloor n}f^n(a)\right\}\ldots(2).$$

Suppose R to have this value which we observe is independent of x.

The expression (1) also vanishes when $x=0$.

Hence, by Art. 91 the differential coefficient of (1) with respect to x must vanish for some value of x between 0 and h; suppose x_1 that value, then

$$f'(a+x)-f'(a)-xf''(a)-\ldots-\frac{x^{n-1}}{\lfloor n-1}f^n(a)-\frac{x^n}{\lfloor n}R\ldots\ldots(3),$$

vanishes when $x=x_1$. But (3) also vanishes when $x=0$; hence there is some value of x between 0 and x_1 for which the differential coefficient of (3) vanishes.

Continuing this process to $n+1$ differentiations of (1) we find that $f^{n+1}(a+x)-R$ is zero for some value of x between 0 and h; let this value of x be θh, where θ is some proper fraction, therefore

$$R=f^{n+1}(a+\theta h).$$

Substitute this value of R in (2) and we have

$$f(a+h)=f(a)+hf'(a)+\frac{h^2}{\lfloor 2}f''(a)+\ldots+\frac{h^n}{\lfloor n}f^n(a)$$

$$+\frac{h^{n+1}}{\lfloor n+1}f^{n+1}(a+\theta h).$$

We may now put x for a in this equation, since there has been no restriction in the value of a, except that all the quantities are to be *finite*, thus we obtain

$$f(x+h) = f(x) + hf'(x) + \frac{h^2}{\lfloor 2}f''(x) + \ldots + \frac{h^n}{\lfloor n}f'^n(x)$$
$$+ \frac{h^{n+1}}{\lfloor n+1}f^{n+1}(x+\theta h)\ldots(4).$$

If the function $f^{n+1}(x+\theta h)$ be such that by making n sufficiently great the term $\dfrac{h^{n+1}}{\lfloor n+1}f^{n+1}(x+\theta h)$ can be made as small as we please, then by carrying on the series

$$f(x) + hf'(x) + \frac{h^2}{\lfloor 2}f''(x) + \frac{h^3}{\lfloor 3}f'''(x) + \ldots,$$

to as many terms as we please, we obtain a result differing as little as we please from $f(x+h)$. Under these circumstances then we may assert the truth of Taylor's Theorem.

93.   Taylor's Theorem is so called from its discoverer Dr Brook Taylor; it was first published in 1715. The theorem contained in equation (4) of Art. 92 is called *Lagrange's Theorem on the limits of Taylor's Theorem*. It gives us an expression for the difference between $f(x+h)$ and the first $n+1$ terms of its expansion by Taylor's Theorem, or as it is called "the remainder after $n+1$ terms."

94.   To the expression $f^{n+1}(x+\theta h)$ which occurs in Art. 92, we must assign the following meaning. "Let $f(x)$ be differentiated $n+1$ times with respect to x, and in the final result change x into $x+\theta h$." We do not know anything of θ, except that it lies between 0 and 1; it will generally be a function of x and h, and hence, to differentiate $f(x+\theta h)$ with respect to x, is not the same thing as to differentiate $f(x)$ with respect to x and then to change x into $x+\theta h$.

95.   *Maclaurin's Theorem.*

In the equation

$$f(x+h) = f(x) + hf'(x) + \frac{h^2}{\lfloor 2}f''(x) + \ldots + \frac{h^n f^n(x)}{\lfloor n}$$
$$+ \frac{h^{n+1}}{\lfloor n+1}f^{n+1}(x+\theta h),$$

put $x=0$, we have then

$$f(h) = f(0) + hf'(0) + \ldots + \frac{h^n f^n(0)}{\lfloor n} + \frac{h^{n+1}}{\lfloor n+1}f^{n+1}(\theta h).$$

We may, if we please, change h into x, and since the quantities $f(0), f'(0), \ldots\ldots f^n(0)$, do not contain x or h, no change is made in any of them: hence

$$f(x) = f(0) + xf'(0) + \ldots + \frac{x^n f^n(0)}{\lfloor n} + \frac{x^{n+1}}{\lfloor n+1} f^{n+1}(\theta x).$$

When the last term, by taking n large enough, can be made as small as we please, we have for $f(x)$ an infinite series proceeding according to powers of x. This series is usually called Maclaurin's, having been published by him in 1742; though, as it had been given a few years previously by Stirling, it sometimes bears the name of the latter.

96. Assuming that any function of x can be expanded in a series of positive integral powers of x, the following method has been given for proving Maclaurin's Theorem.

Let $f(x) = A_0 + A_1 x + A_2 x^2 + \ldots\ldots + A_n x^n + \ldots\ldots$

where $A_0, A_1, A_2 \ldots$ do not contain x.

Differentiate successively, then

$$f'(x) = A_1 + 2A_2 x + \ldots + nA_n x^{n-1} + \ldots\ldots$$
$$f''(x) = 2A_2 + 2.3A_3 x + \ldots + n(n-1)A_n x^{n-2} + \ldots\ldots$$
$$f'''(x) = 2.3A_3 + \ldots + n(n-1)(n-2)A_n x^{n-3} + \ldots\ldots$$
$$\ldots\ldots\ldots\ldots\ldots\ldots\ldots\ldots\ldots\ldots\ldots$$

Now suppose $x = 0$ in each of these equations, we have

$$A_0 = f(0),$$
$$A_1 = f'(0),$$
$$A_2 = \frac{1}{1.2} f''(0),$$
$$A_3 = \frac{1}{1.2.3} f'''(0),$$
$$\ldots\ldots\ldots\ldots\ldots\ldots\ldots\ldots$$

Substitute the values of $A_0, A_1, \ldots$ and we obtain

$$f(x) = f(0) + xf'(0) + \frac{x^2}{\lfloor 2} f''(0) + \ldots + \frac{x^n}{\lfloor n} f^n(0) + \ldots$$

97. The demonstration given of equation (4) in Art. 92, which equation involves Taylor's Theorem, and may even speaking loosely be called Taylor's Theorem, will probably disappoint the reader. Though he may be unable to discover any flaw in the reasoning, he will complain of the artificial and tentative character of the whole, and he will urge the same objection with respect to Cauchy's method of proof which we shall presently give. Without denying the justice of these objections, we may reply that the highly general character of the theorem may be some excuse for the complexity and indirect nature of the investigation. But with respect particularly to the dissatisfaction felt in being compelled to assent to a number of propositions without knowing beforehand the general course which the demonstration might be expected to take, we may remind the student that he must not while engaged in the elements of a subject expect to be able, as it were, to *rediscover the theorems for himself*. Instead of asking, "what suggested this or that step?" he must frequently be contented with the simpler question, "is the reasoning correct?" To this of course he has already, perhaps unconsciously, been accustomed; for example, if a complicated construction occurred in Euclid, he merely confined himself, at least for some time, to an examination of the consistency of the construction, and the truth of the deductions from it, without attempting to retrace the steps by which Euclid arrived at his construction.

98. On account of the importance of Taylor's Theorem we shall add another demonstration; this demonstration is due in substance to Cauchy.

Let $F(x)$ and $f(x)$ be two functions of x which remain continuous, as also their differential coefficients, between the values a and $a + h$ of the variable x. Suppose also that between these same values the derived function $f'(x)$ does not vanish. Then the fraction $\dfrac{F(a + h) - F(a)}{f(a + h) - f(a)}$ shall be equal to the value of $\dfrac{F'(x)}{f'(x)}$, when in the latter x has some value included between the specified values; that is, θ denoting some proper fraction, we shall have

$$\frac{F(a+h)-F(a)}{f(a+h)-f(a)} = \frac{F'(a+\theta h)}{f'(a+\theta h)}.$$

Let
$$R = \frac{F(a+h)-F(a)}{f(a+h)-f(a)};$$

then since $f'(x)$ is continuous and does not vanish between the values a and $a+h$ of x, it retains the same sign; and thus $f(x)$ continually increases or continually decreases: see Art. 89. Hence $f(a+h)-f(a)$ cannot be zero, and we may therefore multiply by it; so that

$$F(a+h) - F(a) - R\{f(a+h) - f(a)\} = 0.$$

Let $\phi(x)$ denote the function

$$F(a+h) - F(x) - R\{f(a+h) - f(x)\}:$$

then $\phi(x)$ is continuous while x lies between a and $a+h$; and so also is the differential coefficient $\phi'(x)$, that is $-F'(x) + Rf'(x)$. Moreover $\phi(x)$ vanishes, by hypothesis, when $x = a$; and $\phi(x)$ obviously vanishes when $x = a+h$. Hence, by Art. 91, it follows that $\phi'(x)$ must vanish for some value of x between a and $a+h$; this value may be denoted by $a+\theta h$, where θ is some proper fraction. Thus

$$-F'(a+\theta h) + Rf'(a+\theta h) = 0;$$

and, by hypothesis, $f'(a+\theta h)$ is not zero, so that we may divide by it: therefore

$$R = \frac{F'(a+\theta h)}{f'(a+\theta h)}.$$

Thus the required result is obtained.

99. The result of the preceding Article has been obtained on the assumption that the functions are continuous and that $f'(x)$ does not vanish between the values a and $a+h$ of the variable x. The result however is true if the functions are continuous and *either* of the two $F'(x)$ and $f'(x)$ does not vanish. For if $F'(x)$ does not vanish we may prove as in the preceding Article that

$$\frac{f(a+h)-f(a)}{F(a+h)-F(a)} = \frac{f'(a+\theta h)}{F'(a+\theta h)},$$

and from this it follows of course that

$$\frac{F(a+h)-F(a)}{f(a+h)-f(a)} = \frac{F'(a+\theta h)}{f'(a+\theta h)}.$$

The reader who wishes to see the application of this result to the establishment of Taylor's Theorem, may pass on to Art. 106 at once, and then return to the consideration of the omitted Articles, in which we shall give another proof of the result, and also some geometrical illustrations.

100.   The enunciation of Art. 98 being supposed, we may arrange the proof thus:

Divide h into a number of equal parts, and let α denote one of these parts.   Consider the fractions

$$\frac{F(a+\alpha)-F(a)}{f(a+\alpha)-f(a)}, \quad \frac{F(a+2\alpha)-F(a+\alpha)}{f(a+2\alpha)-f(a+\alpha)}, \quad \frac{F(a+3\alpha)-F(a+2\alpha)}{f(a+3\alpha)-f(a+2\alpha)},$$

$$\cdots \frac{F(a+h)-F(a+h-\alpha)}{f(a+h)-f(a+h-\alpha)} \cdots\cdots\cdots\cdots(1).$$

Form a new fraction by adding together all the numerators in (1) for a new numerator, and all the denominators in (1) for a new denominator.   We thus obtain

$$\frac{F(a+h)-F(a)}{f(a+h)-f(a)} \cdots\cdots\cdots\cdots\cdots (2).$$

Since the denominators which occur in (1) have by hypothesis all the same sign, we know from algebra that the fraction (2) *lies in value between the greatest and least of those in* (1).   Now

$$\frac{F(a+\alpha)-F(a)}{f(a+\alpha)-f(a)} = \frac{\dfrac{F(a+\alpha)-F(a)}{\alpha}}{\dfrac{f(a+\alpha)-f(a)}{\alpha}};$$

if then we put this fraction equal to $\dfrac{F'(a)}{f'(a)}+\beta$, we know that β diminishes without limit when α does so.

Similarly,

$$\frac{F(a+2\alpha)-F(a+\alpha)}{f(a+2\alpha)-f(a+\alpha)} = \frac{F'(a+\alpha)}{f'(a+\alpha)}+\gamma,$$

$$\frac{F(a+3\alpha)-F(a+2\alpha)}{f(a+3\alpha)-f(a+2\alpha)} = \frac{F'(a+2\alpha)}{f'(a+2\alpha)} + \delta,$$

. . . .

$$\frac{F(a+h)-F(a+h-\alpha)}{f(a+h)-f(a+h-\alpha)} = \frac{F'(a+h-\alpha)}{f'(a+h-\alpha)} + \mu,$$

where γ, δ, ... μ, all diminish without limit when α does so.

Since the fraction in (2) always lies between the greatest and least of the series

$$\frac{F'(a)}{f'(a)} + \beta, \quad \frac{F'(a+\alpha)}{f'(a+\alpha)} + \gamma, \quad \frac{F'(a+2\alpha)}{f'(a+2\alpha)} + \delta,$$

$$\cdots\cdots\cdots\cdots \quad \frac{F'(a+h-\alpha)}{f'(a+h-\alpha)} + \mu,$$

it must lie between the greatest and least limits towards which these tend; that is, it must lie between the greatest and least values which $\dfrac{F'(x)}{f'(x)}$ can assume between a and $a+h$.

But as $\dfrac{F'(x)}{f'(x)}$, in passing from its greatest to its least value passes through all intermediate values, there must be some proper fraction θ, such that

$$\frac{F(a+h)-F(a)}{f(a+h)-f(a)} = \frac{F'(a+\theta h)}{f'(a+\theta h)}.$$

101.   Suppose $f(x) = x-a$; therefore $f'(x) = 1$.

The conditions required to be satisfied by $f(x)$ in the enunciation of Art. 98 are satisfied. And as $f(a+h) = h$, and $f(a) = 0$,

we have $\qquad F(a+h) - F(a) = hF'(a+\theta h).$

This simple case of Art. 98 might of course be proved in the same manner as the general proposition was established.

102.   The result of Art. 101 may be applied to shew that an expression *independent* of x is the only one of which the differential coefficient with respect to x is always zero. For suppose $F(x)$ a function, such that $F'(x)$ is always zero;

then, from the last equation in Art. 101 it follows, whatever be the value of a and $a + h$, that $F(a + h) - F(a) = 0$,

therefore $\qquad\qquad F(a + h) = F(a)$.

Hence the function $F(x)$ has always the same value whatever be the value of the variable; that is, it is constant with respect to x, or in other words does not depend on x.

From this it follows, that two functions which have the same differential coefficient with respect to any variable can only differ by a constant. For the differential coefficient of the difference of these functions being always zero, it follows from what we have just proved that this difference is a constant.

103. The result of Art. 101 admits of the following simple geometrical verification.

We have already shewn, Art. 43, that if u represent the area contained between the axes of x and y, the ordinate y, and any curve, then

$$\frac{du}{dx} = y.$$

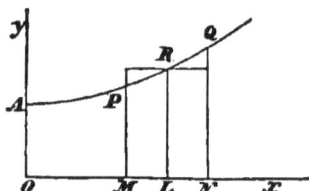

Let $u = F(x)$, and therefore $y = F'(x)$ is the equation to the curve; let $OM = a$, $MN = h$;

then $\qquad\qquad$ area $OAPM = F(a)$,

$\qquad\qquad$ area $OAQN = F(a + h)$,

therefore $\qquad$ area $PQNM = F(a + h) - F(a)$.

Now it is obvious that a point R must exist between P and Q, such that, drawing the ordinate RL,

$\qquad$ the rectangle $RL . MN =$ the area $PQNM$.

But $\qquad\qquad RL = F'(a + \theta h)$,

where θ is some proper fraction; therefore

$$h F'(a + \theta h) = F(a + h) - F(a).$$

104. The following is another geometrical illustration of Art. 101.

If $y = F(x)$ be the equation to a curve, then $F'(x)$ is the trigonometrical tangent of the angle between the axis of x and the tangent to the curve at the point (x, y). See Art. 38.

Let $\quad OM = a, \quad MN = h,$

then $\quad \dfrac{F(a+h) - F'(a)}{h}$

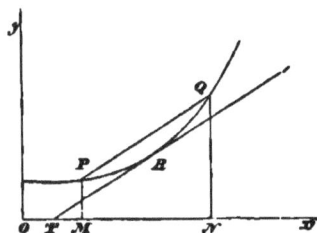

is the tangent of the inclination of the chord PQ to the axis of x. Hence Art. 101 amounts to asserting that *at some point R between P and Q the tangent RT to the curve is parallel to PQ.*

We call this an *illustration.* When, however, the student has sufficiently considered the nature of the tangent to a curve, it may amount to a *proof* of the proposition in question.

105. The following is an *illustration* of the general proposition in Art. 98.

Let there be two curves APQ and apq. . Let $F(x)$ denote the area contained between the first curve, the axes of x and y, and an ordinate to the abscissa x; then $y = F'(x)$ is the equation to this curve. Let $f(x)$ denote a similar area with respect to the second curve; then $y = f'(x)$ is the equation to this curve.

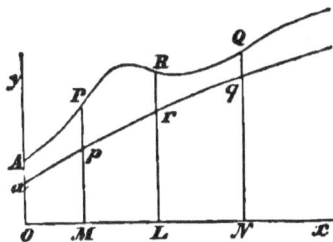

Let $\quad OM = a, \quad MN = h.$

Then $\qquad F(a+h) - F(a) = \text{area } PMNQ,$

$\qquad\qquad f(a+h) - f(a) = \text{area } pMNq.$

Hence the equation

$$\frac{F(a+h) - F(a)}{f(a+h) - f(a)} = \frac{F'(a+\theta h)}{f'(a+\theta h)}$$

amounts to the assertion that there must exist some point R between P and Q, such that

$$\frac{\text{area } PMNQ}{\text{area } pMNq} = \frac{RL}{rL}.$$

106. Suppose now that $F(x)$ and $f(x)$ and all their differential coefficients up to the $(n+1)^{\text{th}}$ inclusive, are continuous between the values a and $a+h$ of the variable x; moreover suppose that one of the two $F'(x)$ and $f'(x)$ does not vanish between the same values, also one of the two $F'''(x)$ and $f''(x)$, and so on up to $F^{n+1}(x)$ and $f^{n+1}(x)$. Then, by Art. 99,

$$\frac{F(a+h) - F(a)}{f(a+h) - f(a)} = \frac{F'(a + \theta_1 h)}{f'(a + \theta_1 h)},$$

$$\frac{F'(a + \theta_1 h) - F'(a)}{f'(a + \theta_1 h) - f'(a)} = \frac{F''(a + \theta_2 h)}{f''(a + \theta_2 h)},$$

$$\frac{F''(a + \theta_2 h) - F''(a)}{f''(a + \theta_2 h) - f''(a)} = \frac{F''''(a + \theta_3 h)}{f'''(a + \theta_3 h)},$$

$$\cdots\cdots\cdots\cdots\cdots$$

$$\frac{F^n(a + \theta_n h) - F^n(a)}{f^n(a + \theta_n h) - f^n(a)} = \frac{F^{n+1}(a + \theta h)}{f^{n+1}(a + \theta h)};$$

where $\theta_1, \theta_2, \ldots\ldots \theta_n, \theta$, are all proper fractions.

Let us now suppose that $F'(x)$, $F'''(x)$, ... $F^n(x)$, $f'(x)$, $f''(x)$, ... $f^n(x)$ all vanish when $x = a$; then from the above equations

$$\frac{F(a+h) - F(a)}{f(a+h) - f(a)} = \frac{F^{n+1}(a + \theta h)}{f^{n+1}(a + \theta h)}.$$

107. If we take $f(x) = (x - a)^{n+1}$ we find that the requisite conditions are all satisfied; that is, $f(x)$ and its differential coefficients are continuous, and the differential coefficients do not vanish between the values a and $a+h$ of the variable; also all the differential coefficients up to the nth inclusive vanish when $x = a$. And

$$f^{n+1}(x) = \lfloor n+1, \quad f(a) = 0, \quad f(a+h) = h^{n+1}.$$

Suppose then that $F(x)$ and all its differential coefficients are continuous between the values a and $a+h$ of the variable, and that all the differential coefficients up to the nth inclusive vanish when $x = a$; we have by Art. 106,

$$F(a+h) - F(a) = \frac{h^{n+1}}{\lfloor n+1} F^{n+1}(a + \theta h).$$

Suppose $a = 0$ and $F(a) = 0$, then

$$F(h) = \frac{h^{n+1}}{\lfloor n+1} F^{n+1}(\theta h).$$

108. *Application to Taylor's Theorem.*

Let $\phi(x+h)$ be a function which is to be expanded in a series of ascending positive integral powers of h.   Let

$$\phi(x+h) - \phi(x) - h\phi'(x) - \frac{h^2}{\lfloor 2} \phi''(x) - \dots - \frac{h^n}{\lfloor n} \phi^n(x) = F(h).$$

Then $F(h)$ and its differential coefficients with respect to h, up to the nth inclusive, vanish when $h = 0$.   Also

$$F^{n+1}(h) = \phi^{n+1}(x + h).$$

Hence, by the last equation of Art. 107,

$$F(h) = \frac{h^{n+1}}{\lfloor n+1} F^{n+1}(\theta h) = \frac{h^{n+1}}{\lfloor n+1} \phi^{n+1}(x + \theta h),$$

and therefore

$$\phi(x+h) = \phi(x) + h\phi'(x) + \frac{h^2}{\lfloor 2} \phi''(x) + \dots + \frac{h^n}{\lfloor n} \phi^n(x)$$

$$+ \frac{h^{n+1}}{\lfloor n+1} \phi^{n+1}(x + \theta h).$$

From this Taylor's Theorem follows whenever the function is such that, by sufficiently increasing n, the term

$$\frac{h^{n+1}}{\lfloor n+1} \phi^{n+1}(x + \theta h)$$

can be made as small as we please.

109. The following proof of Taylor's Theorem deserves notice, as it depends only on the equation which is proved geometrically in Art. 103. Let

$$\phi(z) - \phi(x) - (z-x)\phi'(x) - \frac{(z-x)^2}{\lfloor 2} \phi''(x) - \ldots - \frac{(z-x)^n}{\lfloor n} \phi^n(x)$$

be called $F(x)$, then $F'(x) = -\frac{(z-x)^n}{\lfloor n} \phi^{n+1}(x)$.

Now, by Art. 103, $F(x) = F(z) + (x-z) F'\{z + \theta(x-z)\}$.

Also $\qquad\qquad\qquad F(z) = 0,$

and $\quad F'\{z + \theta(x-z)\} = -\frac{\theta^n(z-x)^n}{\lfloor n} \phi^{n+1}\{z + \theta(x-z)\};$

therefore $\quad \phi(z) - \phi(x) - (z-x)\phi'(x) - \frac{(z-x)^2}{\lfloor 2} \phi''(x) -$

$$\ldots\ldots - \frac{(z-x)^n}{\lfloor n} \phi^n(x)$$

$$= \frac{\theta^n(z-x)^{n+1}}{\lfloor n} \phi^{n+1}\{z + \theta(x-z)\}.$$

Put h for $z - x$, then

$$\phi(x+h) = \phi(x) + h\phi'(x) + \frac{h^2}{\lfloor 2} \phi''(x) + \ldots\ldots + \frac{h^n}{\lfloor n} \phi^n(x)$$

$$+ \frac{\theta^n h^{n+1}}{\lfloor n} \phi^{n+1}(x + h - \theta h).$$

110. The result of the preceding Article gives us an expression for the remainder after $n+1$ terms of the expansion of $\phi(x+h)$, differing in form from that we found before. If we assume $\theta = 1 - \theta_1$, the remainder becomes

$$\frac{(1-\theta_1)^n h^{n+1}}{\lfloor n} \phi^{n+1}(x + \theta_1 h).$$

111. In the proofs given of Taylor's Theorem, we have supposed all the functions that occur to be continuous. If the function we wish to expand, or any of its differential coefficients up to the $(n+1)^{\text{th}}$ inclusive, be infinite for values

of the variable lying between certain values, the demonstration given of the theorem

$$f(x+h) = f(x) + hf'(x) + \ldots\ldots + \frac{h^n}{\underline{|n}} f^n(x) + \frac{h^{n+1}}{\underline{|n+1}} f^{n+1}(x+\theta h),$$

is no longer valid. It is usual to speak of the cases where an infinite value enters as "instances of the failure of Taylor's Theorem." The phrase is connected with the imperfect mode of demonstration given in Arts. 86 and 87, in which it was not settled beforehand when the theorem supposed to be demonstrated was really true and when it was not. For example, suppose

$$f(x) = \sqrt{(x-a)},$$
so that $$f(x+h) = \sqrt{(x-a+h)}.$$

Then it would be said that $f(x+h)$ can *always* be expanded in a series of whole positive powers of h, *except* when $x = a$.

When $x = a$, $f'(x)$, $f''(x)$, ... all become infinite, and $f(x+h)$ becomes $\sqrt{h}$.

112. It was usual in that system of treating the Differential Calculus referred to in Art. 85, to express, or imply, two propositions with respect to the "failure of Taylor's Theorem."

(1) If the *true* expansion of $f(a+h)$ in powers of h contain only integral positive powers of h, then none of the quantities $f(a), f'(a), f''(a)$, ... can be infinite.

(2) If the *true* expansion of $f(a+h)$ in powers of h involve negative or fractional powers of h, then some *one* of the quantities $f(a), f'(a), f''(a)$, ... is infinite, as well as all which succeed it.

By the *true* expansion of $f(a+h)$ is meant the expansion obtained by some legitimate algebraical process, applicable to the example in question, as the Binomial Theorem for example. The proof of the above two propositions was given thus.

Suppose $f(a+h) = A_0 + A_1 h^\alpha + A_2 h^\beta + A_3 h^\gamma + \ldots\ldots$

to be the true expansion, $A_0, A_1, \ldots$, not containing h. Then to obtain $f'(a), f''(a)$, ... we may differentiate $f(a+h)$ successively with respect to h, and put $h = 0$ in the result.

If then a, β, γ,...... be all positive integers, we shall never have *negative* powers of h introduced by successive differentiation of $f(a+h)$. Hence, by putting $h = 0$, we introduce no *infinite* values.

But if any one of the exponents a, β, γ, ... be negative, $f(a+h)$ and all its differential coefficients contain negative powers of h, and therefore $f(a)$, $f'(a)$, $f''(a)$, ... are all infinite.

If none of the exponents be negative, but one or more of them be positive fractions, suppose that γ is the smallest of such fractions, and that it lies between the integers n and $n+1$. Then $f(a+h)$ and all its differential coefficients up to the nth inclusive are free from negative powers of h; but $f^{n+1}(a+h)$ and all the subsequent differential coefficients contain negative powers of h. Hence $f^{n+1}(a)$ is the first differential coefficient that becomes infinite, and all the following differential coefficients are infinite.

113. It will be of use hereafter to remark that if for a finite value of the variable any function becomes infinite, so also does the differential coefficient of the function. In proof of this, it is sufficient to notice the different cases that may arise. An Algebraical function can only become infinite, for a finite value of the variable, by having the form of a fraction the denominator of which vanishes. Now when we differentiate a fraction we never remove the denominator, so that the differential coefficient also has a vanishing denominator, and therefore becomes infinite. Similarly, the second, third, ... differential coefficients are also infinite.

The transcendental functions $\log x$ and $a^{\frac{1}{x}}$, which both become infinite when $x=0$, have their differential coefficients, namely $\dfrac{1}{x}$ and $-\dfrac{\log a}{x^2} a^{\frac{1}{x}}$, also infinite when $x = 0$.

The trigonometrical functions, such as $\tan x$ and $\sec x$, which can become infinite, are fractional forms, and fall under the observations already made.

The proposition is not necessarily true for functions which become infinite for an infinite value of the variable, as may be seen in the case of $\log x$, which is infinite when x is infinite, while its differential coefficient $\dfrac{1}{x}$ vanishes.

MISCELLANEOUS EXAMPLES.

1. If $y = \tan^{-1}\dfrac{a+bx}{b-ax}$, $\dfrac{dy}{dx} = \dfrac{1}{1+x^2}$.

2. If $y = x\tan^{-1}\dfrac{1}{x}$, $\dfrac{dy}{dx} = \tan^{-1}\dfrac{1}{x} - \dfrac{x}{1+x^2}$.

3. If $y = \log\dfrac{\sqrt{(x^2+a^2)}+\sqrt{(x^2+b^2)}}{\sqrt{(x^2+a^2)}-\sqrt{(x^2+b^2)}}$,

$$\dfrac{dy}{dx} = \dfrac{2x}{\sqrt{(x^2+a^2)}\sqrt{(x^2+b^2)}}.$$

4. If $y = \dfrac{\sqrt{(1-x^2)}\,e^{\sin^{-1}x}}{\sqrt{(1-x^2)}+x}$, $\dfrac{dy}{dx} = \dfrac{x\,e^{\sin^{-1}x}}{\sqrt{(1-x^2)}} \cdot \dfrac{\sqrt{(1-x^2)}-x}{\{\sqrt{(1-x^2)}+x\}^2}$.

5. If $y = \left(\dfrac{\sin x}{x}\right)^{\frac{x}{\sin x}}$, $\dfrac{dy}{dx} = \dfrac{y\,(x\cos x - \sin x)}{\sin^2 x}\log\dfrac{ex}{\sin x}$.

6. If $f(x) = \left(\dfrac{a+x}{b+x}\right)^{a+b+2x}$, $f'(0) = \left\{2\log\dfrac{a}{b} + \dfrac{b^2-a^2}{ab}\right\}\left(\dfrac{a}{b}\right)^{a+b}$.

7. If $y = \sqrt[3]{\{(x-a)^2(x-c)\}}$, $\dfrac{d^2y}{dx^2} = \dfrac{-2(a-c)^2 y}{9(x-a)^2(x-c)^2}$.

8. If $x = a\cos\theta + b\sin\theta$, and $y = a\sin\theta - b\cos\theta$, then $\dfrac{d^m x}{d\theta^m}\dfrac{d^n y}{d\theta^n} - \dfrac{d^n x}{d\theta^n}\dfrac{d^m y}{d\theta^m}$ is independent of θ.

9. If $\cos^{-1}\dfrac{y}{a} = \log\left(\dfrac{x}{b}\right)^n$. then

$$x^2\dfrac{d^{n+2}y}{dx^{n+2}} + (2n+1)\,x\dfrac{d^{n+1}y}{dx^{n+1}} + 2n^2\dfrac{d^n y}{dx^n} = 0.$$

10. Shew that $(x-2)\,e^x + x + 2$ is positive for all positive values of x.

CHAPTER VII.

114. WE shall first apply the formulæ of the preceding Chapter to expand certain functions.

Required the expansion of $(1 + x)^m$, m not being assumed to be a positive integer.

If $\qquad f(x) = (1 + x)^m$,

we have $\quad f'(x) = m(1 + x)^{m-1}$,

$\qquad\qquad f''(x) = m(m-1)(1+x)^{m-2}$,

$$\cdots\cdots\cdots\cdots\cdots\cdots\cdots\cdots$$

$\qquad\qquad f^n(x) = m(m-1)\dots(m-n+1)(1+x)^{m-n}$,

$\qquad\qquad f^{n+1}(x) = m(m-1)\dots(m-n)(1+x)^{m-n-1}$;

hence $\qquad f(0) = 1, \qquad f'(0) = m, \qquad f''(0) = m(m-1), \dots$

Therefore, by Art. 95,

$$(1+x)^m = 1 + mx + \frac{m(m-1)}{\lfloor 2} x^2 + \dots + \frac{m(m-1)\dots(m-n+1)}{\lfloor n} x^n$$
$$+ \frac{x^{n+1}}{\lfloor n+1} m(m-1)\dots(m-n)(1+\theta x)^{m-n-1}.$$

If x be less than 1 the last term can be made as small as we please by sufficiently increasing n, and in that case the infinite series

$$1 + mx + \frac{m(m-1)}{\lfloor 2} x^2 + \dots$$

can, by taking a sufficient number of terms, be brought as near as we please to $(1+x)^m$.

115. Let $\qquad f(x) = a^x.$

By Arts. 95 and 79, we have

$$a^x = 1 + x \log a + \frac{x^2}{\lfloor 2} (\log a)^2 + \ldots + \frac{x^n}{\lfloor n} (\log a)^n$$

$$+ \frac{x^{n+1} a^{\theta x} (\log a)^{n+1}}{\lfloor n+1}.$$

Hence, changing a to e, and remembering that $\log e = 1$,

$$e^x = 1 + x + \frac{x^2}{\lfloor 2} + \frac{x^3}{\lfloor 3} + \ldots + \frac{x^n}{\lfloor n} + \frac{x^{n+1} e^{\theta x}}{\lfloor n+1}.$$

The term $\dfrac{x^{n+1} e^{\theta x}}{\lfloor n+1}$ may be made as small as we please by sufficiently increasing n. Hence we obtain an infinite series for e^x, namely,

$$e^x = 1 + x + \frac{x^2}{\lfloor 2} + \frac{x^3}{\lfloor 3} + \ldots$$

Put $x = 1$, and we have

$$e = 1 + 1 + \frac{1}{\lfloor 2} + \frac{1}{\lfloor 3} + \frac{1}{\lfloor 4} + \ldots$$

This series may be used for calculating the approximate value of e, and we may shew from it that e must be an *incommensurable* number. See *Plane Trigonometry*, Chap. x.

It is found that $e = 2 \cdot 718281828 \ldots$

116. Let $\qquad f(x) = \sin x.$

By Arts. 95 and 78,

$$\sin x = x - \frac{x^3}{\lfloor 3} + \frac{x^5}{\lfloor 5} - \ldots\ldots$$

$$+ \frac{x^n}{\lfloor n} \sin\left(\frac{n\pi}{2}\right) + \frac{x^{n+1}}{\lfloor n+1} \sin\left(\frac{n+1}{2}\pi + \theta x\right).$$

Similarly $\quad \cos x = 1 - \frac{x^2}{\lfloor 2} + \frac{x^4}{\lfloor 4} - \ldots$

$$+ \frac{x^n}{\lfloor n} \cos\left(\frac{n\pi}{2}\right) + \frac{x^{n+1}}{\lfloor n+1} \cos\left(\frac{n+1}{2}\pi + \theta x\right).$$

In Arts. 115 and 116, the student will see that the last term can be made as small as we please, whatever be the value of x, if n be taken large enough.

117. Let $f(x) = \log(1 + x)$;

therefore $f'(x) = \dfrac{1}{1 + x}$ and $f'(0) = 1$,

$$f''(x) = -\frac{1}{(1 + x)^2}, \text{ and } f''(0) = -1,$$

$$\cdots\cdots\cdots\cdots\cdots\cdots$$

$$f^n(x) = \frac{(-1)^{n-1}\lfloor n-1}{(1+x)^n} \text{ and } f^n(0) = (-1)^{n-1}\lfloor n-1;$$

$$\cdots\cdots\cdots\cdots\cdots\cdots$$

therefore, by Art. 95,

$$\log(1 + x) = x - \frac{x^2}{2} + \frac{x^3}{3} - \ldots + \frac{(-1)^{n-1}}{n}x^n$$

$$+ \frac{(-1)^n x^{n+1}}{(n+1)(1+\theta x)^{n+1}}.$$

In this series, if we suppose x positive and not greater than unity, then, as $\left(\dfrac{x}{1 + \theta x}\right)^{n+1}$ can not be greater than unity, the error we commit, if we stop at the term $\dfrac{(-1)^{n-1}x^n}{n}$, is not greater than $\dfrac{1}{n+1}$; that is, the error can be made as small as we please by increasing n sufficiently.

If we change the sign of x, we have

$$\log(1 - x) = -x - \frac{x^2}{2} - \frac{x^3}{3} - \ldots - \frac{x^n}{n} - \frac{x^{n+1}}{(n+1)(1-\theta x)^{n+1}},$$

which does not give a very convenient form to the *remainder*. But by Art. 110, we may also write

$$\log(1 - x) = -x - \frac{x^2}{2} - \frac{x^3}{3} - \ldots - \frac{x^n}{n} - \frac{(1-\theta)^n x^{n+1}}{(1-\theta x)^{n+1}},$$

where θ is between 0 and 1;

now $$\frac{(1-\theta)^n x^{n+1}}{(1-\theta x)^{n+1}} = \left(\frac{x-\theta x}{1-\theta x}\right)^n \cdot \frac{x}{1-\theta x}.$$

If x be less than unity, so also is $\dfrac{x-\theta x}{1-\theta x}$, and $\left(\dfrac{x-\theta x}{1-\theta x}\right)^n$ can be made as small as we please by taking n large enough.

Hence, if n be taken large enough, the *remainder* can be made as small as we please.

118.   In the preceding Examples, we have been able to write down the general term of the series, and the *remainder* after $n+1$ terms.   But if $f(x)$ be a complicated function, the expression for $f^n(x)$ will be generally too unwieldy for us to employ.   It is, therefore, not unusual to propose such questions as "expand $e^x \log(1+x)$, by Maclaurin's Theorem, as far as the term involving x^5."   Here we are not required to ascertain the *general term*, or the *remainder*, or to shew when, for the purpose of numerical computation, the remainder may be neglected.   We proceed thus:

$f(x) = e^x \log(1+x),$

therefore $f(0) = 0.$

By Art. 80,

$$f'(x) = e^x \log(1+x) + \frac{e^x}{1+x},$$

therefore $f'(0) = 1;$

$$f''(x) = e^x \log(1+x) + \frac{2e^x}{1+x} - \frac{e^x}{(1+x)^2},$$

therefore $f''(0) = 1;$

$$f'''(x) = e^x \log(1+x) + \frac{3e^x}{1+x} - \frac{3e^x}{(1+x)^2} + \frac{2e^x}{(1+x)^3},$$

therefore $f'''(0) = 2;$

$$f^{iv}(x) = e^x \log(1+x) + \frac{4e^x}{1+x} - \frac{6e^x}{(1+x)^2} + \frac{8e^x}{(1+x)^3} - \frac{6e^x}{(1+x)^4},$$

therefore $f^{iv}(0) = 0;$

$$f^v(x) = e^x \log(1+x) + \frac{5e^x}{1+x} - \frac{10e^x}{(1+x)^2} + \frac{20e^x}{(1+x)^3} - \frac{30e^x}{(1+x)^4} + \frac{24e^x}{(1+x)^5},$$

therefore $f^v(0) = 9.$

Hence $\quad e^x \log (1 + x) = x + \dfrac{x^2}{\lfloor 2} + \dfrac{2x^3}{\lfloor 3} + \dfrac{9x^5}{\lfloor 5} + \dots.$

This may be verified by multiplying the expansion for e^x by that for $\log (1 + x)$.

119. Methods of expansion of more or less rigour are often adopted in special cases of which we will proceed to give examples. We do not lay any stress upon them as exact investigations, but they may serve as exercises in differentiation.

Expand $\tan^{-1} x$ in powers of x.

Assume $\quad \tan^{-1} x = A_0 + A_1 x + A_2 x^2 + \dots + A_n x^n + \dots \dots (1).$

Differentiate both sides with respect to x,

then $\quad \dfrac{1}{1 + x^2} = A_1 + 2A_2 x + \dots + nA_n x^{n-1} + \dots \dots (2).$

But $\quad \dfrac{1}{1 + x^2} = 1 - x^2 + x^4 - x^6 + x^8 - \dots \dots (3),$

by simple division, or by the binomial theorem.

Equating coefficients of like powers of x in (2) and (3), we have

$$A_1 = 1, \ A_2 = 0, \ A_3 = -\tfrac{1}{3}, \ A_4 = 0, \dots$$

and putting $x = 0$ in (1), we get $A_0 = 0$; therefore

$$\tan^{-1} x = x - \dfrac{x^3}{3} + \dfrac{x^5}{5} - \dfrac{x^7}{7} + \dots$$

This example may also be easily treated by the rigorous method already used in Arts. 114...117. It appears from Example 18, page 65, that the nth differential coefficient of $\tan^{-1} x$ with respect to x is

$$\dfrac{(-1)^{n-1} \lfloor n-1}{(1 + x^2)^{\frac{n}{2}}} \sin \left(\dfrac{n\pi}{2} - n \tan^{-1} x \right).$$

Hence we have

$$\tan^{-1}x = x - \frac{x^3}{3} + \frac{x^5}{5} - \ldots + \frac{x^n}{n}(-1)^{n-1}\sin\frac{n\pi}{2}$$

$$+ \frac{(-1)^n x^{n+1}}{(n+1)(1+\theta^2 x^2)^{\frac{n+1}{2}}}\sin\left\{\frac{(n+1)\pi}{2} - (n+1)\tan^{-1}\theta x\right\}.$$

And if x be numerically less than 1, the last term can be made as small as we please by sufficiently increasing n; so that the infinite series

$$x - \frac{x^3}{3} + \frac{x^5}{5} - \frac{x^7}{7} + \ldots$$

can by taking a sufficient number of terms be brought as near as we please to $\tan^{-1}x$.

120.   Expand $\sin^{-1}x$ in powers of x.

Assume $\sin^{-1}x = A_0 + A_1 x + A_2 x^2 + \ldots + A_n x^n + \ldots\ldots(1)$.

Differentiate both sides; thus

$$\frac{1}{\sqrt{(1-x^2)}} = A_1 + 2A_2 x + 3A_3 x^2 + \ldots + nA_n x^{n-1} + \ldots (2).$$

But $\dfrac{1}{\sqrt{(1-x^2)}} = 1 + \frac{1}{2}x^2 + \frac{1.3}{2.4}x^4 + \frac{1.3.5}{2.4.6}x^6 + \ldots\ldots\ldots (3)$,

by the Binomial Theorem.

Hence, comparing the coefficients in (2) and (3), we determine A_1, A_2, ..., and putting $x = 0$ in (1) we get $A_0 = 0$. Substituting in (1), we have

$$\sin^{-1}x = x + \frac{1}{2}\cdot\frac{x^3}{3} + \frac{1.3}{2.4}\cdot\frac{x^5}{5} + \ldots.$$

It should be remarked that there are two considerations which limit the generality of this investigation. We take $\dfrac{1}{\sqrt{(1-x^2)}}$ as the differential coefficient of $\sin^{-1}x$, whereas the radical ought strictly to have the double sign: see Art. 65. And we take $\sin^{-1}x$ to vanish with x, whereas we know, by Trigonometry, that $\sin^{-1}x$ might be any multiple of π when x vanishes.

Similar remarks apply to the expansions in the next two Articles.

121. Expand $e^{a \sin^{-1}x}$ in powers of x.

Put
$$e^{a \sin^{-1}x} = y \dots (1),$$

then
$$\frac{dy}{dx} = e^{a \sin^{-1}x} \frac{a}{\sqrt{(1-x^2)}} \dots (2),$$

$$\frac{d^2y}{dx^2} = e^{a \sin^{-1}x} \frac{a^2}{1-x^2} + \frac{xae^{a\sin^{-1}x}}{(1-x^2)^{\frac{3}{2}}} \dots (3);$$

therefore
$$(1-x^2)\frac{d^2y}{dx^2} - x\frac{dy}{dx} = a^2y \dots (4).$$

Assume
$$y = A_0 + A_1x + A_2x^2 + A_3x^3 + \dots + A_nx^n + \dots (5);$$

therefore $\dfrac{dy}{dx} = A_1 + 2A_2x + \dots + nA_nx^{n-1} + \dots$

$$\frac{d^2y}{dx^2} = 2A_2 + \dots + n(n-1)A_nx^{n-2} + \dots$$

Substitute these values of y, $\dfrac{dy}{dx}$, and $\dfrac{d^2y}{dx^2}$, in (4), then equate the coefficients of like powers of x on both sides, and we obtain

$$A_{n+2} = \frac{a^2 + n^2}{(n+1)(n+2)} A_n \dots (6).$$

Equation (6) will enable us to determine A_2, A_3, A_4, ... as soon as we know A_0 and A_1.

But A_0 is the value of y or $e^{a \sin^{-1}x}$ when $x = 0$, and

A_1 is the value of $\dfrac{dy}{dx}$ or $e^{a \sin^{-1}x}\dfrac{a}{\sqrt{(1-x^2)}}$, when $x = 0$;

therefore $A_0 = 1$, and $A_1 = a$.

Hence, by (6),

$$A_2 = \frac{a^2}{1.2}A_0 = \frac{a^2}{\lfloor 2},$$

$$A_2 = \frac{a^2+1}{2.3} A_1 = \frac{(a^2+1)\,a}{\underline{|3}},$$

and so on ;

therefore $e^{a\sin^{-1}x} = 1 + ax + \frac{a^2x^2}{\underline{|2}} + \frac{a\,(a^2+1)}{\underline{|3}}\,x^3 + \frac{a^2\,(a^2+2^2)}{\underline{|4}}\,x^4$

$$+ \frac{a\,(a^2+1)\,(a^2+3^2)}{\underline{|5}}\,x^5 + \dots$$

Since $e^{a\sin^{-1}x} = 1 + a\sin^{-1}x + \frac{a^2}{\underline{|2}}\,(\sin^{-1}x)^2 + \dots$

we have, by equating the coefficients of a in this series, and in the result just found,

$$\sin^{-1}x = x + \frac{1}{2}\cdot\frac{x^3}{3} + \frac{1.3}{2.4}\cdot\frac{x^5}{5} + \dots$$

as already found.

Also equating the coefficients of a^2, we have

$$(\sin^{-1}x)^2 = x^2 + \frac{2^2}{3.4}\,x^4 + \frac{2^2.4^2}{3.4.5.6}\,x^6 + \frac{2^2.4^2.6^2}{3.4.5.6.7.8}\,x^8 + \dots$$

And equating the coefficients of a^3 we have

$$(\sin^{-1}x)^3 = x^3 + \frac{\underline{|3}}{\underline{|5}}\,3^2\left(1+\frac{1}{3^2}\right)x^5 + \frac{\underline{|3}}{\underline{|7}}\,3^2.5^2\left(1+\frac{1}{3^2}+\frac{1}{5^2}\right)x^7$$

$$+ \dots$$

122. Expand $\sin(m\sin^{-1}x)$ in powers of x.

Putting y for the function, we may shew that

$$(1-x^2)\frac{d^2y}{dx^2} = x\frac{dy}{dx} - m^2y.$$

Proceeding as in Art. 121, we find that

$$(n+1)\,(n+2)\,A_{n+2} = (n^2-m^2)\,A_n; \text{ and thus}$$

$$\sin(m\sin^{-1}x) = \frac{m}{1}x + \frac{m\,(1^2-m^2)}{\underline{|3}}\,x^3 + \frac{m\,(1^2-m^2)\,(3^2-m^2)}{\underline{|5}}\,x^5 + \dots$$

Similarly $\cos(m\sin^{-1}x)$

$$= 1 - \frac{m^2}{\underline{|2}}\,x^2 - \frac{m^2\,(2^2-m^2)}{\underline{|4}}\,x^4 - \frac{m^2\,(2^2-m^2)\,(4^2-m^2)}{\underline{|6}}\,x^6 - \dots$$

123. Expand $\dfrac{x}{e^x - 1}$ in powers of x.

We shall first shew that no *odd* power of x except the first can occur in the expansion. Denote the function by $\phi(x)$.

Then
$$\phi(x) - \phi(-x) = \frac{x}{e^x - 1} - \frac{-x}{e^{-x} - 1}$$

$$= \frac{x}{e^x - 1} + \frac{xe^x}{1 - e^x} = \frac{x(1 - e^x)}{e^x - 1} = -x.$$

This shews that no odd power of x except the first can occur in $\phi(x)$; for every odd power of x which occurs in $\phi(x)$ must also occur in $\phi(x) - \phi(-x)$.

We have $\phi(x)(e^x - 1) = x$; therefore $e^x \phi(x) = x + \phi(x)$.

Differentiate successively with respect to x; thus

$e^x \{\phi'(x) + \phi(x)\} = 1 + \phi'(x),$

$e^x \{\phi''(x) + 2\phi'(x) + \phi(x)\} = \phi''(x),$

$e^x \{\phi'''(x) + 3\phi''(x) + 3\phi'(x) + \phi(x)\} = \phi'''(x),$

$e^x \{\phi''''(x) + 4\phi'''(x) + 6\phi''(x) + 4\phi'(x) + \phi(x)\} = \phi''''(x),$

and so on.

Put $x = 0$ in these equations; thus

$$\phi(0) = 1,$$
$$2\phi'(0) + \phi(0) = 0,$$
$$3\phi''(0) + 3\phi'(0) + \phi(0) = 0,$$
$$4\phi'''(0) + 6\phi''(0) + 4\phi'(0) + \phi(0) = 0,$$

and so on.

Hence we find in succession

$$\phi'(0) = -\frac{1}{2}, \quad \phi''(0) = \frac{1}{6}, \quad \phi'''(0) = 0, \quad \phi''''(0) = -\frac{1}{30}, \dots$$

It is usual to denote the expansion thus:

$$\frac{x}{e^x - 1} = 1 - \tfrac{1}{2}x + \frac{B_1}{\lfloor 2}x^2 - \frac{B_3}{\lfloor 4}x^4 + \frac{B_5}{\lfloor 6}x^6 - \frac{B_7}{\lfloor 8}x^8 + \dots;$$

the coefficients $B_1, B_3, B_5, B_7, \dots$ are called the *numbers of*

Bernoulli, having been first noticed by James Bernoulli. It will be found that

$$B_1 = \frac{1}{6}, \ B_3 = \frac{1}{30}, \ B_5 = \frac{1}{42}, \ B_7 = \frac{1}{30}, \ B_9 = \frac{5}{66}, \ \dots$$

EXAMPLES.

1. If $e^{2x}(3-x) - 4xe^x - x - 3$ be expanded by Maclaurin's Theorem, the first term is $-\dfrac{4x^5}{\lfloor 5}$.

2. Expand $\log(1 + e^x)$ in powers of x.

$$\text{Result.} \quad \log 2 + \frac{x}{2} + \frac{x^2}{2^3} - \frac{x^4}{2^3 \lfloor 4} + \dots$$

3. Expand $e^{x \sin x}$ in powers of x.

$$\text{Result.} \quad 1 + x^2 + \frac{x^4}{3} + \dots$$

4. $e^x \sec x = 1 + x + x^2 + \dfrac{2x^3}{3} + \dots$

5. $\left(\dfrac{1 + e^x}{2}\right)^n = 1 + \dfrac{nx}{2} + \dfrac{n(n+1)}{2 \cdot 2} \dfrac{x^2}{2} + \dots$

6. $\sqrt{(1 + 4x + 12x^2)} = 1 + 2x + 4x^2 + \dots$

7. $(e^x + e^{-x})^n = 2^n \left\{1 + \dfrac{n}{\lfloor 2} x^2 + \dfrac{3n^2 - 2n}{\lfloor 4} x^4 + \dots \right\}.$

8. $(\cos x)^n = 1 - \dfrac{nx^2}{\lfloor 2} + \dfrac{n(3n-2)x^4}{\lfloor 4} - \dfrac{n\{15(n-1)^2 + 1\}x^6}{\lfloor 6}$
$+ \dots$

9. $-\log \cos x = \dfrac{x^2}{\lfloor 2} + \dfrac{2x^4}{\lfloor 4} + \dfrac{16x^6}{\lfloor 6} + \dfrac{16 \times 17x^8}{\lfloor 8} + \dots$

10. $e^{\cos x} = e\left\{1 - \dfrac{x^2}{\lfloor 2} + \dfrac{4x^4}{\lfloor 4} - \dfrac{31x^6}{\lfloor 6} \dots \right\}.$

11. $\sin^{-1}(x + h) = \sin^{-1}x + \dfrac{h}{\sqrt{(1 - x^2)}} + \dfrac{x}{(1 - x^2)^{\frac{3}{2}}} \dfrac{h^2}{\lfloor 2}$

$+ \dfrac{1 + 2x^2}{(1 - x^2)^{\frac{5}{2}}} \dfrac{h^3}{\lfloor 3} + \dfrac{3x(3 + 2x^2)}{(1 - x^2)^{\frac{7}{2}}} \dfrac{h^4}{\lfloor 4} + \dots$

12. $\log(1 - x + x^2) = -x + \dfrac{x^2}{2} + \dfrac{2x^3}{3} + \dfrac{x^4}{4} - \dfrac{x^5}{5} \ldots$

13. $\log\{x + \sqrt{(a^2 + x^2)}\} = \log a + \dfrac{x}{a} - \dfrac{1}{2}\dfrac{x^3}{3a^3} + \dfrac{1.3}{2.4}\dfrac{x^5}{5a^5} - \ldots$

14. $\log(1 + \sin x) = x - \dfrac{x^2}{2} + \dfrac{x^3}{6} \ldots$

15. $e^{\tan^{-1}x} = 1 + x + \dfrac{x^2}{2} - \dfrac{x^3}{6} - \dfrac{7x^4}{24} \ldots$

16. For what values of x does Taylor's Theorem fail, if $y = \sqrt[5]{\left\{\dfrac{(x-a)^7(x-b)^{10}}{(x-c)^2}\right\}}$, and which is the first differential coefficient that becomes infinite ?

17. Shew that

$$\tan^{-1}(x + h) = \tan^{-1}x + h\sin^2\theta - \dfrac{h^2}{2}\sin^2\theta \sin 2\theta$$
$$+ \dfrac{h^3}{3}\sin^3\theta \sin 3\theta - \dfrac{h^4}{4}\sin^4\theta \sin 4\theta + \ldots$$

where $\theta = \dfrac{\pi}{2} - \tan^{-1}x$. See Example 18 of Chapter v.

18. By putting $h = -x$ in Example 17, shew that

$$\dfrac{\pi}{2} - \theta = \sin\theta \cos\theta + \dfrac{\cos^2\theta \sin 2\theta}{2} + \dfrac{\cos^3\theta \sin 3\theta}{3}$$
$$+ \dfrac{\cos^4\theta \sin 4\theta}{4} + \ldots$$

19. By putting $h = -x - \dfrac{1}{x}$ in Example 17, shew that

$$\dfrac{\pi}{2} = \dfrac{\sin\theta}{\cos\theta} + \dfrac{\sin 2\theta}{2\cos^2\theta} + \dfrac{\sin 3\theta}{3\cos^3\theta} + \dfrac{\sin 4\theta}{4\cos^4\theta} + \ldots$$

20. By putting $h = -\sqrt{(1 + x^2)}$ in Example 17, shew that

$$\dfrac{1}{2}(\pi - \theta) = \sin\theta + \dfrac{1}{2}\sin 2\theta + \dfrac{1}{3}\sin 3\theta + \dfrac{1}{4}\sin 4\theta + \ldots$$

CHAPTER VIII.

SUCCESSIVE DIFFERENTIATION. DIFFERENTIATION OF A FUNCTION OF TWO VARIABLES.

124. WE have, in Art. 77, defined the *second* differential coefficient of a function to be the differential coefficient of the differential coefficient of that function. The differential coefficient of the second differential coefficient has been called the *third* differential coefficient, and so on. We are now about to give another view of these successive differential coefficients.

125. Let
$$y = f(x),$$
$$y + \Delta y = f(x + h),$$
therefore
$$\Delta y = f(x + h) - f(x).$$

In the right-hand member of the last equation change x into $x + h$ and subtract the original value; we thus obtain

$$f(x + 2h) - f(x + h) - \{f(x + h) - f(x)\},$$
or
$$f(x + 2h) - 2f(x + h) + f(x).$$

This result, agreeably to our previous notation, may be denoted by $\Delta(\Delta y)$, which we abbreviate into $\Delta^2 y$. Hence

$$\Delta^2 y = f(x + 2h) - 2f(x + h) + f(x).$$

Similarly $\Delta(\Delta^2 y)$ or $\Delta^3 y$ will be equal to

$$f(x + 3h) - 2f(x + 2h) + f(x + h)$$
$$- \{f(x + 2h) - 2f(x + h) + f(x)\},$$

that is, $\Delta^3 y = f(x + 3h) - 3f(x + 2h) + 3f(x + h) - f(x).$

126. By pursuing the method of the last Article we find expressions for $\Delta^4 y$, $\Delta^5 y$, ... We shall not for our purpose require the general expression for $\Delta^n y$. It will, however, be easy for the reader to shew, by an inductive proof, that

$$\Delta^n y = f(x + nh) - nf\{x + (n-1)h\} + \frac{n(n-1)}{\lfloor 2} f\{x + (n-2)h\} - \dots$$
$$\dots \dots \pm nf(x + h) \mp f(x).$$

127. *To shew that the limit of* $\dfrac{\Delta^2 y}{(\Delta x)^2}$ *is* $\dfrac{d^2 y}{dx^2}$.

We have, by Art. 125,

$$\Delta^2 y = f(x + 2h) - 2f(x + h) + f(x).$$

But, by Art. 92,

$$f(x + 2h) = f(x) + 2hf'(x) + \frac{(2h)^2}{\lfloor 2} f''(x) + \frac{(2h)^3}{\lfloor 3} f'''(x + 2\theta h),$$

$$f(x + h) = f(x) + hf'(x) + \frac{h^2}{\lfloor 2} f''(x) + \frac{h^3}{\lfloor 3} f'''(x + \theta_1 h),$$

θ and θ_1 being proper fractions. Hence

$$\Delta^2 y = h^2 f''(x) + \frac{h^3}{3}\{4f'''(x + 2\theta h) - f'''(x + \theta_1 h)\}.$$

Divide both sides by h^2, that is $(\Delta x)^2$, and then let h be diminished indefinitely. Hence we obtain

$$\text{the limit of } \frac{\Delta^2 y}{(\Delta x)^2} = f''(x);$$

that is, the limit of $\dfrac{\Delta^2 y}{(\Delta x)^2}$ is $\dfrac{d^2 y}{dx^2}$.

128. The result of the last Article may be generalized by the inductive method of proof. Assume

$$\Delta^n y = h^n f^n(x) + h^{n+1}\psi(x) \dots\dots\dots\dots (1),$$

where $\psi(x)$ is a function of x and h, which remains finite when h is made $= 0$. From (1) we have

$$\Delta^{n+1} y = h^n f^n(x + h) + h^{n+1}\psi(x + h) - \{h^n f^n(x) + h^{n+1}\psi(x)\}$$
$$= h^n\{f^n(x + h) - f^n(x)\} + h^{n+1}\{\psi(x + h) - \psi(x)\}.$$

H 2

Now, by Art. 92,

$$f^n(x+h) = f^n(x) + hf^{n+1}(x) + \frac{h^2}{2}f^{n+2}(x+\theta h),$$

$$\psi(x+h) = \psi(x) + h\psi'(x+\theta_1 h),$$

therefore

$$\Delta^{n+1}y = h^{n+1}f^{n+1}(x) + h^{n+2}\{\tfrac{1}{2}f^{n+2}(x+\theta h) + \psi'(x+\theta_1 h)\}$$

$$= h^{n+1}f^{n+1}(x) + h^{n+2}\psi_1(x) \dots\dots\dots (2).$$

Equation (2) shews us that, granting the truth of (1), we can deduce for $\Delta^{n+1}y$ a value of the same form as that we assumed for $\Delta^n y$. But Art. 127 gives for $\Delta^x y$ an expression of the assumed form; hence $\Delta^3 y$ has the same form, and so also has $\Delta^4 y$, and generally $\Delta^n y$.

From equation (1), by dividing both sides by h^n and then diminishing h indefinitely, we have

$$\text{the limit of } \frac{\Delta^n y}{(\Delta x)^n} = f^n(x) ;$$

that is, the limit of $\dfrac{\Delta^n y}{(\Delta x)^n}$ is $\dfrac{d^n y}{dx^n}$.

129. Hitherto we have only considered functions of *one independent variable;* that is, we have supposed in the equation $y = f(x)$, although quantities denoted by such symbols as a, b, ... might occur in $f(x)$, yet they were not susceptible of any change. Suppose now we have the equation

$$u = x^2 + xy + y^2,$$

and let y denote some constant quantity and x a variable, we have

$$\frac{du}{dx} = 2x + y.$$

From the same equation, if x be a constant quantity and y a variable, we obtain

$$\frac{du}{dy} = 2y + x.$$

Of course we cannot simultaneously consider x both constant and variable; but there will be no inconsistency if on one occasion and for one purpose we consider x constant, and on another occasion and for another purpose we consider it variable.

130.   If x and y denote quantities such that either of them may be considered to change without affecting the other, they are called *independent variables*, and any quantity u, the value of which depends on the values of x and y, is called a "function of the independent variables x and y;"

$\dfrac{du}{dx}$, $\dfrac{d^2u}{dx^2}$, $\dfrac{d^3u}{dx^3}$, ..., denote the successive differential co-efficients of u, taken on the supposition that x alone varies;

$\dfrac{du}{dy}$, $\dfrac{d^2u}{dy^2}$, $\dfrac{d^3u}{dy^3}$, ..., denote the successive differential co-efficients of u, taken on the supposition that y alone varies.

131.   If u be a function of the independent variables x and y, then $\dfrac{du}{dx}$ will also be generally a function of x and y. Hence we may have occasion for its differential coefficient with respect to x or y.   The former is denoted by

$$\frac{a^2u}{dx^2},$$

as already stated; the latter is denoted by

$$\frac{d\,\dfrac{du}{dx}}{dy},$$

which is abbreviated into $\dfrac{d^2u}{dy\,dx}$.

Again, both $\dfrac{d^2u}{dx^2}$ and $\dfrac{d^2u}{dy\,dx}$ will be generally functions of both x and y. These may require to be differentiated with respect to x or y.   Hence we use such symbols as

$$\frac{d^3u}{dy\,dx^2}, \quad \frac{d^3u}{dx\,dy\,dx}, \quad \text{and } \frac{d^3u}{dy^2\,dx};$$

the meaning of which may be gathered from the preceding remarks.   For example, $\dfrac{d^3u}{dx\,dy\,dx}$ implies the performance of three operations: we are to differentiate u with respect to x, supposing y constant; the resulting function is to be differentiated with respect to y, supposing x constant; this last result is to be differentiated with respect to x, supposing y constant.

132. In considering the equation $y = f(x)$, where we have *one* independent variable, the student could be referred to analytical geometry of two dimensions for illustrations of the nature of a dependent variable and of a differential coefficient. See Arts. 35...43. In like manner, if he is acquainted with the elements of analytical geometry of three dimensions, he will be assisted in the present Chapter of the Differential Calculus. For instance, the equation

$$z = ax + by + c$$

represents a plane; x and y are *two independent* variables, of which z is a function. Here

$$\frac{dz}{dx} = a, \qquad \frac{dz}{dy} = b,$$

and all the higher differential coefficients, $\dfrac{d^2z}{dx^2}, \dfrac{d^3z}{dx^3}, \ldots,$ vanish.

Again,
$$z = \sqrt{(r^2 - x^2 - y^2)} \quad \ldots\ldots\ldots\ldots (1),$$

is the equation to a sphere. If we pass from a point on the sphere, whose co-ordinates are x and y, to a point whose co-ordinates are $x + \Delta x$ and y, we vary x *without varying* y. If in this case the value of the third co-ordinate be $z + \Delta z$, we have

$$z + \Delta z = \sqrt{\{r^2 - y^2 - (x + \Delta x)^2\}} \quad \ldots\ldots\ldots\ldots (2).$$

From (1) and (2) we can of course find $\dfrac{\Delta z}{\Delta x}$; and its limit, which we denote by $\dfrac{dz}{dx}$, will be $\dfrac{-x}{\sqrt{(r^2 - x^2 - y^2)}}$.

The process is the same as if we had given

$$z = \sqrt{(a^2 - x^2)},$$

where a is a constant; from which we deduce

$$\frac{dz}{dx} = \frac{-x}{\sqrt{(a^2 - x^2)}},$$

and finally put $r^2 - y^2$ for a^2.

On the other hand, if we pass from the point (x, y) to a point having x and $y + \Delta y$ for its co-ordinates, we have, as before,

$$z + \Delta z = \sqrt{\{r^2 - x^2 - (y + \Delta y)^2\}} \dots\dots\dots\dots(3).$$

Now, in (2) and (3) we have used Δz; but we do not mean that *the value attached to the symbol is the same in both cases.* If there were any risk of error by confounding them, we could use $\Delta' z$ in (3), or something similar. But in fact we only use (3) to assist us in forming a conception of $\dfrac{dz}{dy}$; and since we look on $\dfrac{dz}{dx}$ and $\dfrac{dz}{dy}$ as *whole symbols* not admitting of decomposition, the question can never occur, "Is the dz in $\dfrac{dz}{dx}$ the same as the dz in $\dfrac{dz}{dy}$?"

133.   When u is a function of two independent variables, the differential coefficients $\dfrac{du}{dx}$, $\dfrac{du}{dy}$, $\dfrac{d^2u}{dx^2}$, $\dfrac{d^2u}{dx\,dy}$, ... are often called "*partial* differential coefficients." Each of these differential coefficients is obtained by one or more operations, every operation being conducted on the supposition that only *one* of the possible variables x and y is *actually* variable.

Let us suppose for example that $u = \tan^{-1} \dfrac{x}{y}$; then

$$\frac{du}{dx} = \frac{y}{x^2 + y^2}, \qquad\qquad \frac{du}{dy} = -\frac{x}{x^2 + y^2},$$

$$\frac{d^2u}{dx^2} = -\frac{2xy}{(x^2 + y^2)^2}, \qquad\qquad \frac{d^2u}{dy^2} = \frac{2xy}{(x^2 + y^2)^2},$$

and so on.

By differentiating $\dfrac{du}{dx}$ with respect to y we obtain

$$\frac{d\dfrac{du}{dx}}{dy} = \frac{x^2 - y^2}{(x^2 + y^2)^2};$$

and by differentiating $\dfrac{du}{dy}$ with respect to x we obtain

$$\frac{d\dfrac{du}{dy}}{dx} = \frac{x^2 - y^2}{(x^2 + y^2)^2}.$$

Thus we see that in this example

$$\frac{d\dfrac{du}{dx}}{dy} = \frac{d\dfrac{du}{dy}}{dx} \quad\dotfill \quad (1),$$

or, as we may write it,

$$\frac{d^2u}{dy\,dx} = \frac{d^2u}{dx\,dy} \quad\dotfill \quad (2).$$

We shall prove in the next Article that this result is universally true. Of the two modes of writing the result given in (1) and (2) the second is the more commodious, but it has the disadvantage of making the theorem which we have to prove *appear* obvious to the student, because it suggests to him that he is merely comparing two *fractions*. But as we have already remarked, a symbol for a differential coefficient is defined as a whole, and is not to be decomposed into a numerator and a denominator. See Arts. 26 and 77.

134. *If* u *be any function of the independent variables* x *and* y,

$$\frac{d\dfrac{du}{dx}}{dy} = \frac{d\dfrac{du}{dy}}{dx}.$$

Let $u = \phi(x, y)$; change x into $x + h$, then by Art. 92,

$$\phi(x + h, y) = \phi(x, y) + h\frac{du}{dx} + \frac{h^2}{2}\phi''(x + \theta h, y);$$

we may therefore write

$$\phi(x + h, y) - \phi(x, y) = h\frac{du}{dx} + h^2v \quad\dotfill \quad (1),$$

where v is a certain function of x and y, which remains finite when $h = 0$. In (1) write $y + k$ for y; then the left-hand member becomes $\phi(x+h, y+k) - \phi(x, y+k)$; by Art. 92

$\dfrac{du}{dx}$ becomes $\dfrac{du}{dx} + k\dfrac{d\dfrac{du}{dx}}{dy} + k^2\beta$, where β remains finite when $k = 0$; and v becomes $v + k\alpha$, where α is a quantity which remains finite when $k = 0$, for it tends to $\dfrac{dv}{dy}$ as its limit. Thus

$$\phi(x+h,\ y+k) - \phi(x, y+k) = h\dfrac{du}{dx} + hk\dfrac{d\dfrac{du}{dx}}{dy} + hk^2\beta$$
$$+ h^2v + h^2k\alpha \dotfill (2).$$

Subtract (1) from (2); thus

$$\phi(x+h, y+k) - \phi(x+h, y) - \phi(x, y+k) + \phi(x, y)$$
$$= hk\dfrac{d\dfrac{du}{dx}}{dy} + h^2k\alpha + hk^2\beta.$$

Divide by hk, and then suppose h and k to diminish indefinitely; therefore

$$\dfrac{d\dfrac{du}{dx}}{dy} = \text{the limit when } h \text{ and } k \text{ vanish of}$$

$$\dfrac{\phi(x+h,\ y+k) - \phi(x+h,\ y) - \phi(x.\ y+k) + \phi(x, y)}{hk}.$$

In a similar way, by *first* changing y into $y + k$, and *afterwards* x into $x + h$, we can prove that $\dfrac{d\dfrac{du}{dy}}{dx}$ is also equal to the above limit.

Hence $\qquad \dfrac{d\dfrac{du}{dx}}{dy} = \dfrac{d\dfrac{du}{dy}}{dx}.$

135. The object of the preceding Article is to prove that $\dfrac{d^2u}{dy\,dx} = \dfrac{d^2u}{dx\,dy}$; this is done by shewing that each of these quantities is equal to the limit of a certain expression. It is

comparatively unimportant what that expression is, but it is
of some interest to notice the analogy of the result to those
in Arts. 127 and 128.

Proofs of the proposition in the preceding Article have
sometimes been given which appear simpler than that here
adopted, but they are deficient in strictness. In particular
an *assumption* has sometimes been made which deserves to
be noticed. The following is substantially a proof that has
been given. To obtain $\dfrac{d\frac{du}{dx}}{dy}$ involves, according to the definition of the symbol, the following operations. (1) In the
function u we put $x + h$ for x, subtract the original value
from the new value, and then divide by h. (2) We find the
limit of the result when $h = 0$. (3) We now put $y + k$ for y,
subtract the original value from the new value, and then
divide by k. (4) We find the limit of the result when $k = 0$.
All this is immediately derived from first principles; the
next step however is the *assumption* that we may perform
the *third* of the above operations *before* the second instead of
after it. With this *assumption* the required result is readily
obtained; for from the first operation we get

$$. \; \frac{\phi(x + h, y) - \phi(x, y)}{h} \; ;$$

then from the third we get

$$\frac{\phi(x + h, y + k) - \phi(x + h, y) - \phi(x, y + k) + \phi(x, y)}{hk},$$

and according to our *assumption*, the limit of this is $\dfrac{d\frac{du}{dx}}{dy}$.

And by a similar *assumption* it is found that $\dfrac{d\frac{du}{dy}}{dx}$ is also
equal to the same limit.

One more remark must be made to guard against a possible
error. In the proof of Art. 134 we have used v for $\frac{1}{2}\phi''(x + \theta h, y)$;
in this expression all that is known of θ is that it is a
proper fraction, and it must not be assumed to be a function

of x *only*. Thus when y is changed into $y + k$ the value of θ will generally change. This does not affect the preceding proof, because it was not necessary there actually to find the value of $\dfrac{dv}{dy}$; but the *assumption* that θ does not change when y changes has rendered some proofs unsound which have been given of the proposition in Art. 134.

136. The important principle proved in Art. 134 is enunciated thus : " The order of independent differentiations is indifferent ;" or it is referred to as the principle of the "convertibility of independent differentiations." It may be extended to any number of differentiations; so that *if a function of two independent variables, x and y, is to be differentiated m times with respect to x, and n times with respect to y, the result will be the same in whatever order the differentiations be performed.* In proof of this we have only to apply the theorem of Art. 134 repeatedly in the manner shewn in the following example.

To prove that
$$\frac{d^3 u}{dy^2\, dx} = \frac{d^3 u}{dx\, dy^2} ;$$

$$\frac{d^3 u}{dy^2\, dx} = \frac{d\,\dfrac{d^2 u}{dy\, dx}}{dy}, \text{ by definition,}$$

$$= \frac{d\,\dfrac{d^2 u}{dx\, dy}}{dy}, \text{ by Art. 134,}$$

$$= \frac{d^3 u}{dy\, dx\, dy}, \text{ by definition,}$$

$$= \frac{d^2 v}{dy\, dx}, \text{ if } v = \frac{du}{dy},$$

$$= \frac{d^2 v}{dx\, dy}, \text{ by Art. 134,}$$

$$= \frac{d^3 u}{dx\, dy^2}.$$

137. If u be a function of the three independent variables, x, y, z, we have in a similar manner

$$\frac{d^2u}{dy\,dz} = \frac{d^2u}{dz\,dy},$$

$$\frac{d^2u}{dx\,dz} = \frac{d^2u}{dz\,dx},$$

$$\frac{d^2u}{dx\,dy} = \frac{d^2u}{dy\,dx},$$

$$\frac{d^3u}{dx\,dy\,dz} = \frac{d^3u}{dx\,dz\,dy} = \frac{d^3u}{dz\,dx\,dy},$$

and so on.

EXAMPLES.

1. If $u = \dfrac{x^2y}{a^2 - z^2}$, find $\dfrac{d^2u}{dx\,dy}$ and $\dfrac{d^2u}{dy\,dz}$.

2. Verify in the following cases the equation

$$\frac{d^2u}{dx\,dy} = \frac{d^2u}{dy\,dx}:$$

$$u = x \sin y + y \sin x,$$

$$u = x \log y,$$

$$u = x^y,$$

$$u = \log \tan \frac{y}{x},$$

$$u = \frac{ay - bx}{by - ax},$$

$$u = y \log (1 + xy).$$

3. If $u = Ax^a y^{a'} + Bx^\beta y^{\beta'} + Cx^\gamma y^{\gamma'} + \dots$

where $\qquad a + a' = \beta + \beta' = \gamma + \gamma' = \dots = n,$

shew that $\qquad\qquad x\dfrac{du}{dx} + y\dfrac{du}{dy} = nu.$

In this example u is called a *homogeneous function of n dimensions*.

4. If u be a homogeneous function of n dimensions, shew that

$$x\frac{d^2u}{dx^2} + y\frac{d^2u}{dx\,dy} = (n-1)\frac{du}{dx}, \qquad x\frac{d^2u}{dx\,dy} + y\frac{d^2u}{dy^2} = (n-1)\frac{du}{dy}.$$

5. If u be a homogeneous function of n dimensions, shew that

$$x^2\frac{d^2u}{dx^2} + 2xy\frac{d^2u}{dx\,dy} + y^2\frac{d^2u}{dy^2} = n(n-1)u.$$

6. Verify the theorems in Examples 3 and 4 in the following cases :

$$u = (x+y)^2,$$

$$u = \frac{xy}{x+y},$$

$$u = \sqrt[3]{(x^2 + y^2)}.$$

7. If $u = x^3 z^4 + e^x y^2 z^3 + x^4 y^2 z^2$, shew that

$$\frac{d^4u}{dx^2\,dy\,dz} = 6e^x yz^2 + 8yz.$$

8. If $u = e^{xyz}$, shew that

$$\frac{d^3u}{dx\,dy\,dz} = (1 + 3xyz + x^2 y^2 z^2)e^{xyz}.$$

9. If $u = y\sqrt{(a^2 - x^2)} + x\sqrt{(a^2 - y^2)}$, shew that

$$\frac{du}{dx}\frac{du}{dy} + \sqrt{(a^2 - x^2)}\sqrt{(a^2 - y^2)}\left(\frac{d^2u}{dx\,dy}\right)^2 = \frac{a^4}{\sqrt{(a^2 - x^2)}\sqrt{(a^2 - y^2)}}.$$

10.   If $u = \tan^{-1} \dfrac{xy}{\sqrt{(1 + x^2 + y^2)}}$, shew that

$$\frac{d^2u}{dx\,dy} = \frac{1}{(1 + x^2 + y^2)^{\frac{3}{2}}}, \quad \frac{d^4u}{dx^2\,dy^2} = \frac{15xy}{(1 + x^2 + y^2)^{\frac{7}{2}}}.$$

11.   If $u = x\sqrt{(a^2 - y^2)}\sqrt{(a^2 - z^2)} + y\sqrt{(a^2 - z^2)}\sqrt{(a^2 - x^2)}$
$$+ z\sqrt{(a^2 - x^2)}\sqrt{(a^2 - y^2)} - xyz,$$

shew that

$$- \sqrt{(a^2 - x^2)}\sqrt{(a^2 - y^2)}\sqrt{(a^2 - z^2)}\frac{d^3u}{dx\,dy\,dz} = \sqrt{(a^2 - x^2)}\frac{du}{dx}$$

$$= \sqrt{(a^2 - y^2)}\frac{du}{dy} = \sqrt{(a^2 - z^2)}\frac{du}{dz}.$$

12.   If $u = \log(x^3 + y^3 + z^3 - 3xyz)$, shew that

$$\frac{1}{6}\frac{d^3u}{dx\,dy\,dz} - \frac{1}{3}\frac{du}{dx}\frac{du}{dy}\frac{du}{dz} = e^{-u},$$

$$\frac{du}{dx} + \frac{du}{dy} + \frac{du}{dz} = \frac{3}{x + y + z},$$

$$\frac{d^2u}{dx^2} + \frac{d^2u}{dy^2} + \frac{d^2u}{dz^2} + 2\frac{d^2u}{dx\,dy} + 2\frac{d^2u}{dy\,dz} + 2\frac{d^2u}{dz\,dx}$$

$$= -\frac{9}{(x + y + z)^2},$$

$$\frac{d^6u}{dx^2\,dy^2\,dz^2} + \frac{d^6u}{dx^3\,dy^2\,dz} + \frac{d^6u}{dx^2\,dy^3\,dz} = -\frac{360}{(x + y + z)^6},$$

$$\frac{d^2u}{dx^2} + \frac{d^2u}{dy^2} + \frac{d^2u}{dz^2} = -\frac{3}{(x + y + z)^2},$$

$$\frac{d^5u}{dx^3\,dy\,dz} + \frac{d^5u}{dx\,dy^3\,dz} + \frac{d^5u}{dx\,dy\,dz^3} = \frac{72}{(x + y + z)^5}.$$

CHAPTER IX

138.   SUPPOSE $y = z + x\phi(y)$ ....(1),

where z and x are independent, and it is required to expand $f(y)$ according to ascending powers of x. Put u for $f(y)$, then, by Maclaurin's theorem, we have

$$u = u_0 + x\frac{du_0}{dx} + \frac{x^2}{1.2}\frac{d^2u_0}{dx^2} + \frac{x^3}{\underline{|3}}\frac{d^3u_0}{dx^3} + \cdots$$

where $u_0, \dfrac{du_0}{dx}, \dfrac{d^2u_0}{dx^2}, \ldots$ denote the values of $u, \dfrac{du}{dx}, \dfrac{d^2u}{dx^2}, \ldots$

when x is put $= 0$ after differentiation.   We proceed to transform these differential coefficients of u with respect to x into a more convenient form in order to ascertain their values when $x = 0$.   We shall first shew that

$$\frac{d}{dx}\left\{F(v)\frac{dv}{dz}\right\} = \frac{d}{dz}\left\{F(v)\frac{dv}{dx}\right\} \quad \ldots\ldots\ldots\ldots (2),$$

supposing that v is any function of the independent quantities x and z, and $F(v)$ any function of v.

To establish (2) we need only observe that the left-hand member is

$$F'(v)\frac{dv}{dx}\frac{dv}{dz} + F(v)\frac{d^2v}{dx\,dz},$$

and the right-hand member is

$$F'(v)\frac{dv}{dx}\frac{dv}{dz} + F(v)\frac{d^2v}{dz\,dx};$$

and these two expressions are equal by Art. 134.

From (1) we have

$$\frac{dy}{dx} = \phi(y) + x\phi'(y)\frac{dy}{dx},$$

therefore

$$\frac{dy}{dx} = \frac{\phi(y)}{1 - x\phi'(y)}.$$

Also

$$\frac{dy}{dz} = 1 + x\phi'(y)\frac{dy}{dz},$$

therefore

$$\frac{dy}{dz} = \frac{1}{1 - x\phi'(y)}.$$

Hence

$$\frac{dy}{dx} = \phi(y)\frac{dy}{dz}.$$

Also

$$\frac{du}{dx} = \frac{du}{dy}\frac{dy}{dx} \text{ and } \frac{du}{dz} = \frac{du}{dy}\frac{dy}{dz},$$

therefore

$$\frac{du}{dx} = \phi(y)\frac{du}{dz} \quad .... (3).$$

Hence

$$\frac{d^2u}{dx^2} = \frac{d}{dx}\left\{\phi(y)\frac{du}{dz}\right\}$$

$$= \frac{d}{dx}\left\{\phi(y)f'(y)\frac{dy}{dz}\right\}, \text{ since } u = f(y),$$

$$= \frac{d}{dz}\left\{\phi(y)f'(y)\frac{dy}{dx}\right\} \text{ by (2)},$$

$$= \frac{d}{dz}\left\{\phi(y)\frac{du}{dx}\right\},$$

$$= \frac{d}{dz}\left\{\overline{\phi(y)}|^2\frac{du}{dz}\right\} \text{ by (3)}.$$

Again

$$\frac{d^3u}{dx^3} = \frac{d^2}{dx\,dz}\left\{\overline{\phi(y)}|^2\frac{du}{dz}\right\}$$

$$= \frac{d^2}{dz\,dx}\left\{\overline{\phi(y)}|^2\frac{du}{dz}\right\}, \text{ by Art. 134},$$

$$= \frac{d^2}{dz^2}\left\{\overline{\phi(y)}|^2\frac{du}{dx}\right\}, \text{ by (2)},$$

$$= \frac{d^2}{dz^2}\left\{\overline{\phi(y)}|^3\frac{du}{dz}, \text{ by (3)}.\right.$$

Suppose, according to this law, that

$$\frac{d^n u}{dx^n} = \frac{d^{n-1}}{dz^{n-1}} \left\{ \overline{[\phi(y)]}^n \frac{du}{dz} \right\};$$

then

$$\frac{d^{n+1} u}{dx^{n+1}} = \frac{d^n}{dx\, dz^{n-1}} \left\{ \overline{[\phi(y)]}^n \frac{du}{dz} \right\}$$

$$= \frac{d^n}{dz^{n-1} dx} \left\{ \overline{[\phi(y)]}^n \frac{du}{dz} \right\}, \text{ by Art. 134,}$$

$$= \frac{d^n}{dz^n} \left\{ \overline{[\phi(y)]}^n \frac{du}{dx} \right\}, \text{ by (2),}$$

$$= \frac{d^n}{dz^n} \left\{ \overline{[\phi(y)]}^{n+1} \frac{du}{dz} \right\},$$

which shews us that the expression for $\dfrac{d^{n+1} u}{dx^{n+1}}$ follows the same law as that for $\dfrac{d^n u}{dx^n}$. Hence, since the law has been proved to hold for $\dfrac{d^2 u}{dx^2}$ and $\dfrac{d^3 u}{dx^3}$, it holds universally.

In $\dfrac{d^n u_0}{dx^n}$ we are to make $x = 0$ *after* the differentiation has been performed; but when we transform $\dfrac{d^n u}{dx^n}$, by the above formula, into an expression involving only differential co-efficients *taken with respect to z*, we may put $x = 0$ *before* the differentiation, since x is to be considered as a constant in differentiating with respect to z. When $x = 0$,

$$y = z,$$
$$\phi(y) = \phi(z),$$

therefore

$$\frac{du_0}{dz} = \frac{df(z)}{dz} = f'(z),$$

$$\frac{du_0}{dx} = \phi(z) f'(z),$$

$$\frac{d^2 u_0}{dx^2} = \frac{d}{dz} \left\{ \overline{[\phi(z)]}^2 f'(z) \right\},$$

....

and thus

$$f(y) = f(z) + x\phi(z)f'(z) + \frac{x^2}{\underline{2}}\frac{d}{dz}\left\{\overline{\phi(z)|}^2 f'(z)\right\}$$

$$+ \frac{x^3}{\underline{3}}\frac{d^2}{dz^2}\left\{\overline{\phi(z)|}^3 f'(z)\right\}$$

$$+ \ldots\ldots + \frac{x^n}{\underline{n}}\frac{d^{n-1}}{dz^{n-1}}\left\{\overline{\phi(z)|}^n f'(z)\right\} + \ldots$$

This result is called Lagrange's Theorem.

139. Suppose $y = F\{z + x\phi(y)\}$;
required the expansion of $f(y)$ in powers of x.

Let t stand for $z + x\phi(y)$; then

$$\frac{dy}{dx} = \frac{dF}{dt}\frac{dt}{dx} = \frac{dF}{dt}\left\{\phi(y) + x\phi'(y)\frac{dy}{dx}\right\},$$

therefore

$$\frac{dy}{dx} = \frac{\phi(y)\dfrac{dF}{dt}}{1 - x\phi'(y)\dfrac{dF}{dt}};$$

also

$$\frac{dy}{dz} = \frac{dF}{dt}\frac{dt}{dz} = \frac{dF}{dt}\left\{1 + x\phi'(y)\frac{dy}{dz}\right\},$$

therefore

$$\frac{dy}{dz} = \frac{\dfrac{dF}{dt}}{1 - x\phi'(y)\dfrac{dF}{dt}}.$$

Hence

$$\frac{dy}{dx} = \phi(y)\frac{dy}{dz}.$$

From this, in the same way as in Art. 138, we deduce that

$$\frac{d^n u}{dx^n} = \frac{d^{n-1}}{dz^{n-1}}\left\{\overline{\phi(y)|}^n\frac{du}{dz}\right\},$$

where $u = f(y)$.

If we make $x = 0$ in the equation

$$y = F\{z + x\phi(y)\},$$

we deduce
$$y = F(z),$$
$$\phi(y) = \phi\{F(z)\},$$
$$\frac{du}{dz} = \frac{df\{F(z)\}}{dz}, \quad \text{-- -- --}$$

and finally,

$$f(y) = f\{F(z)\} + x\phi\{F(z)\}\frac{df\{F(z)\}}{dz} + \frac{x^2}{\underline{2}}\frac{d}{dz}\left[\overline{\phi\{F(z)\}|^2}\frac{df\{F(z)\}}{dz}\right]$$

$$+ \ldots\ldots + \frac{x^n}{\underline{n}}\frac{d^{n-1}}{dz^{n-1}}\left[\overline{\phi\{F(z)\}|}^n\frac{df\{F(z)\}}{dz}\right] + \ldots$$

This is called Laplace's Theorem.

140.   Lagrange's Theorem may of course be deduced from Laplace's, by putting $F(z) = z$.   But Laplace's theorem may also be deduced from Lagrange's, thus:

In the equation $y = F\{z + x\phi(y)\}$ .... (1),

put $z + x\phi(y) = y'$,

then $y = F(y')$,

thus $y' = z + x\phi\{F(y')\}$................... (2),

and $f(y)$ becomes $f\{F(y')\}$.

Thus we are required to expand $f\{F(y')\}$ in powers of x, by means of equation (2).   But this is precisely what Lagrange's Theorem effects, the complex functions $f\{F(y')\}$ and $\phi\{F(y')\}$ taking the place of the simple functions $f(y')$ and $\phi(y')$.

141.   It must be remembered, that in quoting Maclaurin's Theorem, which serves as the foundation for those of Lagrange and Laplace, we ought strictly to have used it in the form given in Art. 95, with an expression for the remainder after $n + 1$ terms. That expression for the remainder however, becomes so complicated in this case, that we have not referred to it.   The investigation of Lagrange's and Laplace's Theorems must be confessed to be imperfect, since the tests of the convergence of these series, which alone can justify our use of them as arithmetical equivalents for the functions they profess to represent, are of too difficult a character for an elementary work.   The advanced student may consult Moigno's *Leçons*

de Calcul Différentiel, 18me *Leçon*, and Liouville's *Journal de Mathématiques*, tom. XI. p. 129 and 313.

142.   If $x = a + y\phi(x)$, we have by Lagrange's Theorem

$$f(x) = f(a) + y\left\{\phi(x)f'(x)\right\} + \frac{y^2}{\lfloor 2}\frac{d}{dx}\left\{\overline{\phi(x)}|^2 f'(x)\right\}$$

$$+ \frac{y^3}{\lfloor 3}\frac{d^2}{dx^2}\left\{\overline{\phi(x)}|^3 f'(x)\right\} + \dots$$

where in the coefficients of the different powers of y, we are to make $x = a$ after the differentiations have been performed. Let y or $\dfrac{x-a}{\phi(x)} = \psi(x)$, so that $x = a$ is a root of $\psi(x) = 0$; then

$$f(x) = f(a) + \psi(x)\left[\frac{f'(x)(x-a)}{\psi(x)}\right] + \frac{\{\psi(x)\}^2}{\lfloor 2}\frac{d}{dx}\left[\frac{f'(x)(x-a)^2}{\{\psi(x)\}^2}\right]$$

$$+ \dots$$

where, in the coefficients of the different powers of $\psi(x)$ after the differentiations, x is to be made $= a$. This series for $f(x)$ in powers of $\psi(x)$ is called Burmann's Theorem.

143.   Let $\psi^{-1}(x)$ denote the inverse function of $\psi(x)$, so that if $u = \psi(x)$ we have $\psi^{-1}(u) = x$, and therefore $\psi\{\psi^{-1}(u)\} = u$. If we write $\psi^{-1}x$ for x in Burmann's Theorem, we have

$$f\{\psi^{-1}(x)\} = f(a) + x\left[\frac{f'(x)(x-a)}{\psi(x)}\right] + \frac{x^2}{\lfloor 2}\frac{d}{dx}\left[\frac{f'(x)(x-a)^2}{\{\psi(x)\}^2}\right]$$

$$+ \frac{x^3}{\lfloor 3}\frac{d^2}{dx^2}\left[\frac{f'(x)(x-a)^3}{\{\psi(x)\}^3}\right] + \dots$$

No change is made in the quantities in the square brackets, for they do not contain x when the operations indicated are completely performed.

If $f(u) = u$, we have

$$\psi^{-1}(x) = a + x\left[\frac{x-a}{\psi(x)}\right] + \frac{x^2}{\lfloor 2}\frac{d}{dx}\left[\frac{(x-a)^2}{\{\psi(x)\}^2}\right]$$

$$+ \frac{x^3}{\lfloor 3}\frac{d^2}{dx^2}\left[\frac{(x-a)^3}{\{\psi(x)\}^3}\right] + \dots$$

and if $a = 0$, so that $\psi(x)$ vanishes with x,

$$\psi^{-1}(x) = x\left[\frac{x}{\psi(x)}\right] + \frac{x^2}{\lfloor 2}\frac{d}{dx}\left[\frac{x^3}{\{\psi(x)\}^2}\right]$$

$$+ \frac{x^3}{\lfloor 3}\frac{d^2}{dx^2}\left[\frac{x^4}{\{\psi(x)\}^3}\right] + \dots$$

EXAMPLES.

1. Given $y = z + xe^y$, expand y in powers of x.

Here
$$\phi(y) = e^y,$$
$$f(y) = y;$$

therefore
$$\frac{d^{n-1}}{dz^{n-1}}\left\{\overline{\phi(z)}|^n f'(z)\right\} = \frac{d^{n-1}}{dz^{n-1}}e^{nz} = n^{n-1}e^{nz}.$$

Thus $y = z + xe^z + \dfrac{x^2}{\lfloor 2}2e^{2z} + \dfrac{x^3}{\lfloor 3}3^2e^{3z} + \dots + \dfrac{x^n}{\lfloor n}n^{n-1}e^{nz} + \dots$

2. Given $y = z + x\dfrac{y^2 - 1}{2}$, expand y in powers of x.

Here
$$\phi(y) = \frac{y^2 - 1}{2},$$
$$f(y) = y;$$

therefore
$$\frac{d^{n-1}}{dz^{n-1}}\left\{\overline{\phi(z)}|^n f'(z)\right\} = \frac{1}{2^n}\frac{d^{n-1}}{dz^{n-1}}(z^2 - 1)^n.$$

Hence $y = z + x\dfrac{1}{2}(z^2 - 1) + \dfrac{x^2}{\lfloor 2}\cdot\dfrac{1}{2^2}\dfrac{d}{dz}(z^2 - 1)^2 + \dots$

$$+ \frac{x^n}{\lfloor n}\cdot\frac{1}{2^n}\frac{d^{n-1}}{dz^{n-1}}(z^2 - 1)^n + \dots$$

3. Given $xy - \log y = 0$, expand y in powers of x. From the given equation

$$y = e^{xy};$$

therefore
$$yx = xe^{xy},$$

say
$$y' = xe^{y'}.$$

If then we put $z = 0$ in the result of the first Example, we deduce

$$y' = x + x^2 + \frac{x^3}{\lfloor 2} 3 + \dots + \frac{x^n}{\lfloor n-1} n^{n-2} + \dots ;$$

restore yx for y' and divide by x; then

$$y = 1 + x + \frac{x^2}{\lfloor 2} 3 + \dots + \frac{x^{n-1}}{\lfloor n-1} n^{n-2} + \dots$$

4. If $y = \dfrac{x}{1 + \sqrt{(1 - x^2)}}$, expand y^n in ascending powers of x.

Since $\qquad\qquad y = \dfrac{x}{1 + \sqrt{(1 - x^2)}}$,

we have $\qquad\qquad y\sqrt{(1 - x^2)} = x - y$;

therefore $\qquad y^2(1 - x^2) = x^2 - 2xy + y^2$ .... (1),

and $\qquad\qquad\qquad y = \dfrac{x}{2} + \dfrac{y^2}{2} x.$

We must then put $y = z + \dfrac{y^2}{2} x,$

so that $\phi(y) = \dfrac{y^2}{2}$, and $f(y) = y^n$.

Thus $y^n = z^n + x \dfrac{n}{2} z^{n+1} + \dots + \dfrac{x^r}{\lfloor r} \dfrac{n}{2^r} \dfrac{d^{r-1}}{dz^{r-1}} (z^{2r+n-1}) + \dots (2),$

and *after* the differentiations are performed, we must put $\dfrac{x}{2}$ for z.

The quadratic equation (1) which we have employed gives two values for y, namely $\dfrac{x}{1 \pm \sqrt{(1 - x^2)}}$; the series which we have obtained in (2) applies to the value with the *upper* sign. For $\dfrac{x}{1 + \sqrt{(1 - x^2)}} = \dfrac{x}{2 - \dfrac{x^2}{2} - \dfrac{x^4}{8} - \dots}$; and if the nth power of this be expanded in ascending powers of x the first term is obviously $\left(\dfrac{x}{2}\right)^n$: whereas the first term of the expansion with the *lower* sign would be $\left(\dfrac{2}{x}\right)^n$, that is $\left(\dfrac{x}{2}\right)^{-n}$.

Now $y = \dfrac{x}{1 + \sqrt{(1 - x^2)}} = \dfrac{1 - \sqrt{(1 - x^2)}}{x}$; thus

$$\left\{\dfrac{1 - \sqrt{(1 - x^2)}}{x}\right\}^n = \left(\dfrac{x}{2}\right)^n + n\left(\dfrac{x}{2}\right)^{n+2} + \dfrac{n(n+3)}{1 \cdot 2}\left(\dfrac{x}{2}\right)^{n+4}$$

$$+ \dfrac{n(n+4)(n+5)}{1 \cdot 2 \cdot 3}\left(\dfrac{x}{2}\right)^{n+6} + \ldots$$

Let $x^2 = 4t$; thus we obtain

$$\left\{\dfrac{1 - \sqrt{(1 - 4t)}}{2t}\right\}^n = 1 + nt + \dfrac{n(n+3)}{1 \cdot 2}t^2$$

$$+ \dfrac{n(n+4)(n+5)}{1 \cdot 2 \cdot 3}t^3 + \ldots$$

Change the sign of n; thus we obtain the expansion in powers of t of $\left\{\dfrac{1 - \sqrt{(1 - 4t)}}{2t}\right\}^{-n}$, that is of $\left\{\dfrac{2t}{1 - \sqrt{(1 - 4t)}}\right\}^n$, that is of $\left\{\dfrac{1 + \sqrt{(1 - 4t)}}{2}\right\}^n$.

Hence $\left\{\dfrac{1 + \sqrt{(1 - 4t)}}{2}\right\}^n = 1 - nt + \dfrac{n(n-3)}{1 \cdot 2}t^2$

$$- \dfrac{n(n-4)(n-5)}{1 \cdot 2 \cdot 3}t^3 + \ldots$$

Hitherto we have put no restriction on the value of n; but let us now suppose that n is a positive integer.

If we expand $\{1 + \sqrt{(1 - 4t)}\}^n$ and $\{1 - \sqrt{(1 - 4t)}\}^n$ by the Binomial Theorem, we see that the sum of the two expressions will be a rational function of t which will be of the degree $\dfrac{n}{2}$ if n be even, and of the degree $\dfrac{n-1}{2}$ if n be odd.

By adding the expansions we have found above we obtain

$$\left\{\frac{1+\sqrt{(1-4t)}}{2}\right\}^n + \left\{\frac{1-\sqrt{(1-4t)}}{2}\right\}^n$$

$$= 1 - nt + \frac{n(n-3)}{1.2} t^2 - \frac{n(n-4)(n-5)}{1.2.3} t^3 + \dots;$$

and by what we have just shewn the series on the right hand extends to $\frac{n}{2}+1$ terms if n be even, and to $\frac{n+1}{2}$ terms if n be odd, so that the remaining terms in the two expansions must disappear; that is, the terms arising from one expansion are cancelled by similar terms arising from the other.

In the same manner as we deduced the expansion of y^n from the equation $y = \frac{x}{1+\sqrt{(1-x^2)}}$ we may deduce the expansion of any other function of y; for example take $\log y$. Thus

$$\log y = \log z + x \frac{1}{2} z + \dots + \frac{x^r}{\underline{|r}} \frac{1}{2^r} \frac{d^{r-1}}{dz^{r-1}} (z^{2r-1}) + \dots$$

where after the differentiations are performed we must put $\frac{x}{2}$ for z. Therefore

$$\log y = \log \frac{x}{2} + \left(\frac{x}{2}\right)^2 + \frac{3}{2}\left(\frac{x}{2}\right)^4 + \frac{4.5}{2.3}\left(\frac{x}{2}\right)^6 + \frac{5.6.7}{2.3.4}\left(\frac{x}{2}\right)^8 + \dots;$$

and $\qquad y = \dfrac{x}{1+\sqrt{(1-x^2)}} = \dfrac{1-\sqrt{(1-x^2)}}{x}.$

Let $x^2 = 4t$, and we shall obtain

$$\log \frac{1-\sqrt{(1-4t)}}{2t} = t + \frac{3}{2}t^2 + \frac{4.5}{2.3}t^3 + \frac{5.6.7}{2.3.4}t^4 + \dots$$

The expansions which this example has furnished are of some importance in mathematics.

5.   If $x = ye^y$, expand $\sin(a+y)$ in powers of x.

We have given $y = xe^{-y}$. Suppose then $y = z + xe^{-y}$, so that $\phi(y) = e^{-y}$, and $f(y) = \sin(a+y)$.

The general term given by Lagrange's Theorem is

$$\frac{x^n}{\underline{|n}} \frac{d^{n-1}}{dz^{n-1}} \{e^{-nz} \cos(a+z)\},$$

which becomes

$$\frac{x^n}{\lfloor n} (-1)^{n-1} (1+n^2)^{\frac{n-1}{2}} e^{-n\alpha} \cos\{\alpha + z - (n-1)\phi\},$$

where $\cot \phi = n$, by a process similar to that in Art. 81.

Putting $z = 0$ in this, we have for the required expansion

$$\sin(\alpha + y) = \sin\alpha + x\cos\alpha + \ldots$$

$$+ \frac{x^n}{\lfloor n} (-1)^{n-1} (1+n^2)^{\frac{n-1}{2}} \cos\{\alpha - (n-1)\cot^{-1}n\} + \ldots$$

6. Given $a - y + x\log y = 0$, find $\sin y$ in powers of x.

7. Given $y = z + xy^p e^{qy}$, expand $y^m e^{ny}$ in powers of x.

8. Given $y = z + x\sin y$, expand $\sin y$ and $\sin 2y$ in powers of x.

9. Given $y = \log(z + x\cos y)$, expand e^y in powers of x.

10. From the equation $xy^4 + 2xy^3 + 3xy^2 + 2y + 1 = 0$ determine y in ascending powers of x.

Result $\quad y = -\dfrac{1}{2} - \dfrac{9}{32} x - \dfrac{9}{32} x^2 - \dfrac{1395}{4096} x^3 \ldots$

11. If $y = e^{\frac{\pi}{4} + z\sin\log y}$, find the first four terms of the expansion of $\cos\log y$ in powers of x.

Result $\quad \dfrac{1}{\sqrt{2}} - \dfrac{x}{2} - \dfrac{3x^2}{4\sqrt{2}} - \dfrac{x^3}{3}.$

12. If $y^3 + my^2 + ny = x$, shew that one value of y is

$$\frac{x}{n} - \frac{m}{n}\left(\frac{x}{n}\right)^2 + \frac{2m^2 - n}{n^2}\left(\frac{x}{n}\right)^3 - \frac{5m^3 - 5mn}{n^3}\left(\frac{x}{n}\right)^4 + \ldots$$

CHAPTER X.

144. In the statement, the limit of $\dfrac{\sin\theta}{\theta} = 1$ when θ diminishes indefinitely, we have an example of a fraction which approaches a finite limit when the numerator and denominator each tend to the limit zero. The object of this Chapter is to find the limit of any fraction of which the numerator and denominator ultimately vanish, and also the limiting value of some other *indeterminate forms*.

145. Form $\dfrac{0}{0}$.

Suppose
$$\frac{\phi(x)}{\psi(x)}$$

such a fraction that both numerator and denominator vanish when $x = a$; it is required to find the limit towards which the above fraction tends as x approaches the limit a.

We have proved in Art. 92 that
$$\phi(a + h) - \phi(a) = h\phi'(a + \theta h),$$
$$\psi(a + h) - \psi(a) = h\psi'(a + \theta_1 h).$$

If then $\phi(a) = 0$ and $\psi(a) = 0$, we have, by division,
$$\frac{\phi(a + h)}{\psi(a + h)} = \frac{\phi'(a + \theta h)}{\psi'(a + \theta_1 h)}.$$

Let h diminish indefinitely; then

the limit when $x = a$ of $\dfrac{\phi(x)}{\psi(x)}$ is $\dfrac{\phi'(a)}{\psi'(a)}$.

146. Suppose that not only
$$\phi(a) = 0, \text{ and } \psi(a) = 0,$$
but also $\quad \phi'(a) = 0, \quad \phi''(a) = 0, \quad \ldots \phi^n(a) = 0,$
and $\quad \psi'(a) = 0, \quad \psi''(a) = 0, \quad \ldots \psi^n(a) = 0.$

By Art. 92,
$$\phi(a+h) - \phi(a) - h\phi'(a) \ldots - \frac{h^n}{\lfloor n} \phi^n(a) = \frac{h^{n+1}}{\lfloor n+1} \phi^{n+1}(a+\theta h),$$

$$\psi(a+h) - \psi(a) - h\psi'(a) \ldots - \frac{h^n}{\lfloor n} \psi^n(a) = \frac{h^{n+1}}{\lfloor n+1} \psi^{n+1}(a+\theta_1 h).$$

Hence, by division, we have
$$\frac{\phi(a+h)}{\psi(a+h)} = \frac{\phi^{n+1}(a+\theta h)}{\psi^{n+1}(a+\theta_1 h)}.$$

Diminish h indefinitely, and we have

the limit when $x = a$ of $\dfrac{\phi(x)}{\psi(x)}$ is $\dfrac{\phi^{n+1}(a)}{\psi^{n+1}(a)}$.

147. In Art. 145, if
$$\psi'(a) = 0,$$
and $\quad \phi'(a) = \text{some finite quantity,}$

we have the limit when $x = a$ of $\dfrac{\phi(x)}{\psi(x)}$ is infinity:

if $\quad \phi'(a) = 0,$

and $\quad \psi'(a) = \text{some finite quantity,}$

we have the limit when $x = a$ of $\dfrac{\phi(x)}{\psi(x)}$ is zero.

And in the same manner, we may shew that if the first of the differential coefficients $\phi'(a), \phi''(a), \ldots$ which does not vanish, is of a *lower* order than the first which does not vanish of the series $\psi'(a), \psi''(a), \ldots$, the limit of $\dfrac{\phi(x)}{\psi(x)}$ when $x = a$, is infinity; if of a *higher* order the limit is zero.

These results may also be obtained without the use of Taylor's Theorem.

If $\phi(a) = 0$ and $\psi(a) = 0$, we have

$$\frac{\phi(a+h)}{\psi(a+h)} = \frac{\phi(a+h) - \phi(a)}{\psi(a+h) - \psi(a)} = \frac{\dfrac{\phi(a+h) - \phi(a)}{h}}{\dfrac{\psi(a+h) - \psi(a)}{h}}.$$

Now diminish h indefinitely, and we have

the limit when $x = a$ of $\dfrac{\phi(x)}{\psi(x)}$ is $\dfrac{\phi'(a)}{\psi'(a)}$.

If $\phi'(a) = 0$ and $\psi'(a) = 0$, we have in the same way

the limit when $x = a$ of $\dfrac{\phi'(x)}{\psi'(x)}$ is $\dfrac{\phi''(a)}{\psi''(a)}$.

Hence, the limit when $x = a$ of $\dfrac{\phi(x)}{\psi(x)}$ is $\dfrac{\phi''(a)}{\psi''(a)}$.

This process may be extended, giving the same result as in Art. 146.

148. Form $\dfrac{\infty}{\infty}$.

Let $\phi(x)$ and $\psi(x)$ be functions which both become infinite when $x = a$; it is required to find the limit of the fraction $\dfrac{\phi(x)}{\psi(x)}$.

$$\frac{\phi(x)}{\psi(x)} = \frac{\dfrac{1}{\psi(x)}}{\dfrac{1}{\phi(x)}},$$

and the fraction on the right-hand side takes the form $\dfrac{0}{0}$ when $x = a$; hence, by the previous rules its limit is

$$-\frac{\dfrac{\psi'(a)}{\{\psi(a)\}^2}}{\dfrac{\phi'(a)}{\{\phi(a)\}^2}} \quad \text{or} \quad \left\{\frac{\phi(a)}{\psi(a)}\right\}^2 \frac{\psi'(a)}{\phi'(a)}.$$

Hence

$$\frac{\phi(a)}{\psi(a)} = \left\{\frac{\phi(a)}{\psi(a)}\right\}^2 \frac{\psi'(a)}{\phi'(a)};$$

therefore

$$\frac{\phi(a)}{\psi(a)} = \frac{\phi'(a)}{\psi'(a)}.$$

149. From the last Article it would appear that the limit of a fraction which tends to the form $\frac{\infty}{\infty}$, may be found by considering the ratio of the differential coefficient of the numerator to the differential coefficient of the denominator. But, by Art. 113, when for a *finite value* of the variable a function becomes infinite, so does its differential coefficient. Hence, if

$$\frac{\phi(a)}{\psi(a)} \text{ takes the form } \frac{\infty}{\infty},$$

$$\frac{\phi'(a)}{\psi'(a)} \text{ takes the same form,}$$

and thus the result of Art. 148 would appear to be of no practical value. It *may*, however, happen that the limit of the fraction $\frac{\phi'(x)}{\psi'(x)}$ is more easy to settle than that of $\frac{\phi(x)}{\psi(x)}$.

For example
$$\frac{\log x}{\frac{1}{x}}$$

when $x = 0$, takes the form $\frac{\infty}{\infty}$.

Here
$$\frac{\phi'(x)}{\psi'(x)} = \frac{\frac{1}{x}}{-\frac{1}{x^2}} = -x,$$

the limit of which is 0.

Hence, the limit of $\dfrac{\log x}{\frac{1}{x}}$, when $x = 0$, is 0.

150. The demonstration in Art. 148, which is that usually given, is satisfactory only in the case in which $\frac{\phi(x)}{\psi(x)}$ *really* has a finite limit. For we divided both sides of an equation by this limit which tacitly assumes that the limit is not zero or infinite.

But the demonstration may be completed thus:

Suppose the limit of $\dfrac{\phi(x)}{\psi(x)}$ is really zero; then the limit

of $\dfrac{\psi(x) + \phi(x)}{\psi(x)}$ is really finite, namely, unity. Hence, it has

been proved that

     the limit of $\dfrac{\psi(x) + \phi(x)}{\psi(x)}$ when $x = a$ is $\dfrac{\psi'(a) + \phi'(a)}{\psi'(a)}$,

that is $1 +$ the limit of $\dfrac{\phi(x)}{\psi(x)} = 1 + \dfrac{\phi'(a)}{\psi'(a)}$;

therefore   the limit of $\dfrac{\phi(x)}{\psi(x)} = \dfrac{\phi'(a)}{\psi'(a)}$.

If the limit of $\dfrac{\phi(x)}{\psi(x)}$ be really infinity, then the limit of

$\dfrac{\psi(x)}{\phi(x)}$ is really zero, and therefore, as just shewn, the limit of

$\dfrac{\psi'(x)}{\phi'(x)}$ will be zero. Hence, the limit of $\dfrac{\phi'(x)}{\psi'(x)}$ will be infinity.

Combining then this Article with Art. 148, we can assert that if $\phi(x)$ and $\psi(x)$ both become infinite when $x = a$, the limit of $\dfrac{\phi(x)}{\psi(x)}$ will be the same as the limit of $\dfrac{\phi'(x)}{\psi'(x)}$.

151. The two Articles 148 and 150 may be replaced by the following mode of exhibiting the proposition.

Suppose $\phi(a) = \infty$, and $\psi(a) = \infty$.

Then $\dfrac{1}{\phi(a)} = 0$ and $\dfrac{1}{\psi(a)} = 0$;

now $\dfrac{\phi(a+h)}{\psi(a+h)} = \dfrac{\dfrac{1}{\psi(a+h)}}{\dfrac{1}{\phi(a+h)}} = \dfrac{\dfrac{\psi'(a+\theta h)}{\{\psi(a+\theta h)\}^2}}{\dfrac{\phi'(a+\theta h)}{\{\phi(a+\theta h)\}^2}}$, (Art. 106);

therefore $\dfrac{\phi'(a+\theta h)}{\psi'(a+\theta h)} = \dfrac{\phi(a+\theta h)}{\psi(a+\theta h)} \cdot \dfrac{\dfrac{\phi(a+\theta h)}{\psi(a+\theta h)}}{\dfrac{\phi(a+h)}{\psi(a+h)}}$.

If $\dfrac{\phi(x)}{\psi(x)}$ has a finite limit when $x = a$, the limit of the second factor on the right-hand side of the equation is unity. Hence

$$\text{the limit of } \frac{\phi(x)}{\psi(x)} = \text{the limit of } \frac{\phi'(x)}{\psi'(x)}.$$

But if $\dfrac{\phi(x)}{\psi(x)}$ tends to 0 or ∞ as x approaches a, it will in general finish by approaching the limit in such a manner that the second factor will in the first case be less than unity, and in the second case greater. Hence, $\dfrac{\phi'(x)}{\psi'(x)}$ becomes zero or infinity at the same time that $\dfrac{\phi(x)}{\psi(x)}$ does.

152. In the preceding rules for finding the limit of a function which takes the form $\dfrac{0}{0}$ or $\dfrac{\infty}{\infty}$ when $x = a$, we have made no supposition as to the magnitude of a. Hence the rules are often applied to the case in which a is infinite. But for a direct demonstration of this case we may proceed thus.

Suppose the limit of $\dfrac{\phi(x)}{\psi(x)}$ required, when $x = \infty$; it being known that then either $\phi(x) = 0$ and $\psi(x) = 0$, or $\phi(x) = \infty$ and $\psi(x) = \infty$.

Put $x = \dfrac{1}{y}$, then

$$\frac{\phi(x)}{\psi(x)} = \frac{\phi\left(\dfrac{1}{y}\right)}{\psi\left(\dfrac{1}{y}\right)}.$$

Now as y tends to zero, we have, by preceding rules,

$$\text{the limit of } \frac{\phi\left(\dfrac{1}{y}\right)}{\psi\left(\dfrac{1}{y}\right)} = \text{the limit of } \frac{\dfrac{1}{y^2}\,\phi'\left(\dfrac{1}{y}\right)}{\dfrac{1}{y^2}\,\psi'\left(\dfrac{1}{y}\right)}$$

$$= \text{the limit of } \frac{\phi'\left(\dfrac{1}{y}\right)}{\psi'\left(\dfrac{1}{y}\right)} = \text{the limit of } \frac{\phi'(x)}{\psi'(x)}.$$

153. For example, required the value of

$$\frac{\dfrac{1}{x}}{\cot x} \text{ when } x = 0.$$

Differentiating both numerator and denominator, we find the required limit is the same as that of

$$\frac{-\dfrac{1}{x^2}}{-\dfrac{1}{\sin^2 x}} \text{ or of } \frac{\sin^2 x}{x^2}, \text{ that is, unity.}$$

The same result may be obtained by writing the proposed fraction in the form $\dfrac{0}{0}$; thus

$$\frac{\dfrac{1}{x}}{\cot x} = \frac{\tan x}{x} \text{ or } \frac{1}{\cos x} \frac{\sin x}{x}.$$

The limit of $\dfrac{1}{\cos x}$ is 1, and the limit of $\dfrac{\sin x}{x}$ is 1; therefore the limit of the proposed fraction is 1.

As another example we may find the limit of $\dfrac{x^n}{e^x}$ when x is infinite, n being positive.

The limit of $\dfrac{x^n}{e^x}$ = the limit of $\dfrac{n x^{n-1}}{e^x}$

$$= \text{the limit of } \frac{n(n-1)x^{n-2}}{e^x} \dots$$

Proceeding thus, we shall, if n be a positive integer, arrive at the fraction $\dfrac{\lfloor n}{e^x}$, the limit of which is 0. If n be a fraction, we shall arrive at a fraction having e^x in the denominator and some *negative* power of x in the numerator, which also has 0 for its limit.

Hence the limit of $\dfrac{x^n}{e^x}$, when $x = \infty$, is zero.

154. A remark should be made for the purpose of preventing a misconception of some of the results of this Chapter. Suppose $\phi(x)$ and $\psi(x)$ both to vanish when $x = a$, and that $\phi'(a) = 0$ while $\psi'(a)$ is finite. We say then, that when $x = a$,

$$\text{the limit of } \frac{\phi(x)}{\psi(x)} = \text{the limit of } \frac{\phi'(x)}{\psi'(x)},$$

meaning that each side of the equation vanishes. *It does not follow necessarily that*

$$\frac{\phi(x)}{\psi(x)} \div \frac{\phi'(x)}{\psi'(x)} \text{ has unity for its limit.}$$

For example, let $\quad \phi(x) = x^2, \qquad \psi(x) = \sin x,$

then $\qquad\qquad\qquad \phi'(x) = 2x, \qquad \psi'(x) = \cos x.$

When x approaches the limit zero, we can infer that, since $\dfrac{\phi'(x)}{\psi'(x)}$ approaches zero, so also does $\dfrac{\phi(x)}{\psi(x)}$. But it is obviously not true that the limit of

$$\frac{x^2}{\sin x} \div \frac{2x}{\cos x} \text{ or of } \frac{x^2 \cos x}{2x \sin x} \text{ is unity;}$$

the limit is in fact $\frac{1}{2}$.

155. It should be observed that there are examples which *may* be solved by means of the Differential Calculus, but which can also be solved, and sometimes more simply, by common algebraical transformations. For instance,

$$\frac{(x-a)^{\frac{1}{2}}}{(x^2 - a^2)^{\frac{1}{4}}}$$

when $x = a$ takes the form $\dfrac{0}{0}$. Put $x = a + h$, and the fraction becomes

$$\frac{h^{\frac{1}{2}}}{h^{\frac{1}{4}}(2a + h)^{\frac{1}{4}}} \text{ or } \frac{h^{\frac{1}{4}}}{(2a + h)^{\frac{1}{4}}};$$

and the limit, when $h = 0$, is 0.

T. D. C. $\qquad\qquad\qquad\qquad\qquad\qquad\qquad\qquad\qquad$ K

Again, suppose we have to find the limit of

$$\frac{\sqrt{x} - 1 + \sqrt{(x - 1)}}{\sqrt{(x^2 - 1)}}$$

as x approaches unity; put $x = 1 + h$, and the fraction becomes

$$\frac{\sqrt{(h + 1)} - 1 + \sqrt{h}}{\sqrt{(h^2 + 2h)}}.$$

Multiply both numerator and denominator by

$$\sqrt{(h + 1)} + 1 - \sqrt{h},$$

and we get

$$\frac{2\sqrt{h}}{\sqrt{h}\sqrt{(h + 2)}\{\sqrt{(h + 1)} + 1 - \sqrt{h}\}} \quad \text{or} \quad \frac{2}{\sqrt{(h + 2)}\{\sqrt{(h + 1)} + 1 - \sqrt{h}\}},$$

and the limit of this, when $h = 0$, is $\dfrac{1}{\sqrt{2}}$.

156.   There are cases in which not only $\phi(x)$ and $\psi(x)$ vanish, but all their differential coefficients, and where, consequently, we are not able to ascertain the limit of $\dfrac{\phi(x)}{\psi(x)}$.

For suppose $\phi(x) = a^{-u}$, where u stands for $\dfrac{1}{x^n}$, a and n being positive numbers, and a greater than unity: we have

$$\phi'(x) = \frac{n \log a . a^{-u}}{x^{n+1}},$$

$$\phi''(x) = n \log a . a^{-u} \left\{ \frac{n \log a}{x^{2(n+1)}} - \frac{n + 1}{x^{n+2}} \right\},$$

and so on.

Put $\dfrac{1}{x} = z$, and let t stand for z^n;

then $$\phi'(x) = \frac{n \log a . z^{n+1}}{a^t},$$

$$\phi''(x) = \frac{n \log a \{n \log a . z^{2(n+1)} - (n + 1) z^{n+2}\}}{a^t};$$

also the value $x = 0$ corresponds to $z = \infty$. But it is easy to see that every expression of the form

$$\frac{z^m}{a^i},$$

where a, m, n, are positive numbers, and a greater than unity, is zero when z is infinite. For if we apply to this example the rule for finding the value of a fraction which assumes the form $\frac{\infty}{\infty}$ and differentiate r times successively, r being the integer next above m, we have

$$\text{the limit of } \frac{z^m}{a^i} = \text{the limit of } \frac{k}{\psi(z)},$$

where k is some constant, and $\psi(z)$ a function of z which is infinite when z is infinite. Consequently, all the differential coefficients of $\phi(x)$ vanish when $x = 0$.

If then we have

$$\phi(x) = a^{-u},$$
$$\psi(x) = b^{-v},$$

where v stands for $\frac{1}{x^v}$, and b is a positive number greater than unity, and v also positive, the differential coefficients of all orders of the two terms of the fraction $\frac{\phi(x)}{\psi(x)}$ will vanish when $x = 0$, and the limit cannot be found by this method.

In the case of $v = n$, the fraction becomes

$$\left(\frac{a}{b}\right)^{-u};$$

this, when $x = 0$, will be 0 or ∞, according as a is greater or less than b.

157. The fraction

$$\frac{e^{-\frac{1}{x}}}{x}$$

takes the form $\frac{0}{0}$ when $x = 0$. Put $x = \frac{1}{y}$ and we have $\frac{y}{e^y}$, the limit of which, when y is infinite, is 0, by Art. 153;

$\dfrac{e^{\frac{1}{x}}}{x}$, or $\dfrac{1}{x} \times e^{\frac{1}{x}}$ is of course infinite when $x = 0$.

Hence, $\dfrac{e^{\frac{1}{x}}}{x}$ is 0 or ∞ when x approaches the limit 0, according as we suppose x negative or positive.

158. Form $0 \times \infty$.

Suppose $\phi(x)$ and $\psi(x)$ two functions of x, such that $\phi(a) = 0$, and $\psi(a) = \infty$; it is required to find the limit of $\phi(x)\,\psi(x)$ as x approaches a.

$$\phi(x)\,\psi(x) = \frac{\phi(x)}{\dfrac{1}{\psi(x)}},$$

and as the fraction on the right-hand side takes the form $\dfrac{0}{0}$ when $x = a$, its limiting value may be found by rules already given.

For example, let $\phi(x) = \log\left(2 - \dfrac{x}{a}\right)$, and $\psi(x) = \tan\dfrac{\pi x}{2a}$.

Here $\phi(x)\,\psi(x)$ takes the form $0 \times \infty$ when $x = a$.

Then $\log\left(2 - \dfrac{x}{a}\right)\tan\dfrac{\pi x}{2a} = \dfrac{\log\left(2 - \dfrac{x}{a}\right)}{\cot\dfrac{\pi x}{2a}}$.

The limit of this when $x = a$, is found by making $x = a$ in

$$\frac{-\dfrac{1}{a} \cdot \dfrac{1}{2 - \dfrac{x}{a}}}{-\dfrac{\pi}{2a} \cdot \dfrac{1}{\sin^2\dfrac{\pi x}{2a}}},$$

which gives $\dfrac{2}{\pi}$.

Again, $x^m (\log x)^n$,

where m and n are positive, takes the form $0 \times \infty$, when $x = 0$

Here $\dfrac{x^m}{\dfrac{1}{(\log x)^n}}$ takes the form $\dfrac{0}{0}$

when $x = 0$; its limit is the same as that of

$$\frac{m x^{m-1}}{-\dfrac{n}{x (\log x)^{n+1}}};$$

which does not assist us.

If we assume $x = e^{-y}$, then $x^m (\log x)^n$ becomes

$$(-1)^n \frac{y^n}{e^{my}};$$

the value of this, when y is ∞, is 0. See Art. 153.

The result in this case should be carefully noticed, as it is frequently wanted in mathematical investigations.

159. Forms 0^0, ∞^0, 1^∞.

Let $\phi(x)$ and $\psi(x)$ be two functions of x, such that when $x = a$, the expression

$$\{\phi(x)\}^{\psi(x)}$$

assumes one of the forms 0^0, ∞^0, 1^∞; it is required to find the limiting value of this expression.

Since $\phi(x) = e^{\log \phi(x)}$,

we have $\{\phi(x)\}^{\psi(x)} = e^{\psi(x) \log \phi(x)}$.

Now $\psi(x) \log \phi(x)$ in each of the proposed cases takes the form $0 \times \infty$, and its limiting value can be found by Art. 158, and thus the value of $\{\phi(x)\}^{\psi(x)}$ becomes known.

For example, x^x, when $x = 0$, takes the form 0^0;

$$x^x = e^{x \log x};$$

and $x \log x = 0$, when $x = 0$, (Art. 158);

therefore, $x^x = 1$, when $x = 0$.

Again, $\left(\dfrac{1}{x}\right)^{\sin x}$ takes the form ∞^0 when $x = 0$; also

$$\left(\dfrac{1}{x}\right)^{\sin x} = e^{-\sin x \log x}.$$

Now, $$\sin x \log x = \dfrac{\sin x}{x} . x \log x;$$

when $x = 0$, we have

$$\dfrac{\sin x}{x} = 1,$$

$$x \log x = 0, \text{ (Art. 158),}$$

therefore $$\sin x \log x = 0, \text{ when } x = 0,$$

therefore $$\left(\dfrac{1}{x}\right)^{\sin x} = 1, \text{ when } x = 0.$$

Again, $\left(2 - \dfrac{x}{a}\right)^{\tan \frac{\pi x}{2a}}$ takes the form 1^∞, when $x = a$.

The above expression $= e^{\tan \frac{\pi x}{2a} \log \left(2 - \frac{x}{a}\right)}$

$$= e^{\frac{2}{\pi}} \text{ when } x = a, \text{ (Art. 158).}$$

160.   Form $\infty - \infty$.

Let $\phi(x)$ and $\psi(x)$ be two functions of x which become infinite when $x = a$, then

$$\phi(x) - \psi(x)$$

assumes the form $\infty - \infty$; it is required to find the value of the expression.

Put $$y = \phi(x) - \psi(x),$$

then $$\epsilon^y = e^{\phi(x) - \psi(x)}$$

$$= \dfrac{e^{-\psi(x)}}{e^{-\phi(x)}}.$$

Thus e^y takes the form $\dfrac{0}{0}$ when $x = a$, and its value may be investigated by Art. 145.

Or we may proceed thus,

$$y = \phi(x) \left\{ 1 - \frac{\psi(x)}{\phi(x)} \right\};$$

then y is infinite unless the limit of $\dfrac{\psi(x)}{\phi(x)}$ is unity; if the limit of $\dfrac{\psi(x)}{\phi(x)}$ is unity,

since
$$y = \frac{1 - \dfrac{\psi(x)}{\phi(x)}}{\dfrac{1}{\phi(x)}}$$

it takes the form $\dfrac{0}{0}$.

For example, suppose $y = \tan x - \sec x$;

then y takes the form $\infty - \infty$ when $x = \dfrac{\pi}{2}$.

Also
$$y = \tan x \left(1 - \frac{\sec x}{\tan x} \right)$$
$$= \frac{1 - \operatorname{cosec} x}{\cot x};$$

this takes the form $\dfrac{0}{0}$, and its limiting value is

$$\frac{\operatorname{cosec} x \cot x}{-\operatorname{cosec}^2 x} \quad \text{or } 0.$$

161. The limit of $\dfrac{F(x)}{x}$ when $x = \infty$, supposing $F(x)$ to be then infinite, will be the same as that of $\dfrac{F'(x)}{1}$, or $F'(x)$. See Art. 151.

But,
$$\frac{F(x+h) - F(x)}{h} = F'(x + \theta h).$$

If x be made to increase indefinitely the limit of the second member of the equation is $F''(x)$.

Hence   the limit when $x = \infty$ of $\dfrac{F(x)}{x}$

$= $ the limit when $x = \infty$ of $\dfrac{F(x+h) - F(x)}{h}$.

If for simplicity we make $h = 1$, we have

    the limit of $\dfrac{F(x)}{x} = $ the limit of $\{F(x+1) - F(x)\}$.

162. The limit of $\{F(x)\}^{\frac{1}{x}}$ when x is infinite, is the same as that of $e^{\frac{\log F(x)}{x}}$.

But, by Art. 161, supposing $F(x)$ to become infinite with x, the limit of $\dfrac{\log F(x)}{x}$ is the same as the limit of

$$\log F(x+1) - \log F(x),$$

or of $\log \dfrac{F(x+1)}{F(x)}$.

Hence the limit when $x = \infty$ of $\{F(x)\}^{\frac{1}{x}}$

$= $ the limit of $\dfrac{F(x+1)}{F(x)}$.

Suppose, for example, that we require the limit when x is infinite of $\left\{\dfrac{x^x}{\lfloor x}\right\}^{\frac{1}{x}}$.

By the theorem just proved the required limit

$= $ the limit of $\dfrac{(x+1)^{x+1}}{\lfloor x+1} \dfrac{\lfloor x}{x^x}$

$= $ the limit of $\left(\dfrac{x+1}{x}\right)^x$

$= $ the limit of $\left(1 + \dfrac{1}{x}\right)^x$

$= e$ by Art. 16.

163. A few remarks may be made on indeterminate fractions involving more than one variable.

A function of two variables may take the form $\frac{0}{0}$, either when one of the variables remains undetermined and the other has a particular value, or when both receive particular values.

As an example of the first case, suppose

$$z = \frac{c\,(x^2 - a^2)}{y\,(x - a) + (x - a)^2};$$

if we make $x = a$ we have $z = \frac{0}{0}$, whatever y may be. But by removing the factor $x - a$ from the numerator and denominator of z, we have

$$z = \frac{c\,(x + a)}{y + x - a}.$$

Hence, when $x = a$, we have

$$z = \frac{2ca}{y}.$$

This case is very simple, and whenever it occurs the application of the preceding rules will give the limiting value towards which z approaches as x approaches its limit.

As an example of the second case, suppose

$$z = \frac{c\,(x - a)}{y - b}.$$

This fraction takes the form $\frac{0}{0}$ when $x = a$ and $y = b$, and is really indeterminate. For suppose $y - b = m\,(x - a)$, then

$$z = \frac{c}{m}.$$

Hence the value of z is indeterminate, for x and y being independent m may have any value we please.

164. It may happen that the values which such a function assumes, although infinite in number, are confined within certain limits. For example, suppose

$$z = \frac{c\,(x - a)\,(y - b)}{(x - a)^2 + (y - b)^2}.$$

Assume
$$y - b = m(x - a);$$

therefore
$$z = \frac{cm}{m^2 + 1} = \frac{c}{m + \dfrac{1}{m}}.$$

Here the greatest value of z is when $m = 1$, and z always lies between $\dfrac{c}{2}$ and $-\dfrac{c}{2}$.

165. We give two more examples.

1st.   Let
$$z = \frac{(x - a)^m + c(y - b)^n}{(x - a)^p + c(y - b)^q};$$

this takes the form $\dfrac{0}{0}$ when $x = a$ and $y = b$.

Put
$$x - a = h \text{ and } y - b = k;$$

therefore
$$z = \frac{h^m + ck^n}{h^p + ck^q}.$$

If now we assume $k = Ah^a$, we have
$$z = \frac{h^m + cA^n h^{an}}{h^p + cA^q h^{aq}},$$

and, according to the different hypotheses we make respecting a, m, p, ..., we shall obtain for z finite, infinite, or zero values.

2nd.   Let $z = \dfrac{(x - y) a^n - (a - y) x^n + (a - x) y^n}{(x - y)(a - y)(a - x)}.$

If $x = a$, and $y = a$, this takes the form $\dfrac{0}{0}$. Put $a + h$ and $a + k$ for x and y respectively; we shall have
$$z = \frac{(h - k) a^n + k(a + h)^n - h(a + k)^n}{(h - k) kh}.$$

If we expand $(a + h)^n$ and $(a + k)^n$, and make some reductions, we obtain
$$z = \frac{n(n - 1)}{1 \cdot 2} a^{n-2} + \frac{n(n - 1)(n - 2)}{1 \cdot 2 \cdot 3} a^{n-3}(h + k) + \dots$$

Hence, putting h and k each zero, we have

$$z = \frac{n\,(n-1)}{1\,.\,2}\,a^{n-2}.$$

This result may also be found by examining the limit towards which z tends as x approaches y, and then the limit towards which this result tends as y approaches a.

The next Article must be omitted until the student has read Chapter XI.

166.   Generally, if $z = \dfrac{f\,(x,\,y)}{F\,(x,\,y)}$, and both numerator and denominator of z vanish for certain values of x and y, the value of z is really indeterminate, and in fact depends upon the arbitrary relation we choose to establish between x and y. Suppose that $x = a$, $y = b$, are the values which make z assume the form $\dfrac{0}{0}$; and assume that $y = \psi\,(x)$, where $\psi\,(x)$ is any function the value of which is b when $x = a$.

Thus the numerator and denominator of z become functions of x only; and by previous rules for ascertaining the value of a fraction which takes the form $\dfrac{0}{0}$, we have

$$z = \frac{\left(\dfrac{df}{dx}\right) + \left(\dfrac{df}{dy}\right)\psi'\,(x)}{\left(\dfrac{dF}{dx}\right) + \left(\dfrac{dF}{dy}\right)\psi'\,(x)},$$

x being put $= a$ and $y = b$ after the differentiations are performed. This value is indeterminate, since $\psi'\,(x)$ is a function which is quite arbitrary.

But if $\left(\dfrac{df}{dx}\right)$ and $\left(\dfrac{dF}{dx}\right)$ both vanish,

or if $\left(\dfrac{df}{dy}\right)$ and $\left(\dfrac{dF}{dy}\right)$ both vanish,

then the value of z becomes determinate.

The value of z is also determinate if

$$\frac{\left(\dfrac{df}{dx}\right)}{\left(\dfrac{dF}{dx}\right)} = \frac{\left(\dfrac{df}{dy}\right)}{\left(\dfrac{dF}{dy}\right)}.$$

If $\quad \left(\dfrac{df}{dx}\right) = 0, \quad \left(\dfrac{dF}{dx}\right) = 0, \quad \left(\dfrac{df}{dy}\right) = 0, \quad \left(\dfrac{dF}{dy}\right) = 0,$

then proceeding to a second differentiation we have

$$z = \frac{\left(\dfrac{d^2f}{dx^2}\right) + 2\left(\dfrac{d^2f}{dx\,dy}\right)\psi'(x) + \left(\dfrac{d^2f}{dy^2}\right)\{\psi'(x)\}^2}{\left(\dfrac{d^2F}{dx^2}\right) + 2\left(\dfrac{d^2F}{dx\,dy}\right)\psi'(x) + \left(\dfrac{d^2F}{dy^2}\right)\{\psi'(x)\}^2}, \text{ (Art. 176)},$$

which is generally indeterminate, since $\psi(x)$ is an arbitrary function.

Example 1. Suppose

$$z = \frac{\log x + \log y}{x + 2y - 3}, \quad a = 1, \quad b = 1 ;$$

$$\left(\frac{df}{dx}\right) = \frac{1}{x} = 1, \quad \text{when } x = 1,$$

$$\left(\frac{df}{dy}\right) = \frac{1}{y} = 1, \quad \text{when } y = 1,$$

$$\left(\frac{dF}{dx}\right) = 1, \qquad \left(\frac{dF}{dy}\right) = 2 ;$$

therefore $\qquad z = \dfrac{1 + \psi'(x)}{1 + 2\psi'(x)},$

which is really indeterminate, and may assume any value between $+\infty$ and $-\infty$.

Example 2. Suppose

$$z = \frac{(x-1)^{\frac{1}{2}} + y^{\frac{1}{2}} - 1}{(x^2-1)^{\frac{1}{2}} - y + 1}.$$

Here z takes the form $\dfrac{0}{0}$ when $x = 1$ and $y = 1$.

Also then $\left(\dfrac{df}{dx}\right) = 0$ and $\left(\dfrac{dF}{dx}\right) = 0$.

Hence z has a determinate value, namely, $-\dfrac{3}{2}$.

Example 3. Suppose $z = \dfrac{(x + y)^2}{x^2 + y^2}$.

Here, when $x = 0$ and $y = 0$, we have

$$\left(\dfrac{df}{dx}\right) = 0, \quad \left(\dfrac{df}{dy}\right) = 0, \quad \left(\dfrac{dF}{dx}\right) = 0, \quad \left(\dfrac{dF}{dy}\right) = 0,$$

and $\quad z = \dfrac{1 + 2\psi'(x) + \{\psi'(x)\}^2}{1 + \{\psi'(x)\}^2} = \dfrac{\{1 + \psi'(x)\}^2}{1 + \{\psi'(x)\}^2}$

$$= \dfrac{(1 + u)^2}{1 + u^2} \text{ say.}$$

Here the value of z is indeterminate; but it will be found that it is confined between the limits 0 and 2, as may be shewn by writing the fraction just given in the form $1 + \dfrac{2u}{1 + u^2}$, remembering that $\dfrac{2u}{1 + u^2}$ is never greater than unity.

167. In solving examples on this Chapter there are various considerations which will abbreviate the labour of the operations, as will be seen in the following case.

Find the value of $\quad \dfrac{\log(1 + x + x^2) + \log(1 - x + x^2)}{\sec x - \cos x}$ when $x = 0$.

The proposed expression takes the form $\dfrac{0}{0}$ when $x = 0$. If we proceed in the ordinary way, we shall find after reduction that the differential coefficient of the numerator is

$$\dfrac{2x + 4x^3}{1 + x^2 + x^4},$$

and that the differential coefficient of the denominator is

$$\dfrac{\sin x}{\cos^2 x} + \sin x.$$

Thus we obtain again the form $\dfrac{0}{0}$, and we may continue in the ordinary way the process of evaluation. We may however obtain the result more easily by arranging the fraction we have now to evaluate thus:

$$\frac{2\,(1+2x^2)\cos^2 x}{(1+x^2+x^4)\,(1+\cos^2 x)} \times \frac{x}{\sin x}.$$

Here the first factor is not indeterminate when $x=0$; its value is then unity. The second factor takes the form $\dfrac{0}{0}$, and its limiting value is known to be unity. Thus unity is the required limiting value of the original expression.

Or the original expression may be evaluated in the following manner. It may be put in the form

$$\frac{\cos x \log (1+x^2+x^4)}{\sin^2 x}.$$

Now $\cos x = 1$ when $x=0$; we need not then pay any attention to this factor, but consider that we have to evaluate

$$\frac{\log (1+x^2+x^4)}{\sin^2 x}$$

when $x=0$; and we may proceed in the usual way to differentiate the numerator and denominator. Or if we are allowed to use the results of the expansions of functions we have

$$\frac{\log (1+x^2+x^4)}{\sin^2 x} = \frac{x^2+x^4-\tfrac{1}{2}\,(x^2+x^4)^2+\tfrac{1}{3}\,(x^2+x^4)^3-\ldots}{\left(x-\dfrac{x^3}{6}+\ldots\right)^2}$$

$$= \frac{x^2+\tfrac{1}{2}\,x^4-\ldots}{x^2-\tfrac{1}{3}\,x^4+\ldots}$$

$$= \frac{1+\tfrac{1}{2}\,x^2-\ldots}{1-\tfrac{1}{3}\,x^2+\ldots}$$

$$= 1 \text{ when } x=0.$$

EXAMPLES.

Find the limits of the following functions:

1. $\dfrac{\log x}{x-1}$,   when $x = 1$.   *Result* 1.

2. $\dfrac{x-1}{x^n-1}$,   when $x = 1$.   *Result* $\dfrac{1}{n}$.

3. $\dfrac{e^x - e^{-x}}{\sin x}$, $x = 0$.   *Result* 2.

4. $\dfrac{e^x - e^{-x} - 2x}{x - \sin x}$, $x = 0$.   *Result* 2.

5. $\dfrac{x - \sin^{-1} x}{(\sin x)^3}$, $x = 0$.   *Result* $-\dfrac{1}{6}$.

6. $\dfrac{a^x - b^x}{x}$, $x = 0$.   *Result* $\log \dfrac{a}{b}$.

7. $\dfrac{\tan x - x}{x - \sin x}$, $x = 0$.   *Result* 2.

8. $\dfrac{x - \sin x}{x^3}$, $x = 0$.   *Result* $\dfrac{1}{6}$.

9. $\dfrac{\sin 3x}{x - \dfrac{3}{2}\sin 2x}$, $x = 0$.   *Result* $-\dfrac{3}{2}$.

10. $\dfrac{1 - x + \log x}{1 - \sqrt{(2x - x^2)}}$, $x = 1$.   *Result* -1.

11. $\dfrac{1}{\log x} - \dfrac{x}{\log x}$, $x = 1$.   *Result* -1.

12. $\dfrac{e^x - 2\cos x + e^{-x}}{x \sin x}$, $x = 0$.   *Result* 2.

13. $\dfrac{\sin 2x + 2\sin^2 x - 2\sin x}{\cos x - \cos^2 x}$, $x = 0$.   *Result* 4.

14    $x \tan x - \dfrac{\pi}{2} \sec x,$ $x = \dfrac{\pi}{2}.$ *Result* $- 1.$

15. $\dfrac{(x - 2)\, e^x + x + 2}{(e^x - 1)^3},$ $x = 0.$ *Result* $\dfrac{1}{6}.$

16. $\dfrac{x^4 + 3x^3 - 7x^2 - 27x - 18}{x^4 - 3x^3 - 7x^2 + 27x - 18},$ $x = 3.$ *Result* $10.$

                                     $x = - 3.$ *Result* $\dfrac{1}{10}.$

17. $\dfrac{x \sqrt{(3x - 2x^4)} - x \sqrt[5]{x}}{1 - x^{\frac{2}{3}}},$ $x = 1.$ *Result* $\dfrac{81}{20}.$

18. $\dfrac{x^{\frac{2}{3}} - 1 + (x - 1)^{\frac{2}{3}}}{(x^2 - 1)^{\frac{2}{3}} - x + 1},$ $x = 1.$ *Result* $- \dfrac{3}{2}.$

19. $\dfrac{x^{\frac{2}{3}} - 1 + (x - 1)^{\frac{2}{3}}}{\sqrt{(x^2 - 1)}},$ $x = 1.$ *Result* $0.$

20. $\dfrac{m \sin x - \sin mx}{x\,(\cos x - \cos mx)},$ $x = 0.$ *Result* $\dfrac{m}{3}.$

21. $\dfrac{x^2}{1 - \cos mx},$ $x = 0.$ *Result* $\dfrac{2}{m^2}.$

22. $\dfrac{\sin (\alpha + x) - \sin (\alpha - x)}{\cos (\alpha + x) - \cos (\alpha - x)},$ $x = 0.$ *Result* $- \cot \alpha.$

23. $\dfrac{\tan nx - n \tan x}{n \sin x - \sin nx},$ $x = 0.$ *Result* $2.$

24. $\dfrac{\sqrt{(a^2 - x^2)} + a - x}{\sqrt{\left(a^2 - \dfrac{x^3}{a}\right)} + \sqrt{(ax - x^2)}},$ $x = a.$ *Result* $\dfrac{\sqrt{2}}{\sqrt{3} + 1}.$

25. $\dfrac{\sqrt{x} - \sqrt{a} + \sqrt{(x - a)}}{\sqrt{(x^2 - a^2)}},$ $x = a.$ *Result* $\dfrac{1}{\sqrt{(2a)}}.$

26. $\sqrt{\left(\dfrac{2 + \cos 2x - \sin x}{x \sin 2x + x \cos x}\right) - \left(\dfrac{\pi - 2x}{2 \sin 2x}\right)^2},$ $x = \dfrac{\pi}{2}.$ *Result* $- \dfrac{1}{4}.$

27. $2^x \sin \dfrac{a}{2^x},$ $x = \infty.$ *Result* $a.$

28. $(a^{\frac{1}{x}} - 1) x,$ $\qquad x = \infty .$ $\qquad$ *Result* $\log a.$

29. $\left(\dfrac{a}{x} + 1\right)^{x},$ $\qquad x = \infty .$ $\qquad$ *Result* $e^{a}.$

30. $\dfrac{m^{x} \sin nx - n^{x} \sin mx}{\tan nx - \tan mx},$ $\quad$ (1) $x = 0.$ $\quad$ (1) *Result* 1.
$\qquad\qquad\qquad\qquad\qquad$ (2) $m = n.$

$\qquad\qquad$ (2) *Result* $n^{x-1} (n \cos nx - \sin nx) \cos^{2} nx.$

31. $\left(1 + \dfrac{1}{x^{2}}\right)^{x},$ $\qquad x = \infty .$ $\qquad$ *Result* 1.

32. $\left(\dfrac{\tan x}{x}\right)^{\frac{1}{x}},$ $\qquad x = 0.$ $\qquad$ *Result* 1.

33. $\left(\dfrac{\tan x}{x}\right)^{\frac{1}{x^{2}}},$ $\qquad x = 0.$ $\qquad$ *Result* $e^{\frac{1}{3}}.$

34. $\left(\dfrac{\tan x}{x}\right)^{\frac{1}{x^{3}}},$ $\qquad x = 0.$ $\qquad$ *Result* $\infty .$

35. $(\cos mx)^{\frac{n}{x}},$ $\qquad x = 0.$ $\qquad$ *Result* 1.

36. $(\cos mx)^{\frac{n}{x^{2}}},$ $\qquad x = 0.$ $\qquad$ *Result* $e^{-\frac{nm^{2}}{2}}.$

37. $(\cos mx)^{\frac{n}{x^{3}}},$ $\qquad x = 0.$ $\qquad$ *Result* 0.

38. $\dfrac{x^{3} (\cot x)^{2} + \sin x}{x},$ $\qquad x = 0.$ $\qquad$ *Result* 2.

39. $\dfrac{(e^{x} - e^{-x})^{2} - 2x^{2} (e^{x} + e^{-x})}{x^{4}},$ $\quad x = 0.$ $\qquad$ *Result* $-\frac{2}{3}.$

40. $\dfrac{1 - \sqrt{(1 - x)}}{\sqrt{(1 + x)} - \sqrt{(1 + x^{2})}},$ $\qquad x = 0.$ $\qquad$ *Result* 1.

41. $(\sin x)^{\tan x},$ $\qquad x = \dfrac{\pi}{2}.$ $\qquad$ *Result* 1.

42. $\dfrac{\sqrt{2} - \sin x - \cos x}{\log \sin 2x},$ $\qquad x = \dfrac{\pi}{4}.$ $\qquad$ *Result* $-\dfrac{1}{2\sqrt{2}}.$

T. D. C. $\qquad\qquad\qquad\qquad\qquad\qquad\qquad\qquad\qquad\qquad\qquad$ L

43. $\sqrt{(a^2 - x^2)} . \cot\left\{\dfrac{\pi}{2}\sqrt{\left(\dfrac{a-x}{a+x}\right)}\right\}$, $x = a$.   Result $\dfrac{4a}{\pi}$.

44. $(1 - x)\tan\dfrac{\pi x}{2}$, $x = 1$.   Result $\dfrac{2}{\pi}$.

45. $x^{\frac{1}{1-x}}$, $x = 1$.   Result $\dfrac{1}{e}$.

46. $x^{x^{-a}}$, $x = 0$.   Result 0.

47. $\dfrac{\sec\dfrac{\pi x}{2}}{\log(1-x)^2}$, $x = 1$.   Result ∞.

48. $(Ax^m + Bx^{m-1} \ldots + Mx + N)^{\frac{1}{x}}$, $x = \infty$.   Result 1.

49. $\dfrac{1}{\sqrt{x}}\log\dfrac{2x + b + 2\sqrt{(ax + bx + x^2)}}{b + 2\sqrt{(ax)}}$, $x = 0$.

Result $\dfrac{2}{b}\{\sqrt{(a+b)} - \sqrt{a}\}$.

50. $\sqrt{\left\{\dfrac{1}{x(x-1)} + \dfrac{1}{4x^2}\right\}} - \dfrac{1}{2x}$, $x = 0$.   Result -1.

51. $\dfrac{\cos x\theta - \cos a\theta}{e^{-x^2\theta} - e^{-a^2\theta}}$, $x = a$.   Result $\dfrac{\sin a\theta e^{a^2\theta}}{2a}$.

52. $\dfrac{e^x + \log\left(\dfrac{1-x}{e}\right)}{\tan x - x}$, $x = 0$.   Result $-\frac{1}{2}$.

53. $\dfrac{e^x\sin x - e^a\{\sin a + \sqrt{2}(x - a)\cos(a - \frac{1}{4}\pi)\}}{e^x - e^a(x + 1 - a)}$, $x = a$.

Result $2\cos a$.

54. $\left(\dfrac{a_1^{\frac{1}{x}} + a_2^{\frac{1}{x}} + \ldots + a_n^{\frac{1}{x}}}{n}\right)^{nx}$, $x = \infty$.   Result $a_1 a_2 \ldots a_n$.

55. $\dfrac{(x + \sin x - 4\sin\frac{1}{2}x)^4}{(3 + \cos x - 4\cos\frac{1}{2}x)^3}$, $x = 0$.   Result $\dfrac{128}{81}$.

56. $\left\{\dfrac{\log x}{x}\right\}^{\frac{1}{x}}$, $x = \infty.$ *Result* 1.

57. $\dfrac{(\log x)^{\frac{3}{2}} + (1 - x^2)^{\frac{3}{4}}}{\sin^{\frac{3}{2}} (x - 1)}$, $x = 1.$ *Result* 1.

58. Shew that when x is infinite $\dfrac{a^{x^m}}{b^{x^n}}$ is infinite or zero, according as m is greater or less than n; a and b being both greater than unity.

59. Shew that when x is infinite

$$x - x^2 \log \left(1 + \frac{1}{x}\right) = \frac{1}{2}.$$

60. If $u\sqrt{(xc)} = \tan^{-1} \dfrac{a\sqrt{x}}{\sqrt{c}} + \log \left\{\dfrac{a\sqrt{x}}{\sqrt{c}} + \sqrt{\left(1 + \dfrac{a^2 x}{c}\right)}\right\}$, shew

that $u = \dfrac{2a}{c}$ and $\dfrac{du}{dx} = -\dfrac{a^3}{2c^2}$ when $x = 0$; and that $u = 0$

and $\dfrac{du}{dx} = 0$ when $x = \infty$.

CHAPTER XI.

DIFFERENTIAL COEFFICIENT OF A FUNCTION OF FUNCTIONS AND OF IMPLICIT FUNCTIONS.

168. Suppose u a function of y and z, and y and z themselves functions of x, it is required to find $\dfrac{du}{dx}$. This of course might be obtained by substituting in u for y and z their values in terms of x, by which substitution u becomes an explicit function of x, and $\dfrac{du}{dx}$ can be found by previous methods. But it is often convenient to have a rule which gives $\dfrac{du}{dx}$ without requiring the substitution for y and z. To this rule we proceed.

169. Suppose $u = \phi\,(y,\,z)$,

where y and z are both functions of x. Let x become $x + \Delta x$, and in consequence let y, z, and u, become respectively $y + \Delta y$, $z + \Delta z$, and $u + \Delta u$. Then

$$\Delta u = \phi\,(y + \Delta y,\, z + \Delta z) - \phi\,(y,\, z)$$
$$= \phi\,(y + \Delta y,\, z + \Delta z) - \phi\,(y,\, z + \Delta z) + \phi\,(y,\, z + \Delta z) - \phi\,(y,\, z)\,;$$

therefore
$$\frac{\Delta u}{\Delta x} = \frac{\phi(y + \Delta y,\, z + \Delta z) - \phi(y,\, z + \Delta z)}{\Delta y}\,\frac{\Delta y}{\Delta x}$$
$$+ \frac{\phi(y,\, z + \Delta z) - \phi(y,\, z)}{\Delta z}\,\frac{\Delta z}{\Delta x}\,.$$

Now let Δx and consequently Δy, Δz, and Δu, diminish without limit; then

$$\text{the limit of } \frac{\Delta u}{\Delta x} \text{ is } \frac{du}{dx},$$

$$\text{the limit of } \frac{\Delta y}{\Delta x} \text{ is } \frac{dy}{dx},$$

$$\text{the limit of } \frac{\Delta z}{\Delta x} \text{ is } \frac{dz}{dx}.$$

The limit of $\dfrac{\phi\,(y,\,z + \Delta z) - \phi\,(y,\,z)}{\Delta z}$ is the differential coefficient of $\phi\,(y,\,z)$ or u, with respect to z, taken on the supposition that z is the only variable; and may therefore be denoted by $\dfrac{du}{dz}$.

The limit of $\dfrac{\phi\,(y + \Delta y,\,z + \Delta z) - \phi\,(y,\,z + \Delta z)}{\Delta y}$ would, if Δz did not change, be the differential coefficient of $\phi\,(y,\,z + \Delta z)$ with respect to y. But as Δz diminishes without limit with Δy, the limit is the differential coefficient of $\phi\,(y,\,z)$, with respect to y, *taken on the supposition that y is the only variable.*

We have then finally

$$\frac{du}{dx} = \frac{du}{dy}\frac{dy}{dx} + \frac{du}{dz}\frac{dz}{dx}.$$

170. In this result $\dfrac{du}{dy}$ denotes, as stated, "the differential coefficient of u, taken with respect to y, *supposing y alone to vary.*" It is not impossible that the reader may be inclined to say, "But y and z being both functions of x, if y varies, z *must* vary too, how then *can* I make the supposition that y alone varies?" His own further reflexion will probably remove the difficulty, if such it be. Should he however be unable to satisfy himself, it may be suggested to him that we do not make the supposition that y alone varies as a *final* supposition. We allow for the variation of both y and z, but it is convenient for our purpose to consider these variations *one at a time.*

It is usual to write $\left(\dfrac{du}{dy}\right)$ and $\left(\dfrac{du}{dz}\right)$, instead of $\dfrac{du}{dy}$ and $\dfrac{du}{dz}$, the brackets serving to remind us of the suppositions to be made in finding the values of these differential coefficients. Hence the above equation should be written

$$\frac{du}{dx} = \left(\frac{du}{dy}\right)\frac{dy}{dx} + \left(\frac{du}{dz}\right)\frac{dz}{dx}.$$

Of course the brackets *may* be omitted, and indeed frequently are omitted, provided we can feel certain of remembering the conditions which they are designed to express. The *beginner* will do well to use them, although as he advances in the subject he may be able to dispense with them.

171.   For example, let $u = z^2 + y^3 + zy$,

$$z = \sin x,$$

$$y = e^x;$$

then

$$\left(\frac{du}{dy}\right) = 3y^2 + z,$$

$$\left(\frac{du}{dz}\right) = 2z + y,$$

$$\frac{dy}{dx} = e^x,$$

$$\frac{dz}{dx} = \cos x;$$

therefore $\dfrac{du}{dx} = (3y^2 + z)\,e^x + (2z + y)\cos x$

$$= (3e^{2x} + \sin x)\,e^x + (2\sin x + e^x)\cos x$$

$$= 3e^{3x} + e^x\,(\sin x + \cos x) + \sin 2x;$$

and this value is of course precisely what we obtain if we substitute in u for y and z their values in terms of x, thus obtaining $u = e^{3x} + e^x \sin x + \sin^2 x$, and then differentiate with respect to x.

172.   An important case of the general proposition is obtained by supposing $z = x$ so that $\dfrac{dz}{dx} = 1$.   We have then

$$\frac{du}{dx} = \left(\frac{du}{dy}\right)\frac{dy}{dx} + \left(\frac{du}{dx}\right).$$

Here we cannot dispense with the brackets or some equivalent notation, $\left(\dfrac{du}{dx}\right)$ denoting what *would be* the differential coefficient of u with respect to x, if y were not a function of x, and $\dfrac{du}{dx}$ denoting the actual differential coefficient of u with respect to x, when y *is* a function of x.

173.　For example, let $u = \tan^{-1}(xy)$,

$$y = e^x;$$

then
$$\left(\frac{du}{dx}\right) = \frac{y}{1 + x^2 y^2},$$

$$\left(\frac{du}{dy}\right) = \frac{x}{1 + x^2 y^2},$$

$$\frac{dy}{dx} = e^x;$$

therefore
$$\frac{du}{dx} = \frac{e^x x + y}{1 + x^2 y^2}$$

$$= \frac{e^x (1 + x)}{1 + x^2 e^{2x}},$$

which of course is what we obtain if we differentiate $\tan^{-1}(xe^x)$ with respect to x.

174.　Suppose $u = \phi(v, y, z)$ where v, y, z, are each functions of x. We have, as before,

$$\Delta u = \phi(v + \Delta v, y + \Delta y, z + \Delta z) - \phi(v, y, z)$$
$$= \phi(v + \Delta v, y + \Delta y, z + \Delta z) - \phi(v, y + \Delta y, z + \Delta z)$$
$$+ \phi(v, y + \Delta y, z + \Delta z) - \phi(v, y, z + \Delta z)$$
$$+ \phi(v, y, z + \Delta z) - \phi(v, y, z);$$

$$\frac{\Delta u}{\Delta x} = \frac{\phi(v + \Delta v, y + \Delta y, z + \Delta z) - \phi(v, y + \Delta y, z + \Delta z)}{\Delta v} \frac{\Delta v}{\Delta x}$$
$$+ \frac{\phi(v, y + \Delta y, z + \Delta z) - \phi(v, y, z + \Delta z)}{\Delta y} \frac{\Delta y}{\Delta x}$$
$$+ \frac{\phi(v, y, z + \Delta z) - \phi(v, y, z)}{\Delta z} \frac{\Delta z}{\Delta x}.$$

Proceeding to the limit, we obtain

$$\frac{du}{dx} \quad \left(\frac{du}{dv}\right)\frac{dv}{dx} + \left(\frac{du}{dy}\right)\frac{dy}{dx} + \left(\frac{du}{dz}\right)\frac{dz}{dx}.$$

The process may be extended to the case in which u involves more than three functions of x.

175. Examples may occur more complicated in appearance, but essentially involving the same principles as those of the preceding Articles. Suppose for instance

$$u = \phi\,(v,\ y,\ z,\ x),$$
$$v = \psi\,(y,\ z,\ x),$$
$$y = f\,(x),$$
$$z = F\,(x),$$

so that u could, by performing the requisite substitutions, be made an explicit function of x: it is required to express the differential coefficient of u with respect to x, without previously making these substitutions.

$$\frac{du}{dx} = \left(\frac{du}{dv}\right)\frac{dv}{dx} + \left(\frac{du}{dy}\right)\frac{dy}{dx} + \left(\frac{du}{dz}\right)\frac{dz}{dx} + \left(\frac{du}{dx}\right),$$

$$\frac{dv}{dx} = \left(\frac{dv}{dy}\right)\frac{dy}{dx} + \left(\frac{dv}{dz}\right)\frac{dz}{dx} + \left(\frac{dv}{dx}\right),$$

$$\frac{dy}{dx} = f'\,(x),\qquad \frac{dz}{dx} = F'\,(x).$$

Hence $\dfrac{du}{dx} = \left(\dfrac{du}{dv}\right)\left\{\left(\dfrac{dv}{dy}\right)f'\,(x) + \left(\dfrac{dv}{dz}\right)F'\,(x) + \left(\dfrac{dv}{dx}\right)\right\}$

$$+ \left(\frac{du}{dy}\right)f'\,(x) + \left(\frac{du}{dz}\right)F'\,(x) + \left(\frac{du}{dx}\right).$$

176. The same suppositions being made as in Art. 169, it is required to express $\dfrac{d^2u}{dx^2}$. We have

$$\frac{du}{dx} = \left(\frac{du}{dy}\right)\frac{dy}{dx} + \left(\frac{du}{dz}\right)\frac{dz}{dx}.$$

Now $\left(\dfrac{du}{dy}\right)$ is itself a function of y and z. If we denote it by v its differential coefficient with respect to x will be

$$\left(\frac{dv}{dy}\right)\frac{dy}{dx} + \left(\frac{dv}{dz}\right)\frac{dz}{dx},$$

which may be written

$$\left(\frac{d^2u}{dy^2}\right)\frac{dy}{dx} + \left(\frac{d^2u}{dz\,dy}\right)\frac{dz}{dx}.$$

The differential coefficient of $\dfrac{dy}{dx}$ with respect to x is $\dfrac{d^2y}{dx^2}$.
Proceeding in the same way with the term

$$\left(\frac{du}{dz}\right)\frac{dz}{dx},$$

and remembering, (Art. 134), that

$$\left(\frac{d^2u}{dz\,dy}\right) = \left(\frac{d^2u}{dy\,dz}\right),$$

we have

$$\frac{d^2u}{dx^2} = \left(\frac{d^2u}{dy^2}\right)\left(\frac{dy}{dx}\right)^2 + 2\left(\frac{d^2u}{dy\,dz}\right)\frac{dy}{dx}\frac{dz}{dx} + \left(\frac{d^2u}{dz^2}\right)\left(\frac{dz}{dx}\right)^2$$

$$+ \left(\frac{du}{dy}\right)\frac{d^2y}{dx^2} + \left(\frac{du}{dz}\right)\frac{d^2z}{dx^2}.$$

If $z = x$, we have $\dfrac{dz}{dx} = 1$, $\dfrac{d^2z}{dx^2} = 0$; thus

$$\frac{d^2u}{dx^2} = \left(\frac{d^2u}{dy^2}\right)\left(\frac{dy}{dx}\right)^2 + 2\left(\frac{d^2u}{dy\,dx}\right)\frac{dy}{dx} + \left(\frac{d^2u}{dx^2}\right) + \left(\frac{du}{dy}\right)\frac{d^2y}{dx^2}.$$

177. Hitherto in this Chapter we have given methods which, although often convenient, are not absolutely *necessary*, as in all cases, by effecting the required substitutions, we may obtain an explicit function of x, and differentiate it by known rules. But the case we now consider is one in which a new method is frequently *indispensable*.

Let $\phi(x, y) = 0$ be an equation connecting the variables x and y: it is required to find $\dfrac{dy}{dx}$. If the given equation can be solved so as to give y in terms of x, say $y = \psi(x)$, then the differential coefficient of y with respect to x can be found by previous rules. If x can be expressed in terms of y, we can determine $\dfrac{dx}{dy}$ and then $\dfrac{dy}{dx}$, since $\dfrac{dx}{dy} \times \dfrac{dy}{dx} = 1$. But as it is often difficult, and sometimes impossible, to solve the given equation, it is necessary to investigate a rule for finding $\dfrac{dy}{dx}$ which does not require this operation.

Put u for $\phi(x, y)$. From the given equation y is *some* definite function of x; hence

$$\left(\frac{du}{dy}\right)\frac{dy}{dx} + \left(\frac{du}{dx}\right)$$

is, by Art. 172, the differential coefficient of u with respect to x. But u is always zero, and therefore so also is its differential coefficient with respect to x. Hence

$$0 = \left(\frac{du}{dy}\right)\frac{dy}{dx} + \left(\frac{du}{dx}\right),$$

therefore
$$\frac{dy}{dx} = -\frac{\left(\dfrac{du}{dx}\right)}{\left(\dfrac{du}{dy}\right)}.$$

178. This important result may also be obtained thus, which is in effect combining into one Article portions of the preceding pages. Let

$$\phi(x, y) = 0.$$

Suppose x to become $x + \Delta x$ and y to become $y + \Delta y$, so that

$$\phi(x + \Delta x, y + \Delta y) = 0.$$

Hence $\qquad \phi(x + \Delta x, y + \Delta y) - \phi(x, y) = 0,$

and $\phi(x+\Delta x, y+\Delta y) - \phi(x+\Delta x, y) + \phi(x+\Delta x, y) - \phi(x, y) = 0.$

Divide by Δx, and we have

$$\frac{\phi(x+\Delta x,\, y+\Delta y) - \phi(x+\Delta x,\, y)}{\Delta y} \cdot \frac{\Delta y}{\Delta x} + \frac{\phi(x+\Delta x, y) - \phi(x, y)}{\Delta x} = 0.$$

This equation, being always true, remains so when Δx and Δy are diminished indefinitely.

The limit of $\dfrac{\phi\,(x+\Delta x,\ y)-\phi\,(x,\ y)}{\Delta x}$, when Δx diminishes, is the differential coefficient of $\phi\,(x,\ y)$ *with respect to x, formed on the supposition that x alone varies*, and if we put u for $\phi\,(x,\ y)$, this limit may be denoted by $\left(\dfrac{du}{dx}\right)$.

The limit of $\dfrac{\phi\,(x+\Delta x,\ y+\Delta y)-\phi\,(x+\Delta x,\ y)}{\Delta y}$ would, if Δx remained constant, be the differential coefficient of $\phi\,(x+\Delta x,\ y)$ with respect to y, *formed on the supposition that y alone varies*. But as Δx diminishes without limit when Δy does so, the limit is the differential coefficient of u with respect to y, formed on the supposition that y alone varies. It may be denoted by $\left(\dfrac{du}{dy}\right)$.

The limit of $\dfrac{\Delta y}{\Delta x}$ is $\dfrac{dy}{dx}$. Hence finally

$$\left(\frac{du}{dy}\right)\frac{dy}{dx}+\left(\frac{du}{dx}\right)=0.$$

179. For example, suppose $a^2y^2+b^2x^2-a^2b^2=0$.

Here $\qquad u=a^2y^2+b^2x^2-a^2b^2,$

$$\left(\frac{du}{dx}\right)=2b^2x,$$

$$\left(\frac{du}{dy}\right)=2a^2y\ ;$$

therefore $\qquad a^2y\,\dfrac{dy}{dx}+b^2x=0,$

therefore $\qquad \dfrac{dy}{dx}=-\dfrac{b^2x}{a^2y}$ .... (1).

Since $y=\dfrac{b}{a}\,\sqrt{(a^2-x^2)}$ from the given equation, we obtain directly

$$\frac{dy}{dx}=-\frac{bx}{a\,\sqrt{(a^2-x^2)}}\ \ldots\ldots\ldots\ldots\ldots(2).$$

When in (1) we substitute the value of y in terms of x, the result agrees with (2).

In this case we can verify our new rule, by comparing its results with those previously found. In more complex examples, such as

$$x^5 - ax^3y + bx^2y^2 - y^5 = 0,$$

we can find $\dfrac{dy}{dx}$ only by the new method;

putting u for $x^5 - ax^3y + bx^2y^2 - y^5$, we have

$$\left(\frac{du}{dx}\right) = 5x^4 - 3ax^2y + 2bxy^2,$$

$$\left(\frac{du}{dy}\right) = - ax^3 + 2bx^2y - 5y^4;$$

therefore $\dfrac{dy}{dx} = \dfrac{5x^4 - 3ax^2y + 2bxy^2}{5y^4 - 2bx^2y + ax^3}.$

180. We shall now investigate the *second* differential coefficient of an implicit function.

From the equation

$$u \text{ or } \phi(x, y) = 0,$$

we have deduced $\dfrac{dy}{dx} = - \dfrac{\left(\dfrac{du}{dx}\right)}{\left(\dfrac{du}{dy}\right)}$ .... (1):

it is required to find $\dfrac{d^2y}{dx^2}$.

We observe that $\left(\dfrac{du}{dx}\right)$ being a function of both x and y, its differential coefficient with respect to x must be found by Art. 172. If we put v for $\left(\dfrac{du}{dx}\right)$, the required differential coefficient will be

$$\left(\frac{dv}{dy}\right)\frac{dy}{dx} + \left(\frac{dv}{dx}\right).$$

Similarly, denoting $\left(\dfrac{du}{dy}\right)$ by w, we have for its differential coefficient with respect to x,

$$\left(\frac{dw}{dy}\right)\frac{dy}{dx} + \left(\frac{dw}{dx}\right).$$

Hence, from (1),

$$\frac{d^2y}{dx^2} = -\frac{w\left\{\left(\frac{dv}{dy}\right)\frac{dy}{dx} + \left(\frac{dv}{dx}\right)\right\} - v\left\{\left(\frac{dw}{dy}\right)\frac{dy}{dx} + \left(\frac{dw}{dx}\right)\right\}}{w^2} \quad \ldots\ldots(2).$$

Now

$$\left(\frac{dv}{dx}\right) = \left(\frac{d^2u}{dx^2}\right),$$

the latter symbol denoting that u is to be differentiated twice with respect to x, *on the supposition that x alone varies*; also

$$\left(\frac{dv}{dy}\right) = \left(\frac{d^2u}{dy\,dx}\right),$$

the latter symbol denoting that u is to be differentiated with respect to x, supposing x *alone to vary*, and the result with respect to y, supposing y alone to vary. Similarly

$$\left(\frac{dw}{dx}\right) = \left(\frac{d^2u}{dx\,dy}\right),$$

$$\left(\frac{dw}{dy}\right) = \left(\frac{d^2u}{dy^2}\right).$$

Hence, substituting in (2), we have

$$\frac{d^2y}{dx^2} = -\frac{\left(\frac{du}{dy}\right)\left\{\left(\frac{d^2u}{dy\,dx}\right)\frac{dy}{dx} + \left(\frac{d^2u}{dx^2}\right)\right\} - \left(\frac{du}{dx}\right)\left\{\left(\frac{d^2u}{dy^2}\right)\frac{dy}{dx} + \left(\frac{d^2u}{dx\,dy}\right)\right\}}{\left(\frac{du}{dy}\right)^2}$$

$$\ldots\ldots\ldots\ldots\ldots\ldots (3).$$

If we substitute in (3) the value of $\frac{dy}{dx}$ given by (1), we have, since $\left(\frac{d^2u}{dy\,dx}\right) = \left(\frac{d^2u}{dx\,dy}\right)$ by Art. 134,

$$\frac{d^2y}{dx^2} = -\frac{\left(\frac{d^2u}{dx^2}\right)\left(\frac{du}{dy}\right)^2 - 2\left(\frac{d^2u}{dx\,dy}\right)\left(\frac{du}{dx}\right)\left(\frac{du}{dy}\right) + \left(\frac{d^2u}{dy^2}\right)\left(\frac{du}{dx}\right)^2}{\left(\frac{du}{dy}\right)^3} \quad \ldots(4).$$

181. This result may also be found from Art. 176, by supposing $u = 0$ always, and therefore $\dfrac{d^2u}{dx^2} = 0$; or independently thus.

From
$$u = 0$$

it follows that
$$\left(\frac{du}{dy}\right)\frac{dy}{dx} + \left(\frac{du}{dx}\right) = 0 \dots\dots\dots\dots (1).$$

Denote this result for the sake of shortness by
$$v = 0.$$

Hence
$$\left(\frac{dv}{dy}\right)\frac{dy}{dx} + \left(\frac{dv}{dx}\right) = 0 \dots\dots\dots\dots (2),$$

which result, expressed in terms of u, is

$$\left(\frac{d^2u}{dx^2}\right) + 2\left(\frac{d^2u}{dx\,dy}\right)\frac{dy}{dx} + \left(\frac{d^2u}{dy^2}\right)\left(\frac{dy}{dx}\right)^2 + \left(\frac{du}{dy}\right)\frac{d^2y}{dx^2} = 0 \dots (3);$$

as $\dfrac{dy}{dx}$ is already known, this equation will furnish $\dfrac{d^2y}{dx^2}$.

Equation (1) is frequently called the "first derived equation," or "the differential equation of the first order;" and equation (3) is called "the second derived equation," or the "differential equation of the second order;" the equation $u = 0$ being called the "primitive equation."

182. Should the reader succeed in correctly deducing for himself the important equation (3) of the last Article, he may omit the next two Articles, as it seems unnecessary to direct his attention to difficulties he *might* have felt, or mistakes he *might* have made. If however he has failed in his attempts, he may compare his process with the following.

In (1), put p for $\dfrac{dy}{dx}$, so that v stands for
$$\left(\frac{du}{dy}\right)p + \left(\frac{du}{dx}\right).$$

Hence
$$\left(\frac{dv}{dx}\right) = \left(\frac{d^2u}{dx\,dy}\right)p + \left(\frac{du}{dy}\right)\left(\frac{dp}{dx}\right) + \left(\frac{d^2u}{dx^2}\right),$$

$$\left(\frac{dv}{dy}\right) = \left(\frac{d^2u}{dy^2}\right)p + \left(\frac{du}{dy}\right)\left(\frac{dp}{dy}\right) + \left(\frac{d^2u}{dy\,dx}\right).$$

Thus (2) becomes

$$\left\{ \left(\frac{d^2u}{dy^2}\right) p + \left(\frac{du}{dy}\right)\left(\frac{dp}{dy}\right) + \left(\frac{d^2u}{dy\,dx}\right)\right\} p$$

$$+ \left(\frac{d^2u}{dx\,dy}\right) p + \left(\frac{du}{dy}\right)\left(\frac{dp}{dx}\right) + \left(\frac{d^2u}{dx^2}\right) = 0,$$

or

$$\left(\frac{d^2u}{dx^2}\right) + 2\left(\frac{d^2u}{dx\,dy}\right) p + \left(\frac{d^2u}{dy^2}\right) p^2 + \left(\frac{du}{dy}\right)\left\{\left(\frac{dp}{dy}\right) p + \left(\frac{dp}{dx}\right)\right\} = 0.$$

But $\left(\frac{dp}{dy}\right) p + \left(\frac{dp}{dx}\right) = \frac{dp}{dx}$, that is $\frac{d^2y}{dx^2}$, (Art. 172), and with

this simplification we obtain the required result.

A very common mistake is to omit the brackets in $\left(\frac{dp}{dy}\right) p + \left(\frac{dp}{dx}\right)$, and thus $\left(\frac{dp}{dx}\right)$ is written $\frac{d^2y}{dx^2}$, and there remains a superfluous term, namely $\frac{dp}{dy}$, or as it has perhaps been written by the student, $\frac{d^2y}{dy\,dx}$.

183. In Art. 182 we proceeded very strictly according to the literal requirements of the rule involved in equation (2) of Art. 181. We might have reasoned thus.

We have merely to express symbolically the fact, that the differential coefficient of

$$\left(\frac{du}{dy}\right)\frac{dy}{dx} + \left(\frac{du}{dx}\right)$$

with respect to x is zero.

Now the differential coefficient of $\left(\frac{du}{dy}\right)$ with respect to x

is $\qquad \left(\frac{d^2u}{dx\,dy}\right) + \left(\frac{d^2u}{dy^2}\right)\frac{dy}{dx};$

and the differential coefficient of $\frac{dy}{dx}$ with respect to x is $\frac{d^2y}{dx^2}$.

Hence the differential coefficient of $\left(\dfrac{du}{dy}\right)\dfrac{dy}{dx}$ is

$$\left\{\left(\frac{d^2u}{dx\,dy}\right)+\left(\frac{d^2u}{dy^2}\right)\frac{dy}{dx}\right\}\frac{dy}{dx}+\left(\frac{du}{dy}\right)\frac{d^2y}{dx^2}\dots\dots\dots(1).$$

Also the differential coefficient of $\left(\dfrac{du}{dx}\right)$ is

$$\left(\frac{d^2u}{dy\,dx}\right)\frac{dy}{dx}+\left(\frac{d^2u}{dx^2}\right)\dots\dots\dots\dots(2).$$

Collecting the terms in (1) and (2), we have

$$\left(\frac{d^2u}{dx^2}\right)+2\left(\frac{d^2u}{dx\,dy}\right)\frac{dy}{dx}+\left(\frac{d^2u}{dy^2}\right)\left(\frac{dy}{dx}\right)^2+\left(\frac{du}{dy}\right)\frac{d^2y}{dx^2}=0.$$

184. It is not necessary to proceed further with the successive differential coefficients of implicit functions, as the equations become too complicated to be often used. The reader may, as an exercise, obtain the following result from equation (3) of Art. 181, by either of the methods we have used in Arts. 182 and 183:

$$\left(\frac{d^3u}{dx^3}\right)+3\left(\frac{d^3u}{dx^2dy}\right)\frac{dy}{dx}+3\left(\frac{d^3u}{dxdy^2}\right)\left(\frac{dy}{dx}\right)^2+\left(\frac{d^3u}{dy^3}\right)\left(\frac{dy}{dx}\right)^3$$

$$+3\left\{\left(\frac{d^2u}{dxdy}\right)+\left(\frac{d^2u}{dy^2}\right)\frac{dy}{dx}\right\}\frac{d^2y}{dx^2}+\left(\frac{du}{dy}\right)\frac{d^3y}{dx^3}=0.$$

We may observe that it is often found convenient to use a certain abbreviated notation for partial differential coefficients. Thus if $\phi(x,y)$ be any function of x and y, any partial differential coefficient of the function may be indicated by the letter ϕ, with accents *above* corresponding to the number of differentiations with respect to x, and with accents *below* corresponding to the number of differentiations with respect to y. For example, ϕ'' will indicate $\left(\dfrac{d^2\phi\,(x,\,y)}{dx^2}\right)$, and $\phi_{\prime}'$ will indicate $\left(\dfrac{d^2\phi\,(x,\,y)}{dxdy}\right)$, and so on.

We may also use y' for $\dfrac{dy}{dx}$, and y'' for $\dfrac{d^2y}{dx^2}$, and so on. Thus with the present notation the equations (1) and (3) of

Art. 181, and the equation which may be obtained from (3) will be expressed respectively as follows:

$$\phi' + \phi_{,}y' = 0,$$
$$\phi'' + 2\phi_{,}'y' + \phi_{,,}y'^{2} + \phi_{,}y'' = 0,$$
$$\phi''' + 3\phi_{,}''y' + 3\phi_{,,}'y'^{2} + \phi_{,,,}y'^{3} + 3\left(\phi_{,}' + \phi_{,,}y'\right)y'' + \phi_{,}y''' = 0.$$

185. Suppose the two equations

$$f(x, y, z) = 0,$$
$$F(x, y, z) = 0,$$

exist *simultaneously*, in which x is the independent variable and y and z dependent variables. From the two given equations we may eliminate z, and thus find an equation connecting y and x. Hence $\dfrac{dy}{dx}$ may be determined. Again, from the two given equations we may eliminate y, and thus find an equation connecting z and x, whence $\dfrac{dz}{dx}$ may be determined. In cases where the elimination is tedious or impracticable we may proceed thus.

Let u denote $f(x, y, z)$ and v denote $F(x, y, z)$. Since y and z are functions of x, the differential coefficient of u with respect to x is, by Arts. 172 and 174,

$$\left(\frac{du}{dx}\right) + \left(\frac{du}{dy}\right)\frac{dy}{dx} + \left(\frac{du}{dz}\right)\frac{dz}{dx};$$

and since u always $= 0$, we have

$$0 = \left(\frac{du}{dx}\right) + \left(\frac{du}{dy}\right)\frac{dy}{dx} + \left(\frac{du}{dz}\right)\frac{dz}{dx} \quad\ldots\ldots\ldots\ldots (1).$$

Similarly, $0 = \left(\dfrac{dv}{dx}\right) + \left(\dfrac{dv}{dy}\right)\dfrac{dy}{dx} + \left(\dfrac{dv}{dz}\right)\dfrac{dz}{dx} \quad\ldots\ldots\ldots\ldots (2);$

from which we find

$$\frac{dy}{dx} = -\frac{\left(\dfrac{du}{dx}\right)\left(\dfrac{dv}{dz}\right) - \left(\dfrac{dv}{dx}\right)\left(\dfrac{du}{dz}\right)}{\left(\dfrac{du}{dy}\right)\left(\dfrac{dv}{dz}\right) - \left(\dfrac{dv}{dy}\right)\left(\dfrac{du}{dz}\right)} \quad\ldots\ldots\ldots\ldots (3),$$

$$\frac{dz}{dx} = -\frac{\left(\frac{dv}{dx}\right)\left(\frac{du}{dy}\right) - \left(\frac{du}{dx}\right)\left(\frac{dv}{dy}\right)}{\left(\frac{dv}{dz}\right)\left(\frac{du}{dy}\right) - \left(\frac{du}{dz}\right)\left(\frac{dv}{dy}\right)} \quad \ldots\ldots\ldots\ldots (4).$$

186. By differentiating equations (1), (2) of the last Article with respect to x, we obtain

$$\left(\frac{d^2u}{dx^2}\right) + 2\left(\frac{d^2u}{dx\,dy}\right)\frac{dy}{dx} + 2\left(\frac{d^2u}{dx\,dz}\right)\frac{dz}{dx} + \left(\frac{d^2u}{dy^2}\right)\left(\frac{dy}{dx}\right)^2$$

$$+ 2\left(\frac{d^2u}{dy\,dz}\right)\frac{dy}{dx}\frac{dz}{dx} + \left(\frac{d^2u}{dz^2}\right)\left(\frac{dz}{dx}\right)^2 + \left(\frac{du}{dy}\right)\frac{d^2y}{dx^2} + \left(\frac{du}{dz}\right)\frac{d^2z}{dx^2} = 0,$$

$$\left(\frac{d^2v}{dx^2}\right) + 2\left(\frac{d^2v}{dx\,dy}\right)\frac{dy}{dx} + 2\left(\frac{d^2v}{dx\,dz}\right)\frac{dz}{dx} + \left(\frac{d^2v}{dy^2}\right)\left(\frac{dy}{dx}\right)^2$$

$$+ 2\left(\frac{d^2v}{dy\,dz}\right)\frac{dy}{dx}\frac{dz}{dx} + \left(\frac{d^2v}{dz^2}\right)\left(\frac{dz}{dx}\right)^2 + \left(\frac{dv}{dy}\right)\frac{d^2y}{dx^2} + \left(\frac{dv}{dz}\right)\frac{d^2z}{dx^2} = 0.$$

From these equations we can deduce $\frac{d^2y}{dx^2}$ and $\frac{d^2z}{dx^2}$, which may also be found by differentiating equations (3) and (4) of the preceding Article.

187. Suppose we have n equations connecting $n+1$ variables x, y, z, ... t. Let the equations be

$$F_1(x, y, z, \ldots\ldots t) = 0, \quad \text{say } u_1 = 0,$$

$$F_2(x, y, z, \ldots\ldots t) = 0, \quad \text{say } u_2 = 0,$$

$$\ldots\ldots\ldots\ldots\ldots\ldots\ldots \ldots\ldots\ldots\ldots$$

$$F_n(x, y, z, \ldots\ldots t) = 0, \quad \text{say } u_n = 0.$$

From these equations all the variables but one may be considered functions of that one. If x be the independent variable, we have by differentiation, as in Art. 185,

$$0 = \left(\frac{du_1}{dx}\right) + \left(\frac{du_1}{dy}\right)\frac{dy}{dx} + \left(\frac{du_1}{dz}\right)\frac{dz}{dx} + \ldots\ldots + \left(\frac{du_1}{dt}\right)\frac{dt}{dx},$$

$$0 = \left(\frac{du_2}{dx}\right) + \left(\frac{du_2}{dy}\right)\frac{dy}{dx} + \left(\frac{du_2}{dz}\right)\frac{dz}{dx} + \ldots\ldots + \left(\frac{du_2}{dt}\right)\frac{dt}{dx},$$

$$\ldots\ldots\ldots\ldots\ldots\ldots\ldots\ldots\ldots\ldots$$

$$0 = \left(\frac{du_n}{dx}\right) + \left(\frac{du_n}{dy}\right)\frac{dy}{dx} + \ldots\ldots\ldots \quad + \left(\frac{du_n}{dt}\right)\frac{dt}{dx},$$

from which n equations we can determine the n quantities

$$\frac{dy}{dx}, \quad \frac{dz}{dx}, \quad \ldots\ldots\ldots \quad \frac{dt}{dx}.$$

188. Suppose $\phi(x, y, z) = 0$ to be the only equation connecting three variables, so that z may be considered an implicit function of the two independent variables x and y: it is required to find $\dfrac{dz}{dx}$ and $\dfrac{dz}{dy}$.

By $\dfrac{dz}{dx}$ is meant the differential coefficient of z with respect to x, supposing y constant, and by $\dfrac{dz}{dy}$ the differential coefficient of z with respect to y supposing x constant. Theoretically we may from the given equation find the value of z in terms of x and y and then effect the differentiation by common rules; (see Art. 131). But to avoid the difficulty of solving the given equation we adopt another method. Suppose y constant, so that we have two variables x and z, and let u stand for $\phi(x, y, z)$, then by Art. 178

$$\left(\frac{du}{dx}\right) + \left(\frac{du}{dz}\right)\frac{dz}{dx} = 0 \ldots\ldots\ldots\ldots\ldots (1);$$

where $\left(\dfrac{du}{dx}\right)$ stands for the differential coefficient of u taken on the supposition that x alone varies, and $\left(\dfrac{du}{dz}\right)$ for the differential coefficient of u taken on the supposition that z alone varies. Similarly

$$\left(\frac{du}{dy}\right) + \left(\frac{du}{dz}\right)\frac{dz}{dy} = 0 \ldots\ldots\ldots\ldots\ldots (2).$$

Equations (1) and (2) determine $\dfrac{dz}{dx}$ and $\dfrac{dz}{dy}$.

We may determine $\dfrac{d^2z}{dx^2}$ and $\dfrac{d^2z}{dy^2}$ by the method of Art. 180, or by that of Art. 181. If we adopt the latter method, the two equations we obtain are

$$\left(\frac{d^2u}{dx^2}\right) + 2\left(\frac{d^2u}{dx\,dz}\right)\frac{dz}{dx} + \left(\frac{d^2u}{dz^2}\right)\left(\frac{dz}{dx}\right)^2 + \left(\frac{du}{dz}\right)\frac{d^2z}{dx^2} = 0,$$

$$\left(\frac{d^2u}{dy^2}\right) + 2\left(\frac{d^2u}{dy\,dz}\right)\frac{dz}{dy} + \left(\frac{d^2u}{dz^2}\right)\left(\frac{dz}{dy}\right)^2 + \left(\frac{du}{dz}\right)\frac{d^2z}{dy^2} = 0.$$

We can obtain an equation for finding $\dfrac{d^2z}{dy\,dx}$ by differentiating (1) with respect to y, or by differentiating (2) with respect to x. We thus deduce

$$\left(\frac{d^2u}{dx\,dy}\right) + \left(\frac{d^2u}{dz\,dx}\right)\frac{dz}{dy} + \left(\frac{d^2u}{dz\,dy}\right)\frac{dz}{dx} + \left(\frac{d^2u}{dz^2}\right)\frac{dz}{dy}\frac{dz}{dx}$$
$$+ \left(\frac{du}{dz}\right)\frac{d^2z}{dy\,dx} = 0.$$

189. Suppose we have two equations connecting four variables; for example,

$$f(v, x, y, z) = 0, \quad \text{say } u_1 = 0,$$
$$F(v, x, y, z) = 0, \quad \text{say } u_2 = 0;$$

from these equations v and z may be considered functions of the independent variables x and y. If we eliminate v we obtain an equation connecting z, x, and y, so that $\dfrac{dz}{dx}$ and $\dfrac{dz}{dy}$ may be obtained by the preceding Article; and similarly if we eliminate z we may find $\dfrac{dv}{dx}$ and $\dfrac{dv}{dy}$. Or we may proceed thus: from the equation $u_1 = 0$ we deduce, by Art. 174,

$$\left(\frac{du_1}{dx}\right) + \left(\frac{du_1}{dv}\right)\frac{dv}{dx} + \left(\frac{du_1}{dz}\right)\frac{dz}{dx} = 0,$$

and from the equation $u_2 = 0$ we deduce

$$\left(\frac{du_2}{dx}\right) + \left(\frac{du_2}{dv}\right)\frac{dv}{dx} + \left(\frac{du_2}{dz}\right)\frac{dz}{dx} = 0,$$

from which $\dfrac{dv}{dx}$ and $\dfrac{dz}{dx}$ can be found.

Similarly, from $u_1 = 0$ and $u_2 = 0$ we deduce

$$\left(\frac{du_1}{dy}\right) + \left(\frac{du_1}{dv}\right)\frac{dv}{dy} + \left(\frac{du_1}{dz}\right)\frac{dz}{dy} = 0,$$

and

$$\left(\frac{du_2}{dy}\right) + \left(\frac{du_2}{dv}\right)\frac{dv}{dy} + \left(\frac{du_2}{dz}\right)\frac{dz}{dy} = 0,$$

from which $\dfrac{dz}{dy}$ and $\dfrac{dv}{dy}$ can be found.

In such equations as those in the present Article it is very common to write $\dfrac{df}{dy}$, $\dfrac{df}{dv}$, $\dfrac{dF}{dy}$, ..., to denote $\dfrac{du_1}{dy}$, $\dfrac{du_1}{dv}$, $\dfrac{du_2}{dy}$, ..

190.   If values of x and y which satisfy an equation $u = 0$ involving x and y, also make $\left(\dfrac{du}{dx}\right)$ and $\left(\dfrac{du}{dy}\right)$ vanish, then $\dfrac{dy}{dx}$, which $= -\dfrac{\left(\dfrac{du}{dx}\right)}{\left(\dfrac{du}{dy}\right)}$, assumes the indeterminate form $\dfrac{0}{0}$.

If we apply the method of Art. 145, we have

the limit of $\dfrac{\left(\dfrac{du}{dx}\right)}{\left(\dfrac{du}{dy}\right)} =$ the limit of $\dfrac{\left(\dfrac{d^2u}{dx^2}\right) + \left(\dfrac{d^2u}{dx\,dy}\right)\dfrac{dy}{dx}}{\left(\dfrac{d^2u}{dx\,dy}\right) + \left(\dfrac{d^2u}{dy^2}\right)\dfrac{dy}{dx}}$,

the numerator and denominator of the second fraction being respectively the differential coefficient of $\left(\dfrac{du}{dx}\right)$ and of $\left(\dfrac{du}{dy}\right)$ with respect to x.

We have then

$$\frac{dy}{dx} = -\frac{\left(\dfrac{d^2u}{dx^2}\right) + \left(\dfrac{d^2u}{dx\,dy}\right)\dfrac{dy}{dx}}{\left(\dfrac{d^2u}{dx\,dy}\right) + \left(\dfrac{d^2u}{dy^2}\right)\dfrac{dy}{dx}} \quad\dots\dots\dots\dots\dots (1).$$

In this expression we must substitute in $\left(\dfrac{d^2u}{dx^2}\right)$, $\left(\dfrac{d^2u}{dx\,dy}\right)$, and $\left(\dfrac{d^2u}{dy^2}\right)$, the values of x and y under consideration, and thus we obtain a quadratic for finding $\dfrac{dy}{dx}$. This quadratic is

$$\left(\frac{d^2u}{dy^2}\right)\left(\frac{dy}{dx}\right)^2 + 2\left(\frac{d^2u}{dx\,dy}\right)\frac{dy}{dx} + \left(\frac{d^2u}{dx^2}\right) = 0 \ldots\ldots\ldots(2)\,;$$

equation (2) agrees with equation (3) of Art. 181, remembering that by hypothesis $\left(\dfrac{du}{dy}\right) = 0.$

191. Should the values of x and y we are considering in addition to making $u = 0$, $\left(\dfrac{du}{dx}\right) = 0$, $\left(\dfrac{du}{dy}\right) = 0$, also make $\left(\dfrac{d^2u}{dx^2}\right) = 0$, $\left(\dfrac{d^2u}{dy^2}\right) = 0$, $\left(\dfrac{d^2u}{dx\,dy}\right) = 0$, then the value of $\dfrac{dy}{dx}$ given in equation (1) of the preceding Article also takes the form $\dfrac{0}{0}$. Hence, applying again the rule for finding the limit of such a fraction, we have

$$\frac{dy}{dx} = -\frac{\left(\dfrac{d^3u}{dx^3}\right) + 2\left(\dfrac{d^3u}{dx^2dy}\right)\dfrac{dy}{dx} + \left(\dfrac{d^3u}{dxdy^2}\right)\left(\dfrac{dy}{dx}\right)^2 + \left(\dfrac{d^2u}{dxdy}\right)\dfrac{d^2y}{dx^2}}{\left(\dfrac{d^3u}{dx^2dy}\right) + 2\left(\dfrac{d^3u}{dxdy^2}\right)\dfrac{dy}{dx} + \left(\dfrac{d^3u}{dy^3}\right)\left(\dfrac{dy}{dx}\right)^2 + \left(\dfrac{d^2u}{dy^2}\right)\dfrac{d^2y}{dx^2}} \ldots(1).$$

Since $\left(\dfrac{d^2u}{dx\,dy}\right)$ and $\left(\dfrac{d^2u}{dy^2}\right)$ vanish, we obtain from (1)

$$\left(\frac{d^3u}{dy^3}\right)\left(\frac{dy}{dx}\right)^3 + 3\left(\frac{d^3u}{dx\,dy^2}\right)\left(\frac{dy}{dx}\right)^2 + 3\left(\frac{d^3u}{dx^2dy}\right)\frac{dy}{dx} + \left(\frac{d^3u}{dx^3}\right) = 0 \ldots(2),$$

where in all the differential coefficients of u we must substitute the values of x and y under consideration, giving a cubic equation to determine $\dfrac{dy}{dx}$. Compare Art. 184.

It must be observed that this method is liable to an objection. We *assume* that $\dfrac{d^2u}{dx\,dy}\dfrac{d^2y}{dx^2}$ and $\dfrac{d^2u}{dy^2}\dfrac{d^2y}{dx^2}$ vanish because in each case one factor vanishes; if however $\dfrac{d^2y}{dx^2}$ be *infinite*, it does not follow necessarily that $\dfrac{d^2u}{dx\,dy}\dfrac{d^2y}{dx^2}$ and $\dfrac{d^2u}{dy^2}\dfrac{d^2y}{dx^2}$ vanish. See Art. 380.

192. Example. $y^4 + 3a^2y^2 - 4a^2xy - a^2x^2 = 0$, or $u = 0$.

Here
$$\left(\frac{du}{dx}\right) = -4a^2y - 2a^2x,$$
$$\left(\frac{du}{dy}\right) = 4y^3 + 6a^2y - 4a^2x\,;$$

therefore $\dfrac{dy}{dx} = \dfrac{4a^2y + 2a^2x}{4y^3 + 6a^2y - 4a^2x} = \dfrac{2a^2y + a^2x}{2y^3 + 3a^2y - 2a^2x}$.

Here $x = 0$, $y = 0$, satisfy $u = 0$, and make $\dfrac{dy}{dx}$ assume the form $\dfrac{0}{0}$.

Differentiate both numerator and denominator, and we have

$$\frac{dy}{dx} = \text{the limit of } \frac{2a^2\dfrac{dy}{dx} + a^2}{(6y^2 + 3a^2)\dfrac{dy}{dx} - 2a^2}$$

$$= \frac{2\dfrac{dy}{dx} + 1}{3\dfrac{dy}{dx} - 2} \text{ ultimately.}$$

Hence,
$$\frac{dy}{dx}\left(3\frac{dy}{dx} - 2\right) = 2\frac{dy}{dx} + 1\,;$$

therefore
$$3\left(\frac{dy}{dx}\right)^2 - 4\frac{dy}{dx} - 1 = 0\,;$$

therefore
$$\frac{dy}{dx} = \frac{2 \pm \sqrt{7}}{3}\,.$$

Again, suppose $ay^3 - bx^2y + x^4 = 0$ to be the given equation.

Then

$$\left(\frac{du}{dx}\right) = 4x^3 - 2bxy,$$

$$\left(\frac{du}{dy}\right) = 3ay^2 - bx^2;$$

therefore

$$\frac{dy}{dx} = \frac{4x^3 - 2bxy}{bx^2 - 3ay^2}.$$

This value of $\frac{dy}{dx}$ takes the form $\frac{0}{0}$ when x and y vanish. Hence, differentiating the numerator and denominator, we have

$$\frac{dy}{dx} = \frac{12x^2 - 2by - 2bx\dfrac{dy}{dx}}{2bx - 6ay\dfrac{dy}{dx}},$$

when x and y are made $= 0$.

Again, we have the form $\frac{0}{0}$. Hence, differentiating again,

$$\frac{dy}{dx} = \frac{24x - 4b\dfrac{dy}{dx} - 2bx\dfrac{d^2y}{dx^2}}{2b - 6a\left(\dfrac{dy}{dx}\right)^2 - 6ay\dfrac{d^2y}{dx^2}},$$

x and y being made each $= 0$. Thus *assuming* that $x\dfrac{d^2y}{dx^2}$ and $y\dfrac{d^2y}{dx^2}$ vanish, we have

$$\frac{dy}{dx}\left\{2b - 6a\left(\frac{dy}{dx}\right)^2\right\} = -4b\frac{dy}{dx},$$

from which

$$\frac{dy}{dx} = 0,$$

or

$$\frac{dy}{dx} = \pm\sqrt{\frac{b}{a}}.$$

193. It may be noticed that equation (2) of Art. 190 differs from equation (3) of Art. 181 only in the omission of

the term $\left(\dfrac{du}{dy}\right)\dfrac{d^2y}{dx^2}$. This term would not occur if $\dfrac{dy}{dx}$ were a constant quantity, for then $\dfrac{d^2y}{dx^2}$ would be zero. Hence equation (2) of Art. 190 *may* be derived by differentiating the equation

$$\left(\frac{du}{dx}\right) + \left(\frac{du}{dy}\right)\frac{dy}{dx} = 0,$$

with respect to x *and treating* $\dfrac{dy}{dx}$ *as if it were a constant.*

Similarly, equation (2) of Art. 191 *may* be deduced from equation (2) of Art. 190 by differentiating with respect to x *and treating* $\dfrac{dy}{dx}$ *as if it were a constant.*

194. If in equation (2) of Art. 190 we have $\left(\dfrac{d^2u}{dy^2}\right) = 0$, then

$$\frac{dy}{dx} = -\frac{\left(\dfrac{d^2u}{dx^2}\right)}{2\left(\dfrac{d^2u}{dx\,dy}\right)}$$

as *one* value of $\dfrac{dy}{dx}$. The other value of $\dfrac{dy}{dx}$ will be infinite, for we know from Algebra that if we have a quadratic equation and the coefficient of the highest power of the unknown quantity gradually diminishes without limit, then one of the roots simultaneously increases without limit. See *Algebra*, Chapter XXII.

195. The value of $\dfrac{dy}{dx}$, when the values $x = 0$, $y = 0$, make it assume an indeterminate form, may often be more simply found thus. We have only to seek the limit of $\dfrac{y}{x}$ as x and y diminish without limit; this is obvious from the meaning of $\dfrac{dy}{dx}$, or from Art. 145; it will be seen too if we refer to the geometrical illustration of Art. 38.

Example. $y^4 + 3a^2y^2 - 4a^2xy - a^2x^2 = 0.$

Hence, $y^2 \left(\dfrac{y}{x}\right)^2 + 3a^2 \left(\dfrac{y}{x}\right)^2 - 4a^2 \dfrac{y}{x} - a^2 = 0.$

If now $\dfrac{y}{x}$ have any *finite* limit, the term $y^2 \left(\dfrac{y}{x}\right)^2$ will vanish when $y = 0$, and we have for finding the ultimate value of $\dfrac{y}{x}$ the equation

$$3a^2 \left(\dfrac{y}{x}\right)^2 - 4a^2 \left(\dfrac{y}{x}\right) - a^2 = 0,$$

or

$$3 \left(\dfrac{y}{x}\right)^2 - 4 \left(\dfrac{y}{x}\right) - 1 = 0;$$

therefore

$$\dfrac{y}{x} = \dfrac{2 \pm \sqrt{7}}{3}.$$

If $\dfrac{y}{x}$ have an infinite value, then $\dfrac{x}{y}$ has a value zero: putting the given equation in the form

$$y^2 + 3a^2 - 4a^2 \dfrac{x}{y} - a^2 \left(\dfrac{x}{y}\right)^2 = 0,$$

we see that $\dfrac{x}{y} = 0$ ultimately would *not* satisfy it. Hence $\dfrac{y}{x}$ has not an infinite value.

Again, suppose $ay^3 - bx^2y + x^4 = 0;$

therefore $a \left(\dfrac{y}{x}\right)^3 - b \left(\dfrac{y}{x}\right) + x = 0:$

when x vanishes, we have $\dfrac{y}{x} \left\{ a \left(\dfrac{y}{x}\right)^2 - b \right\} = 0;$

therefore $\dfrac{y}{x} = 0$ ultimately, or $\dfrac{y}{x} = \pm \sqrt{\dfrac{b}{a}}.$

Again, suppose $x^4 + ax^2y + bxy^2 - y^4 = 0;$

therefore $x + a \dfrac{y}{x} + b \left(\dfrac{y}{x}\right)^2 - y \left(\dfrac{y}{x}\right)^3 = 0.$

The *finite* limiting values of $\frac{y}{x}$ are given by

$$a\,\frac{y}{x} + b\left(\frac{y}{x}\right)^2 = 0;$$

therefore $\qquad \frac{y}{x} = 0,\ \text{ or }\ \frac{y}{x} = -\frac{a}{b}.$

And since the given equation may be put in the form

$$x\left(\frac{x}{y}\right)^3 + a\left(\frac{x}{y}\right)^2 + b\left(\frac{x}{y}\right) - y = 0,$$

we see that $\frac{x}{y} = 0$ ultimately satisfies it;

therefore $\qquad \frac{y}{x} = \infty$ ultimately for another value.

Hence the limits of $\frac{y}{x}$ are 0, or $-\frac{a}{b}$, or ∞.

This method is free from the difficulty which is pointed out at the end of Art. 191.

If we wish to ascertain by the method of the present Article the value of $\frac{dy}{dx}$ at a point for which $x = a$, $y = b$, we may put $a + x'$ for x and $b + y'$ for y in the equation which connects x and y. We shall then have to find the value of $\frac{dy'}{dx'}$, when $x' = 0$ and $y' = 0$; and this may be ascertained by the method shewn in the preceding Examples.

EXAMPLES.

1. If $u = \sqrt{\left(\dfrac{z^2 - y^2}{z^2 + y^2}\right)}$, where z and y are functions of x, find $\dfrac{du}{dx}$.

2. If $u = \sin^{-1}\dfrac{z}{y}$, where z and y are functions of x, find $\dfrac{du}{dx}$.

3. If $ye^{ny} = ax^m$,　　$\dfrac{dy}{dx} = \dfrac{my}{x\,(1+ny)}$.

4. If $x^y - y^x = 0$,　　$\dfrac{dy}{dx} = \dfrac{y^2 - xy\log y}{x^2 - xy\log x}$.

5. If $(a+y)^2\,(b^2 - y^2) + (x+a)^2\,y^2 = 0$, find $\dfrac{dy}{dx}$.

6. If $\sin(xy) = mx$, find $\dfrac{dy}{dx}$.

7. Given $y^3 + x^3 - 3axy = 0$, shew that $\dfrac{d^2y}{dx^2} = -\dfrac{2a^3xy}{(y^2 - ax)^3}$

8. Given $x^4 + 2ax^2y = ay^3$, find $\dfrac{dy}{dx}$ and $\dfrac{d^2y}{dx^2}$, and write down the third derived equation.

9. If $y = \phi\,(x,\,y,\,u)$ and $\psi\,(x,\,y,\,u) = 0$, find $\dfrac{du}{dx}$.

Result $\dfrac{du}{dx} = -\dfrac{\dfrac{d\psi}{dx}\dfrac{d\phi}{dy} - \dfrac{d\psi}{dy}\dfrac{d\phi}{dx} - \dfrac{d\psi}{dx}}{\dfrac{d\psi}{du}\dfrac{d\phi}{dy} - \dfrac{d\psi}{dy}\dfrac{d\phi}{du} - \dfrac{d\psi}{du}}$.

10. If $u = \phi\,(x,\,y)$, and $u = \chi\,(x)$, find $\dfrac{du}{dy}$.

Result $\dfrac{du}{dy}\left\{\chi'\,(x) - \left(\dfrac{d\phi}{dx}\right)\right\} = \left(\dfrac{d\phi}{dy}\right)\chi'\,(x)$.

11. If $u = a^{x^y} + \sqrt{(\sec xy)}$, find $\dfrac{du}{dx}$,　(1) when x and y are independent,　(2) when $x + y = a$.

12. If $a^{x^y} + \sqrt{(\sec xy)} = 0$, find $\dfrac{dy}{dx}$.

Result $\dfrac{dy}{dx} = -\dfrac{y\sqrt{(\sec xy)}\tan xy + 2a^{x^y}yx^{y-1}\log a}{x\sqrt{(\sec xy)}\tan xy + 2a^{x^y}x^y\log a\log x}$.

13. If $x^4 + 2ax^3y - ay^3 = 0$, shew that $\dfrac{dy}{dx} = 0$, or $\pm \sqrt{2}$,

when $x = 0$ and $y = 0$.

14. If $x^4 - ay^3 + 2axy^2 + 3ax^2y = 0$, shew that $\dfrac{dy}{dx} = 0$, or -1,

or 3, when $x = 0$ and $y = 0$.

15. If $ax^3 + x^2y - ay^3 = 0$, shew that $\dfrac{dy}{dx} = 1$, when $x = 0$

and $y = 0$.

16. If $x^2y^2 = (a^2 - y^2)(b + y)^2$, shew that $\dfrac{dy}{dx} = \pm \dfrac{b}{\sqrt{(a^2 - b^2)}}$,

when $x = 0$ and $y = -b$.

17. If $(y^2 - x^2)(x - 1)\left(x - \dfrac{3}{2}\right) = 2(y^2 + x^2 - 2x)^2$,

find $\dfrac{dy}{dx}$ when x and y vanish, and when $x = 1$, $y = 1$.

Results $\sqrt{\left(\dfrac{19}{3}\right)}$ and $\dfrac{-1 \pm \sqrt{(33)}}{16}$.

18. If $y^4 - y^2 + 3xy - 2x^2 = 0$, find $\dfrac{dy}{dx}$ when $x = 0$.

Result 1, 2, or $-\dfrac{3}{2}$.

19. Find $\dfrac{du}{dx}$ if $u^2 + x^2 + y^2 + z^2 = c^2$,

$$\log(xy) + \dfrac{y}{x} = a^2,$$

$$\log\left(\dfrac{z}{x}\right) + zx = b^2.$$

Result $u\dfrac{du}{dx} = \dfrac{y^2(x - y)}{x(x + y)} + \dfrac{z^2(xz - 1)}{x(xz + 1)} - x.$

20. If $\dfrac{x^2}{a^2} + \dfrac{y^2}{b^2} + \dfrac{z^2}{c^2} - 1 = 0$, find $\dfrac{d^2z}{dx^2}$, $\dfrac{d^2z}{dx\,dy}$, and $\dfrac{d^2z}{dy^2}$.

CHAPTER XII.

CHANGE OF THE INDEPENDENT VARIABLE.

196. In Art. 60 we have shewn that

$$\frac{dy}{dx} = \frac{1}{\dfrac{dx}{dy}} \quad\dotfill (1),$$

and in Art. 63 we have shewn that

$$\frac{dy}{dx} = \frac{dy}{dz}\frac{dz}{dx} \quad\dotfill (2);$$

and we now proceed to some extensions of these formulæ.

Given x and y, both functions of a third variable z, it is required to express the successive differential coefficients of y with respect to x, in terms of those of y and x with respect to z.

We have
$$\frac{dy}{dx} = \frac{dy}{dz}\frac{dz}{dx} \quad \text{by (2)},$$

$$= \frac{\dfrac{dy}{dz}}{\dfrac{dx}{dz}} \quad \text{by (1)}.$$

Hence
$$\frac{d^2y}{dx^2} = \frac{d}{dx}\frac{\dfrac{dy}{dz}}{\dfrac{dx}{dz}} = \frac{d}{dz}\frac{\dfrac{dy}{dz}}{\dfrac{dx}{dz}}\cdot\frac{dz}{dx} \quad \text{by (2)},$$

$$= \frac{\dfrac{d^2y}{dz^2}\dfrac{dx}{dz} - \dfrac{d^2x}{dz^2}\dfrac{dy}{dz}}{\left(\dfrac{dx}{dz}\right)^2}\cdot\frac{dz}{dx},$$

$$= \frac{\dfrac{d^2y}{dz^2}\dfrac{dx}{dz} - \dfrac{d^2x}{dz^2}\dfrac{dy}{dz}}{\left(\dfrac{dx}{dz}\right)^3} \quad \text{by (1).}$$

Again, $\quad \dfrac{d^3y}{dx^3} = \dfrac{d}{dz}\dfrac{\dfrac{d^2y}{dz^2}\dfrac{dx}{dz} - \dfrac{d^2x}{dz^2}\dfrac{dy}{dz}}{\left(\dfrac{dx}{dz}\right)^3} \cdot \dfrac{dz}{dx}$

$$= \frac{\left(\dfrac{d^3y}{dz^3}\dfrac{dx}{dz} - \dfrac{d^3x}{dz^3}\dfrac{dy}{dz}\right)\left(\dfrac{dx}{dz}\right)^3 - 3\left(\dfrac{dx}{dz}\right)^2\dfrac{d^2x}{dz^2}\left(\dfrac{d^2y}{dz^2}\dfrac{dx}{dz} - \dfrac{d^2x}{dz^2}\dfrac{dy}{dz}\right)}{\left(\dfrac{dx}{dz}\right)^6} \, \dfrac{dz}{dx}$$

$$= \frac{\left(\dfrac{d^3y}{dz^3}\dfrac{dx}{dz} - \dfrac{d^3x}{dz^3}\dfrac{dy}{dz}\right)\dfrac{dx}{dz} - 3\dfrac{d^2x}{dz^2}\left(\dfrac{d^2y}{dz^2}\dfrac{dx}{dz} - \dfrac{d^2x}{dz^2}\dfrac{dy}{dz}\right)}{\left(\dfrac{dx}{dz}\right)^5} \, .$$

Similarly we might express $\dfrac{d^4y}{dx^4}$, $\dfrac{d^5y}{dx^5}$, ...

This process is called "*changing the independent variable from x to z*;" since in $\dfrac{d^2y}{dx^2}$ the independent variable is x, but in the expression $\dfrac{\dfrac{d^2y}{dz^2}\dfrac{dx}{dz} - \dfrac{d^2x}{dz^2}\dfrac{dy}{dz}}{\left(\dfrac{dx}{dz}\right)^3}$ the independent variable is z.

197. Suppose in the preceding Article we put $z = y$.

We have $\quad \dfrac{dy}{dz} = 1, \quad \dfrac{d^2y}{dz^2} = 0, \quad \dfrac{d^3y}{dz^3} = 0, \ ......$

$$\dfrac{dx}{dz} = \dfrac{dx}{dy}, \quad \dfrac{d^2x}{dz^2} = \dfrac{d^2x}{dy^2}, \ ...... ;$$

and thus
$$\frac{dy}{dx} = \frac{1}{\dfrac{dx}{dy}},$$

$$\frac{d^2y}{dx^2} = -\frac{\dfrac{d^2x}{dy^2}}{\left(\dfrac{dx}{dy}\right)^3},$$

$$\frac{d^3y}{dx^3} = -\frac{\dfrac{dx}{dy}\dfrac{d^3x}{dy^3} - 3\left(\dfrac{d^2x}{dy^2}\right)^2}{\left(\dfrac{dx}{dy}\right)^5}.$$

198. The formulæ of Art. 197 may also be obtained directly thus:

$$\frac{dy}{dx} = \frac{1}{\dfrac{dx}{dy}};$$

therefore
$$\frac{d^2y}{dx^2} = \frac{d}{dx}\frac{1}{\dfrac{dx}{dy}}$$

$$= \frac{d}{dy}\frac{1}{\dfrac{dx}{dy}} \cdot \frac{dy}{dx}$$

$$= -\frac{\dfrac{d^2x}{dy^2}}{\left(\dfrac{dx}{dy}\right)^2} \cdot \frac{dy}{dx} = -\frac{\dfrac{d^2x}{dy^2}}{\left(\dfrac{dx}{dy}\right)^3},$$

$$\frac{d^3y}{dx^3} = -\frac{d}{dx}\frac{\dfrac{d^2x}{dy^2}}{\left(\dfrac{dx}{dy}\right)^3} = -\frac{d}{dy}\frac{\dfrac{d^2x}{dy^2}}{\left(\dfrac{dx}{dy}\right)^3} \cdot \frac{dy}{dx}$$

$$= -\frac{\dfrac{d^3x}{dy^3}\left(\dfrac{dx}{dy}\right)^3 - 3\left(\dfrac{dx}{dy}\right)^2\left(\dfrac{d^2x}{dy^2}\right)^2}{\left(\dfrac{dx}{dy}\right)^7}$$

$$= -\frac{\dfrac{d^3x}{dy^3}\dfrac{dx}{dy} - 3\left(\dfrac{d^2x}{dy^2}\right)^2}{\left(\dfrac{dx}{dy}\right)^5}.$$

This process is called *changing the independent variable from x to y*.

199. With respect to the *use* of the preceding Articles we must observe that, as is the case with some other parts of the Differential Calculus, the student is here acquiring materials which will be available in some of his following subjects. Expressions which present themselves can sometimes be much simplified by transforming them in the manner above indicated; of this examples will be seen at the end of this Chapter.

200. The following is an important special case.

Change the independent variable in $x^n \dfrac{d^n y}{dx^n}$ from x to t, where $x = e^t$.

We have $\dfrac{d}{dt}\left(x^n \dfrac{d^n y}{dx^n}\right) = \dfrac{d}{dx}\left(x^n \dfrac{d^n y}{dx^n}\right)\dfrac{dx}{dt}$

$$= \left(nx^{n-1}\dfrac{d^n y}{dx^n} + x^n \dfrac{d^{n+1}y}{dx^{n+1}}\right)x$$

$$= nx^n \dfrac{d^n y}{dx^n} + x^{n+1}\dfrac{d^{n+1}y}{dx^{n+1}};$$

therefore $\dfrac{d}{dt}\left(x^n \dfrac{d^n y}{dx^n}\right) - nx^n \dfrac{d^n y}{dx^n} = x^{n+1}\dfrac{d^{n+1}y}{dx^{n+1}}.$

This result may for the sake of abbreviation be thus expressed,

$$\left(\dfrac{d}{dt} - n\right)x^n \dfrac{d^n y}{dx^n} = x^{n+1}\dfrac{d^{n+1}y}{dx^{n+1}} \dots\dots\dots\dots(1)$$

Put $n = 1$; then

$$\left(\frac{d}{dt} - 1\right) x \frac{dy}{dx} = x^2 \frac{d^2y}{dx^2}.$$

But

$$\frac{dy}{dt} = \frac{dy}{dx}\frac{dx}{dt} = x \frac{dy}{dx};$$

therefore

$$x^2 \frac{d^2y}{dx^2} = \left(\frac{d}{dt} - 1\right) \frac{dy}{dt} \quad \text{.................... (2).}$$

Put $n = 2$ in (1); then

$$\left(\frac{d}{dt} - 2\right) x^2 \frac{d^2y}{dx^2} = x^3 \frac{d^3y}{dx^3};$$

or from (2),

$$x^3 \frac{d^3y}{dx^3} = \left(\frac{d}{dt} - 2\right)\left(\frac{d}{dt} - 1\right) \frac{dy}{dt} \quad \text{................ (3).}$$

Proceeding thus we deduce

$$x^n \frac{d^ny}{dx^n} = \left\{\frac{d}{dt} - (n-1)\right\}\left\{\frac{d}{dt} - (n-2)\right\} \dots \left\{\frac{d}{dt} - 1\right\} \frac{dy}{dt} \quad \text{......(4).}$$

201. It is often useful in geometrical applications of the Differential Calculus to have expressions for $\frac{dy}{dx}$ and $\frac{d^2y}{dx^2}$ in terms of θ, supposing

$$\left.\begin{array}{l} x = r \cos \theta \\ y = r \sin \theta \end{array}\right\} \quad \text{........................... (1).}$$

Since y is by supposition some function of x, it follows from (1) that an equation subsists between r and θ, so that r may be considered some function of θ.

Now

$$\frac{dy}{dx} = \frac{\dfrac{dy}{d\theta}}{\dfrac{dx}{d\theta}} = \frac{\sin\theta \dfrac{dr}{d\theta} + r\cos\theta}{\cos\theta \dfrac{dr}{d\theta} - r\sin\theta}, \text{ from (1),}$$

$$\frac{d^2y}{dx^2} = \frac{d}{d\theta} \frac{\sin\theta \dfrac{dr}{d\theta} + r\cos\theta}{\cos\theta \dfrac{dr}{d\theta} - r\sin\theta} \cdot \frac{d\theta}{dx}.$$

The numerator of this fraction is

$$\left(\sin\theta\frac{d^2r}{d\theta^2}+2\cos\theta\frac{dr}{d\theta}-r\sin\theta\right)\left(\cos\theta\frac{dr}{d\theta}-r\sin\theta\right)$$

$$-\left(\cos\theta\frac{d^2r}{d\theta^2}-2\sin\theta\frac{dr}{d\theta}-r\cos\theta\right)\left(\sin\theta\frac{dr}{d\theta}+r\cos\theta\right)$$

and the denominator is

$$\left(\cos\theta\frac{dr}{d\theta}-r\sin\theta\right)^3.$$

Hence we obtain $\dfrac{d^2y}{dx^2}=\dfrac{r^2+2\left(\dfrac{dr}{d\theta}\right)^2-r\dfrac{d^2r}{d\theta^2}}{\left(\cos\theta\dfrac{dr}{d\theta}-r\sin\theta\right)^3}.$

202. Let u be a function of the independent variables x and y, say $u=f(x,y)$; and suppose x and y functions of two new independent variables r, θ, so that

$$x=F_1(r,\theta),$$
$$y=F_2(r,\theta).$$

It is required to find the values of $\dfrac{du}{dx}$ and $\dfrac{du}{dy}$ in terms of differential coefficients of u taken with respect to the new variables.

If for x and y we substitute their values in terms of r and θ, we make u an explicit function of r and θ. Now, by Art. 169,

$$\frac{du}{dr}=\frac{du}{dx}\frac{dx}{dr}+\frac{du}{dy}\frac{dy}{dr},$$

$$\frac{du}{d\theta}=\frac{du}{dx}\frac{dx}{d\theta}+\frac{du}{dy}\frac{dy}{d\theta}.$$

From these equations $\dfrac{du}{dx}$ and $\dfrac{du}{dy}$ can be found.

203. If the equations which connect x, y, r, θ, instead of those in Art. 202, are given in the form

$$r=F_1(x,y),$$
$$\theta=F_2(x,y),$$

we may use the formulæ

$$\frac{du}{dx} = \frac{du}{dr}\frac{dr}{dx} + \frac{du}{d\theta}\frac{d\theta}{dx},$$

$$\frac{du}{dy} = \frac{du}{dr}\frac{dr}{dy} + \frac{du}{d\theta}\frac{d\theta}{dy}.$$

204. If the equations which connect x, y, r, θ, are given in the form

$$F_1(x, y, r, \theta) = 0 \dotfill (1),$$

$$F_2(x, y, r, \theta) = 0 \dotfill (2),$$

we may, in order to find the values of $\dfrac{dx}{dr}$, $\dfrac{dx}{d\theta}$, $\dfrac{dy}{dr}$, $\dfrac{dy}{d\theta}$, required by the formulæ of Art. 202, by successively eliminating y and x from (1) and (2), obtain explicitly the values of x and y in terms of r and θ. Or, by Art. 189, we may find $\dfrac{dx}{d\theta}$ and $\dfrac{dy}{d\theta}$ from the equations

$$\left(\frac{dF_1}{d\theta}\right) + \left(\frac{dF_1}{dx}\right)\frac{dx}{d\theta} + \left(\frac{dF_1}{dy}\right)\frac{dy}{d\theta} = 0,$$

$$\left(\frac{dF_2}{d\theta}\right) + \left(\frac{dF_2}{dx}\right)\frac{dx}{d\theta} + \left(\frac{dF_2}{dy}\right)\frac{dy}{d\theta} = 0,$$

and use two similar equations for $\dfrac{dx}{dr}$ and $\dfrac{dy}{dr}$.

205. **Example.** $u = f(x, y),$

$$x = r\cos\theta,$$

$$y = r\sin\theta;$$

here $\dfrac{dx}{d\theta} = -r\sin\theta,$ $\dfrac{dy}{d\theta} = r\cos\theta,$

$\dfrac{dx}{dr} = \cos\theta,$ $\dfrac{dy}{dr} = \sin\theta.$

Hence, by Art. 202,

$$\frac{du}{dr} = \cos\theta\,\frac{du}{dx} + \sin\theta\,\frac{du}{dy},$$

$$\frac{du}{d\theta} = -r\sin\theta\,\frac{du}{dx} + r\cos\theta\,\frac{du}{dy};$$

therefore
$$\left.\begin{aligned}
\frac{du}{dx} &= \cos\theta\,\frac{du}{dr} - \frac{1}{r}\sin\theta\,\frac{du}{d\theta}, \\
\frac{du}{dy} &= \sin\theta\,\frac{du}{dr} + \frac{1}{r}\cos\theta\,\frac{du}{d\theta},
\end{aligned}\right\} \quad\ldots\ldots\ldots\ldots (1).$$

If we proceed according to Art. 203, we must put the equations between x, y, r, θ, in the form

$$r = \sqrt{(x^2 + y^2)}, \qquad \theta = \tan^{-1}\frac{y}{x};$$

here
$$\frac{dr}{dx} = \frac{x}{\sqrt{(x^2+y^2)}} = \frac{x}{r}, \qquad \frac{d\theta}{dx} = -\frac{y}{x^2+y^2} = -\frac{y}{r^2},$$

$$\frac{dr}{dy} = \frac{y}{\sqrt{(x^2+y^2)}} = \frac{y}{r}, \qquad \frac{d\theta}{dy} = \frac{x}{x^2+y^2} = \frac{x}{r^2};$$

therefore
$$\left.\begin{aligned}
\frac{du}{dx} &= \frac{x}{r}\frac{du}{dr} - \frac{y}{r^2}\frac{du}{d\theta}, \\
\frac{du}{dy} &= \frac{y}{r}\frac{du}{dr} + \frac{x}{r^2}\frac{du}{d\theta},
\end{aligned}\right\} \quad\ldots\ldots\ldots\ldots (2).$$

Since $\dfrac{x}{r} = \cos\theta$ and $\dfrac{y}{r} = \sin\theta$, the formulæ (1) and (2) agree.

In this branch of the subject beginners are liable to mistakes from not paying sufficient attention to the *precise meaning of the symbols*. Generally speaking mathematical notation is so definite that the meaning of any symbol can be settled without regard to the context; but sometimes instead of using a complex symbol to express our meaning without any possibility of mistake we use a symbol which in itself may be ambiguous, but which is rendered perfectly definite by means of the connexion in which it occurs. Thus, for example, as we have stated in Art 170, the *brackets* expressive of differentiation under certain conditions are sometimes omitted, that is, they are left to be suggested by the context.

In the present case the meaning of the symbols $\frac{du}{dr}$, $\frac{du}{d\theta}$, $\frac{du}{dx}$, $\frac{du}{dy}$ which occur in Arts. 202 and 203 must be carefully observed. We might use a more complex notation, as for example the following; let $\psi(x, y)$ be any function of x and y, and let $\chi(r, \theta)$ be the form which $\psi(x, y)$ takes when for x and y we substitute their values in terms of r and θ; then

$$\frac{d\chi(r, \theta)}{dr} = \left\{ \frac{d\psi(x, y)}{dx} \right\} \frac{dx}{dr} + \left\{ \frac{d\psi(x, y)}{dy} \right\} \frac{dy}{dr},$$

and this is the equation which in Art. 202 is expressed more briefly thus,

$$\frac{du}{dr} = \frac{du}{dx}\frac{dx}{dr} + \frac{du}{dy}\frac{dy}{dr}.$$

The beginner however must remember that the second form is an abbreviation of the first form, and he should recur to the first form if he has any doubt of the meaning of the symbols $\frac{du}{dx}$, $\frac{du}{dy}$, $\frac{du}{dr}$.

It is however with respect to the symbols $\frac{dx}{dr}$, $\frac{dy}{dr}$, $\frac{dx}{d\theta}$, $\frac{dy}{d\theta}$ which occur in Art. 202, and the symbols $\frac{dr}{dx}$, $\frac{d\theta}{dx}$, $\frac{dr}{dy}$, $\frac{d\theta}{dy}$, which occur in Art. 203, that mistakes are most frequently made. For example, beginners sometimes imagine that the $\frac{dx}{dr}$ of Art. 202 and the $\frac{dr}{dx}$ of Art. 203 are *connected by the formula* $\frac{dx}{dr} \times \frac{dr}{dx} = 1$. This formula however is quite inapplicable here; for it implies that there is a single equation involving x and r and no other variable, which is not the case here.

In Art. 202 we suppose that x and y are expressed as functions of r and θ; and $\dfrac{dx}{dr}$ means the differential coefficient of x when r varies but θ does not vary: and as r varies y will also vary, so that on the whole r, x, and y vary, and θ does not vary. In Art. 203 we suppose that r and θ are expressed as functions of x and y; and $\dfrac{dr}{dx}$ means the differential coefficient of r when x varies but y does not vary: and as x varies θ will also vary, so that on the whole x, r, and θ vary, and y does not vary.

Thus the $\dfrac{dx}{dr}$ of Art. 202 and the $\dfrac{dr}{dx}$ of Art. 203 are formed on different suppositions as to the quantities which vary and the quantities which do not vary.

In the example of the present Article we find that the $\dfrac{dx}{dr}$ of Art. 202 $= \cos\theta$, and the $\dfrac{dr}{dx}$ of Art. 203 $= \dfrac{x}{r} = \cos\theta$; and the product of the two is *not* unity.

206. Suppose u a function of the three independent variables x, y, z, and that these are connected by three equations with three new independent variables θ, ϕ, r: it is required to express $\dfrac{du}{dx}$, $\dfrac{du}{dy}$, $\dfrac{du}{dz}$, in terms of differential coefficients of u taken with respect to the new variables.

We have, by Art. 174,

$$\left. \begin{aligned} \frac{du}{dx} &= \frac{du}{d\theta}\frac{d\theta}{dx} + \frac{du}{d\phi}\frac{d\phi}{dx} + \frac{du}{dr}\frac{dr}{dx} \\ \frac{du}{dy} &= \frac{du}{d\theta}\frac{d\theta}{dy} + \frac{du}{d\phi}\frac{d\phi}{dy} + \frac{du}{dr}\frac{dr}{dy} \\ \frac{du}{dz} &= \frac{du}{d\theta}\frac{d\theta}{dz} + \frac{du}{d\phi}\frac{d\phi}{dz} + \frac{du}{dr}\frac{dr}{dz} \end{aligned} \right\} \dotsc\dotsc (1).$$

But by means of the three equations between x, y, z, θ, ϕ, r, we can determine the values of

$$\frac{d\theta}{dx}, \quad \frac{d\theta}{dy}, \quad \frac{d\theta}{dz}, \quad \frac{d\phi}{dx}, \quad \frac{d\phi}{dy}, \quad \frac{d\phi}{dz}, \quad \frac{dr}{dx}, \quad \frac{dr}{dy}, \quad \frac{dr}{dz},$$

and hence the above equations express $\dfrac{du}{dx}$, $\dfrac{du}{dy}$, and $\dfrac{du}{dz}$, in terms of $\dfrac{du}{d\theta}$, $\dfrac{du}{d\phi}$, and $\dfrac{du}{dr}$.

Also by solving the above equations we can express $\dfrac{du}{d\theta}$, $\dfrac{du}{d\phi}$, $\dfrac{du}{dr}$, in terms of $\dfrac{du}{dx}$, $\dfrac{du}{dy}$, and $\dfrac{du}{dz}$, which can also be found by the equations

$$\left. \begin{aligned} \frac{du}{d\theta} &= \frac{du}{dx}\frac{dx}{d\theta} + \frac{du}{dy}\frac{dy}{d\theta} + \frac{du}{dz}\frac{dz}{d\theta} \\ \frac{du}{d\phi} &= \frac{du}{dx}\frac{dx}{d\phi} + \frac{du}{dy}\frac{dy}{d\phi} + \frac{du}{dz}\frac{dz}{d\phi} \\ \frac{du}{dr} &= \frac{du}{dx}\frac{dx}{dr} + \frac{du}{dy}\frac{dy}{dr} + \frac{du}{dz}\frac{dz}{dr} \end{aligned} \right\} \ \dots\dots (2).$$

207. Suppose, to exemplify the above, we put

$$x = r \sin\theta \cos\phi, \quad y = r \sin\theta \sin\phi, \quad z = r \cos\theta.$$

Hence, to apply equations (2) of Art. 206, we have

$$\frac{dx}{d\theta} = r\cos\theta\cos\phi, \qquad \frac{dy}{d\theta} = r\cos\theta\sin\phi, \qquad \frac{dz}{d\theta} = -r\sin\theta,$$

$$\frac{dx}{d\phi} = -r\sin\theta\sin\phi, \qquad \frac{dy}{d\phi} = r\sin\theta\cos\phi, \qquad \frac{dz}{d\phi} = 0,$$

$$\frac{dx}{dr} = \sin\theta\cos\phi, \qquad \frac{dy}{dr} = \sin\theta\sin\phi, \qquad \frac{dz}{dr} = \cos\theta;$$

therefore

$$\left. \begin{aligned} \frac{du}{d\theta} &= r\cos\theta\cos\phi\,\frac{du}{dx} + r\cos\theta\sin\phi\,\frac{du}{dy} - r\sin\theta\,\frac{du}{dz} \\ \frac{du}{d\phi} &= -r\sin\theta\sin\phi\,\frac{du}{dx} + r\sin\theta\cos\phi\,\frac{du}{dy} \\ \frac{du}{dr} &= \sin\theta\cos\phi\,\frac{du}{dx} + \sin\theta\sin\phi\,\frac{du}{dy} + \cos\theta\,\frac{du}{dz} \end{aligned} \right\} \ \dots (1).$$

If we employ equations (1) of Art. 206, we must put the relations between x, y, and z, in the form

$$r = \sqrt{(x^2 + y^2 + z^2)},$$

$$\theta = \tan^{-1} \frac{\sqrt{(x^2 + y^2)}}{z},$$

$$\phi = \tan^{-1} \frac{y}{x};$$

therefore
$$\frac{dr}{dx} = \frac{x}{\sqrt{(x^2 + y^2 + z^2)}} = \frac{x}{r} = \sin\theta\cos\phi,$$

$$\frac{dr}{dy} = \frac{y}{r} = \sin\theta\sin\phi,$$

$$\frac{dr}{dz} = \frac{z}{r} = \cos\theta,$$

$$\frac{d\theta}{dx} = \frac{z}{x^2 + y^2 + z^2} \cdot \frac{x}{\sqrt{(x^2 + y^2)}} = \frac{\cos\theta\cos\phi}{r},$$

$$\frac{d\theta}{dy} = \frac{z}{x^2 + y^2 + z^2} \cdot \frac{y}{\sqrt{(x^2 + y^2)}} = \frac{\cos\theta\sin\phi}{r},$$

$$\frac{d\theta}{dz} = -\frac{\sqrt{(x^2 + y^2)}}{x^2 + y^2 + z^2} = -\frac{\sin\theta}{r},$$

$$\frac{d\phi}{dx} = -\frac{y}{x^2 + y^2} = -\frac{\sin\phi}{r\sin\theta},$$

$$\frac{d\phi}{dy} = \frac{x}{x^2 + y^2} = \frac{\cos\phi}{r\sin\theta},$$

$$\frac{d\phi}{dz} = 0;$$

therefore

$$\left.\begin{aligned}
\frac{du}{dx} &= \sin\theta\cos\phi\,\frac{du}{dr} + \frac{\cos\theta\cos\phi}{r}\frac{du}{d\theta} - \frac{\sin\phi}{r\sin\theta}\frac{du}{d\phi} \\
\frac{du}{dy} &= \sin\theta\sin\phi\,\frac{du}{dr} + \frac{\cos\theta\sin\phi}{r}\frac{du}{d\theta} + \frac{\cos\phi}{r\sin\theta}\frac{du}{d\phi} \\
\frac{du}{dz} &= \cos\theta\,\frac{du}{dr} - \frac{\sin\theta}{r}\frac{du}{d\theta}
\end{aligned}\right\} \quad \ldots\ldots(2),$$

which will be found consistent with (1).

For exercise we give the results arising from differentiating equations (2) of the preceding investigation.

$$\frac{d^2u}{dx\,dy} = \frac{\sin 2\phi}{2}\left\{\sin^2\theta\,\frac{d^2u}{dr^2} + \frac{\cos^2\theta}{r^2}\frac{d^2u}{d\theta^2} - \frac{1}{r^2\sin^2\theta}\frac{d^2u}{d\phi^2}\right.$$

$$+ \frac{\sin 2\theta}{r}\frac{d^2u}{d\theta\,dr} - \frac{\sin^2\theta}{r}\frac{du}{dr} - \frac{\cos\theta}{r^2}\left(2\sin\theta + \frac{1}{\sin\theta}\right)\frac{du}{d\theta}\right\}$$

$$+ \cos 2\phi\left\{\frac{1}{r}\frac{d^2u}{d\phi\,dr} + \frac{\cot\theta}{r^2}\frac{d^2u}{d\phi\,d\theta} - \frac{1}{r^2\sin^2\theta}\frac{du}{d\phi}\right\};$$

$$\frac{d^2u}{dx\,dz} = \frac{\sin 2\theta}{2}\cos\phi\left\{\frac{d^2u}{dr^2} - \frac{1}{r^2}\frac{d^2u}{d\theta^2} - \frac{1}{r}\frac{du}{dr}\right\}$$

$$+ \frac{\cos 2\theta\cos\phi}{r}\left\{\frac{d^2u}{d\theta\,dr} - \frac{1}{r}\frac{du}{d\theta}\right\}$$

$$+ \frac{\sin\phi}{r}\left\{\frac{1}{r}\frac{d^2u}{d\theta\,d\phi} - \frac{\cos\theta}{\sin\theta}\frac{d^2u}{dr\,d\phi}\right\}$$

$$= \frac{\sin 2\theta\cos\phi}{2}A + \frac{\cos 2\theta\cos\phi}{r}B + \frac{\sin\phi}{r}C,\ \text{say};$$

$$\frac{d^2u}{dy\,dz} = \frac{\sin 2\theta\sin\phi}{2}A + \frac{\cos 2\theta\sin\phi}{r}B - \frac{\cos\phi}{r}C;$$

$$\frac{d^2u}{dz^2} = \cos^2\theta\,\frac{d^2u}{dr^2} + \frac{\sin^2\theta}{r}\left(\frac{1}{r}\frac{d^2u}{d\theta^2} + \frac{du}{dr}\right) + \frac{\sin 2\theta}{r}\left(\frac{1}{r}\frac{du}{d\theta} - \frac{d^2u}{dr\,d\theta}\right);$$

$$\frac{d^2u}{dx^2} =$$

$$\cos^2\phi\left\{\sin^2\theta\frac{d^2u}{dr^2} + \frac{\cos^2\theta}{r^2}\frac{d^2u}{d\theta^2} + \frac{\sin 2\theta}{r}\frac{d^2u}{dr\,d\theta} + \frac{\cos^2\theta}{r}\frac{du}{dr} - \frac{\sin 2\theta}{r^2}\frac{du}{d\theta}\right\}$$

$$- \frac{\sin 2\phi}{r}\left\{\frac{d^2u}{d\phi\,dr} + \frac{\cot\theta}{r}\frac{d^2u}{d\theta\,d\phi} - \frac{1}{r\sin^2\theta}\frac{du}{d\phi}\right\}$$

$$+ \frac{\sin^2\phi}{r}\left\{\frac{1}{r\sin^2\theta}\frac{d^2u}{d\phi^2} + \frac{du}{dr} + \frac{\cot\theta}{r}\frac{du}{d\theta}\right\}$$

$$= \cos^2\phi\,L - \frac{\sin 2\phi}{r}M + \frac{\sin^2\phi}{r}N,\ \text{say};$$

$$\frac{d^2u}{dy^2} = \sin^2\phi\,L + \frac{\sin 2\phi}{r}M + \frac{\cos^2\phi}{r}N.$$

By addition we have

$$\frac{d^2u}{dx^2} + \frac{d^2u}{dy^2} + \frac{d^2u}{dz^2} = \frac{d^2u}{dr^2} + \frac{1}{r^2}\frac{d^2u}{d\theta^2} + \frac{1}{r^2\sin^2\theta}\frac{d^2u}{d\phi^2} + \frac{2}{r}\frac{du}{dr} + \frac{\cot\theta}{r^2}\frac{du}{d\theta}.$$

208. The following example for *two* independent variables is analogous to that in Art. 200 for *one* independent variable.

If $x = e^\theta$ and $y = e^\phi$ it is required to change the independent variables from x and y to θ and ϕ in the expression

$$x^n\frac{d^nu}{dx^n} + nx^{n-1}y\frac{d^nu}{dx^{n-1}dy} + \frac{n(n-1)}{\lfloor 2}x^{n-2}y^2\frac{d^nu}{dx^{n-2}dy^2} + \cdots$$

Let this expression be denoted by v_n, and let v_{n+1} denote what it becomes when n is changed into $n+1$; we shall prove that

$$v_{n+1} = \frac{dv_n}{d\theta} + \frac{dv_n}{d\phi} - nv_n \dots\dots\dots\dots (1).$$

For
$$\frac{dv_n}{d\theta} = \frac{dv_n}{dx}\frac{dx}{d\theta} = x\frac{dv_n}{dx},$$

and
$$\frac{dv_n}{d\phi} = \frac{dv_n}{dy}\frac{dy}{d\phi} = y\frac{dv_n}{dy}.$$

Now take any term in the expression represented by v_n and perform the following operations: differentiate the term with respect to x and afterwards multiply by x; differentiate the term with respect to y and afterwards multiply by y; then add the two results. Take for example the $(r+1)^{\text{th}}$ term which is

$$\frac{n(n-1)\dots(n-r+1)}{\lfloor r}x^{n-r}y^r\frac{d^nu}{dx^{n-r}dy^r},$$

and by performing the operations we obtain

$$\frac{n(n-1)\dots(n-r+1)}{\lfloor r}\left\{x^{n+1-r}y^r\frac{d^{n+1}u}{dx^{n+1-r}dy^r} + x^{n-r}y^{r+1}\frac{d^{n+1}u}{dx^{n-r}dy^{r+1}}\right.$$
$$\left. + nx^{n-r}y^r\frac{d^nu}{dx^{n-r}dy^r}\right\}.$$

Hence we infer that $x\dfrac{dv_n}{dx} + y\dfrac{dv_n}{dy}$ is equal to nv_n together with two series; and by uniting like terms in the two series we obtain a single series of which the general term is

$$\frac{(n+1)\,n\,(n-1)\ldots(n+1-r+1)}{\underline{|r}}\,x^{n+1-r}y^r\,\frac{d^{n+1}u}{dx^{n+1-r}dy^r}.$$

Therefore

$$x\frac{dv_n}{dx} + y\frac{dv_n}{dy} = nv_n + v_{n+1};$$

and thus (1) is proved; we may write (1) for abbreviation thus,

$$v_{n+1} = \left\{\frac{d}{d\theta} + \frac{d}{d\phi} - n\right\}v_n\ \ldots\ldots\ldots\ldots\ldots (2).$$

Put $n = 1$ in (2); then

$$v_2 = \left\{\frac{d}{d\theta} + \frac{d}{d\phi} - 1\right\}v_1 = \left\{\frac{d}{d\theta} + \frac{d}{d\phi} - 1\right\}\left\{x\frac{du}{dx} + y\frac{du}{dy}\right\}$$

$$= \left\{\frac{d}{d\theta} + \frac{d}{d\phi} - 1\right\}\left\{\frac{du}{d\theta} + \frac{du}{d\phi}\right\} = \left\{\frac{d}{d\theta} + \frac{d}{d\phi} - 1\right\}\left\{\frac{d}{d\theta} + \frac{d}{d\phi}\right\}u,$$

as we may write it; again put $n = 2$ in (2); then

$$v_3 = \left\{\frac{d}{d\theta} + \frac{d}{d\phi} - 2\right\}v_2 = \left\{\frac{d}{d\theta} + \frac{d}{d\phi} - 2\right\}\left\{\frac{d}{d\theta} + \frac{d}{d\phi} - 1\right\}\left\{\frac{d}{d\theta} + \frac{d}{d\phi}\right\}u.$$

Proceeding in this way we obtain

$$v_n = \left\{\frac{d}{d\theta} + \frac{d}{d\phi} - (n-1)\right\}\ldots\left\{\frac{d}{d\theta} + \frac{d}{d\phi} - 2\right\}\left\{\frac{d}{d\theta} + \frac{d}{d\phi} - 1\right\}\left\{\frac{d}{d\theta} + \frac{d}{d\phi}\right\}u.$$

EXAMPLES.

1. Change the independent variable from x to y in the equation

$$x^2\frac{d^2u}{dx^2} + x\frac{du}{dx} + u = 0,\text{ supposing } y = \log x.$$

$$\textit{Result }\frac{d^2u}{dy^2} + u = 0.$$

2. Transform $\dfrac{d^2y}{dx^2} + \dfrac{2x}{1+x^2}\dfrac{dy}{dx} + \dfrac{y}{(1+x^2)^2} = 0$ into an equation in which θ is the independent variable, where $\theta = \tan^{-1} x$.

$\qquad\qquad\qquad$ Result $\dfrac{d^2y}{d\theta^2} + y = 0.$

3. Transform $\dfrac{d^2y}{dx^2} + \dfrac{1}{x}\dfrac{dy}{dx} + y = 0$, into an equation in which t is the independent variable where $x^2 = 4t$.

$\qquad\qquad\qquad$ Result $t\dfrac{d^2y}{dt^2} + \dfrac{dy}{dt} + y = 0.$

4. If $\dfrac{d^2y}{dx^2} = \dfrac{y}{(e^x + e^{-x})^2}$, and $x = \log \dfrac{t}{\sqrt{(1-t^2)}}$, shew that

$$(t - t^3)\dfrac{d^2y}{dt^2} + (1 - 3t^2)\dfrac{dy}{dt} = ty.$$

5. If $x = \cos t$, then

$$(1 - x^2)\dfrac{d^2y}{dx^2} - x\dfrac{dy}{dx} = 0 \text{ becomes } \dfrac{d^2y}{dt^2} = 0.$$

6. Transform $\dfrac{x\dfrac{dy}{dx} - y}{\sqrt{\left\{1 + \left(\dfrac{dy}{dx}\right)^2\right\}}}$, by assuming $x = r\cos\theta$, $y = r\sin\theta$.

$\qquad\qquad\qquad$ Result $\dfrac{r^2}{\sqrt{\left\{r^2 + \left(\dfrac{dr}{d\theta}\right)^2\right\}}}$.

7. If $x = r\cos\theta$, $y = r\sin\theta$, shew that

$$\dfrac{x + y\dfrac{dy}{dx}}{x\dfrac{dy}{dx} - y} = \dfrac{1}{r}\dfrac{dr}{d\theta}.$$

8. If $x = a(1 - \cos t)$ and $y = a(nt + \sin t)$, express $\dfrac{d^2y}{dx^2}$ in terms of t.

$\qquad\qquad\qquad$ Result $-\dfrac{n\cos t + 1}{a\sin^3 t}$.

9. Suppose u to be a function of r and

$$r^2 = x_1^2 + x_2^2 + x_3^2 + \ldots\ldots + x_n^2 ;$$

then if

$$\frac{d^2u}{dx_1^2} + \frac{d^2u}{dx_2^2} + \frac{d^2u}{dx_3^2} + \ldots\ldots + \frac{d^2u}{dx_n^2} = 0,$$

shew that

$$\frac{d^2u}{dr^2} + \frac{n-1}{r}\frac{du}{dr} = 0.$$

10. Given $x = a \cos \phi$, $y = b \sin \phi$, express

$$\frac{\left\{1 + \left(\frac{dy}{dx}\right)^2\right\}^{\frac{3}{2}}}{-\dfrac{d^2y}{dx^2}} \quad \text{in terms of } \phi.$$

$$Result \quad \frac{(a^2 \sin^2 \phi + b^2 \cos^2 \phi)^{\frac{3}{2}}}{ab}.$$

11. Transform $\dfrac{d^2y}{dx^2} + 2\dfrac{e^{2x} - e^{-2x}}{e^{2x} + e^{-2x}}\dfrac{dy}{dx} + \dfrac{4n^2y}{(e^{2x} + e^{-2x})^2} = 0$ into an equation in which t shall be the independent variable, having given $x = \log \sqrt{(\tan t)}$.

$$Result \quad \frac{d^2y}{dt^2} + n^2y = 0.$$

12. Change the independent variable from y to x in $\dfrac{d^3u}{dy^3} - 4\tan y\dfrac{d^2u}{dy^2} + 2\tan^2 y\dfrac{du}{dy} = 0$, supposing $\tan y = x$.

$$Result \quad (1 + x^2)^2\frac{d^3u}{dx^3} + 2x(1 + x^2)\frac{d^2u}{dx^2} + 2\frac{du}{dx} = 0.$$

13. Transform $\dfrac{d^2y}{dx^2} + \left(\dfrac{dy}{dx}\right)^2$ into an expression in which y is the independent variable.

14. Given $x = t + t^2$, transform $\dfrac{d^2u}{dt^2}$ into an expression in which x is the independent variable.

15. If $z = u - e \sin u$, and $\tan \dfrac{u}{2} = \sqrt{\left(\dfrac{1-e}{1+e}\right)} \tan \dfrac{v}{2}$, shew

that $\dfrac{dz}{dv} = \dfrac{(1-e^2)^{\frac{3}{2}}}{(1 + e \cos v)^2}$.

16. If $(a^2 - x^2)\dfrac{d^2z}{dx^2} - \dfrac{a^2}{x}\dfrac{dz}{dx} - z = 0$, and $x^2 + y^2 = a^2$, shew

that $x^2 \dfrac{d^2z}{dy^2} - z = 0$.

17. Transform

$$(a + bx)^2 \frac{d^2y}{dx^2} + A(a + bx)\frac{dy}{dx} + By = F(x),$$

by assuming $\qquad a + bx = e^t$.

18. If z be a function of the two independent variables x and y, and x and y be connected with two new variables r and θ by means of two equations, shew how to express $\dfrac{d^2z}{dx^2}$, $\dfrac{d^2z}{dx\,dy}$, and $\dfrac{d^2z}{dy^2}$, in terms of the new variables.

For instance, if $x = r \cos \theta$, $y = r \sin \theta$, shew that

$$\frac{d^2z}{dx^2} = A + B \cos 2\theta - C \sin 2\theta,$$

$$\frac{d^2z}{dy^2} = A - B \cos 2\theta + C \sin 2\theta,$$

$$\frac{d^2z}{dx\,dy} = B \sin 2\theta + C \cos 2\theta;$$

where $\quad A + B = \dfrac{d^2z}{dr^2}$, $\quad A - B = \dfrac{1}{r^2}\dfrac{d^2z}{d\theta^2} + \dfrac{1}{r}\dfrac{dz}{dr}$,

$$C = \frac{1}{r}\frac{d^2z}{dr\,d\theta} - \frac{1}{r^2}\frac{dz}{d\theta}.$$

19. If x, y, z, and ξ, η, ζ, be co-ordinates of the same point P referred to two different rectangular systems, shew that

$$\frac{d^2\phi}{dx^2} + \frac{d^2\phi}{dy^2} + \frac{d^2\phi}{dz^2} = \frac{d^2\phi}{d\xi^2} + \frac{d^2\phi}{d\eta^2} + \frac{d^2\phi}{d\zeta^2}.$$

20. If $x\dfrac{d^2y}{dx^2} - \dfrac{x}{y}\left(\dfrac{dy}{dx}\right)^2 + \dfrac{dy}{dx} = 0$, and $x = ye^z$, shew that

$$y\frac{d^2z}{dy^2} + \frac{dz}{dy} = 0.$$

21. Given $u = \left(\dfrac{d^2x}{ds^2}\right)^2 + \left(\dfrac{d^2y}{ds^2}\right)^2$, and $\left(\dfrac{dx}{ds}\right)^2 + \left(\dfrac{dy}{ds}\right)^2 = 1$,

shew that $u\left(\dfrac{ds}{dt}\right)^4 = \left(\dfrac{d^2x}{dt^2}\right)^2 + \left(\dfrac{d^2y}{dt^2}\right)^2 - \left(\dfrac{d^2s}{dt^2}\right)^2.$

22. Transform $\dfrac{d^2y}{d\theta^2} - \sec\theta\,\operatorname{cosec}\theta\,\dfrac{dy}{d\theta} + yn^2\tan^2\theta = 0$, into an equation in which x shall be the independent variable, having given $x = \log(\sec\theta)$.

$$\text{Result } \frac{d^2y}{dx^2} + n^2y = 0$$

23. If $y = e^{-\theta}$ and $x = \sin\theta$, shew that

$$\frac{d^3y}{dx^3} = \frac{e^{-\theta}}{\cos^5\theta}\{3\sin\theta\cos\theta - \sin^2\theta - 2\}.$$

24. Express $\dfrac{d^2u}{dx^2} + 2\dfrac{d^2u}{dx\,dy} + \dfrac{d^2u}{dy^2}$ in terms of $\dfrac{du}{ds}$, $\dfrac{du}{dt}$, ..., where $s = e^x + e^y$, and $t = e^{-x} + e^{-y}$.

$$\text{Result } s^2\frac{d^2u}{ds^2} - 2st\frac{d^2u}{ds\,dt} + t^2\frac{d^2u}{dt^2} + s\frac{du}{ds} + t\frac{du}{dt}.$$

25. If $x = ae^\theta\cos\phi$, and $y = ae^\theta\sin\phi$, shew that

$$y^2\frac{d^2u}{dx^2} - 2xy\frac{d^2u}{dx\,dy} + x^2\frac{d^2u}{dy^2} = \frac{d^2u}{d\phi^2} + \frac{du}{d\theta}.$$

CHAPTER XIII.

MAXIMA AND MINIMA OF A FUNCTION OF ONE VARIABLE.

209. Suppose $\phi(x)$ to denote a certain function of x, and that while the variable x changes gradually from one definite value to another, $\phi(x)$ changes in such a manner that it is sometimes increasing and sometimes decreasing. There must then be certain values of x, for which $\phi(x)$ begins to decrease, having previously been increasing, or begins to increase, having previously been decreasing. In the former case, $\phi(x)$ has a greater value for the particular value of x than it has for adjacent values of x, and is said to have a *maximum* value. In the latter case, $\phi(x)$ has a less value for the particular value of x than it has for adjacent values of x, and is said to have a *minimum* value. Hence, these terms *maximum* and *minimum* are not used to denote the arithmetically greatest and least values which a function can assume; for it appears from the above explanation that a function *may* have several maxima and minima values, and that some particular minimum may be greater than some particular maximum.

210. Definition. If as x increases or decreases from the value a through a finite interval, however small, $\phi(x)$ is always less than $\phi(a)$, then $\phi(a)$ is called a *maximum* value of $\phi(x)$; if $\phi(x)$ is always greater than $\phi(a)$, then $\phi(a)$ is called a *minimum* value of $\phi(x)$.

211. *Rule for discovering maxima and minima values.*

Let $\phi(x)$ denote any function of x. By Art. 92, we have

$$\phi(x+h) = \phi(x) + h\phi'(x) + \frac{h^2}{2}\phi''(x+\theta h).$$

If $\phi'(x)$ be not zero we can give such a value to h that the sign of

$$h\phi'(x) + \frac{h^2}{2}\phi''(x + \theta h)$$

shall for that value of h, and all inferior values of h, be the same as the sign of $h\phi'(x)$, because $\frac{h}{2}\phi''(x + \theta h)$ can always be made less than $\phi'(x)$ by taking h small enough. In this case

$$\phi(x + h) - \phi(x)$$

and $$\phi(x - h) - \phi(x)$$

have *different* signs, and therefore $\phi(x)$ has neither a maximum nor minimum value. ·

Hence, as the first condition for the existence of a maximum or minimum value of $\phi(x)$, we must have

$$\phi'(x) = 0 \ldots\ldots\ldots\ldots\ldots (1).$$

Let a be a value of x deduced from equation (1), so that

$$\phi'(a) = 0.$$

We have now, by Art. 92,

$$\phi(a + h) = \phi(a) + \frac{h^2}{\lfloor 2}\phi''(a) + \frac{h^3}{\lfloor 3}\phi'''(a + \theta h).$$

Suppose $\phi''(a)$ not zero; then by giving to h some value sufficiently small, the sign of

$$\frac{h^2}{\lfloor 2}\phi''(a) + \frac{h^3}{\lfloor 3}\phi'''(a + \theta h)$$

will be the same as that of $\frac{h^2}{\lfloor 2}\phi''(a)$, or of $\phi''(a)$, for that value of h and all inferior values ; .

therefore $$\phi(a + h) - \phi(a)$$

and $$\phi(a - h) - \phi(a)$$

have the *same* signs.

If then $\phi''(a)$ be *positive* $\phi(a)$ is a minimum value of $\phi(x)$; if $\phi''(a)$ be *negative* $\phi(a)$ is a maximum value of $\phi(x)$.

If $\phi''(a)$ vanish as well as $\phi'(a)$ then, by Art. 92,

$$\phi(a+h) = \phi(a) + \frac{h^3}{\lfloor 3} \phi'''(a) + \frac{h^4}{\lfloor 4} \phi''''(a+\theta h).$$

By reasoning similar to that used before, we may shew that unless $\phi'''(a)$ also vanish $\phi(a)$ can be neither a maximum nor minimum value of $\phi(x)$; but that if $\phi'''(a)$ vanish and $\phi''''(a)$ be positive $\phi(a)$ is a minimum value, and if $\phi'''(a)$ vanish and $\phi''''(a)$ be negative $\phi(a)$ is a maximum value.

Since this process may be continued until we arrive at a differential coefficient which does *not* vanish when $x = a$, we have the following result. In order that $\phi(x)$ may have a maximum or minimum value when $x = a$, it is necessary that this value of x should make an *odd* number of the successive differential coefficients of $\phi(x)$ vanish, beginning with the first; when this condition is satisfied $\phi(a)$ is a maximum value if the next differential coefficient be negative and a minimum value if it be positive.

212. It is to be observed that in the above demonstration we have used θ to denote *a fraction less than unity*, and it is not to be assumed that the *same* fraction is denoted whenever the symbol is used. Also we have supposed as usual that none of the functions $\phi'(a)$, $\phi''(a)$, ... are infinite. We shall shew hereafter, that maxima and minima values *may* occur when $\phi'(x) = \infty$, as well as when $\phi'(x) = 0$: see Art. 214.

213. Suppose that when $x = a$, the function $\phi(x)$ has a maximum or minimum value, and that $\phi^n(a)$ is the first differential coefficient that does not vanish, n *being even*. By Art. 92, since $\phi'(a)$, $\phi''(a)$, ... all vanish up to $\phi^{n-1}(a)$ inclusive, we have

$$\phi'(a+h) = \frac{h^{n-1}}{\lfloor n-1} \phi^n(a) + \frac{h^n}{\lfloor n} \phi^{n+1}(a+\theta h),$$

$$\phi'(a-h) = -\frac{h^{n-1}}{\lfloor n-1} \phi^n(a) + \frac{h^n}{\lfloor n} \phi^{n+1}(a-\theta_1 h),$$

where θ and θ_1 are proper fractions.

From these values of $\phi'(a+h)$ and $\phi'(a-h)$ we see that $\phi'(x)$ changes sign as x passes through the value a. If we suppose x to *increase* and pass through the value a, then

$\phi'(x)$ changes from positive to negative if $\phi''(a)$ be negative, that is, if $\phi(a)$ be a maximum ; and $\phi'(x)$ changes from negative to positive if $\phi''(a)$ be positive, that is, if $\phi(a)$ be a minimum. This suggests another form for the definition of maxima and minima values and for the investigation of the conditions of their existence which we give in the next Article.

214. DEFINITION. If as x varies through any finite interval, however small, $\phi(x)$ increase until $x = a$ and then decrease, $\phi(a)$ is called a maximum value of $\phi(x)$; if $\phi(x)$ decrease until $x = a$ and then increase, $\phi(a)$ is called a minimum value.

By Art. 89, if the differential coefficient of a function be positive that function increases with the variable, and if the differential coefficient be negative the function decreases as the variable increases. Hence, as x increases $\phi'(x)$ must change from positive to negative when $x = a$, if $\phi(a)$ be a maximum, and from negative to positive if $\phi(a)$ be a minimum. But a function can only change its sign by passing through zero or infinity. Hence, we must find the values of x that make

$$\phi'(x) = 0,$$
or
$$\phi'(x) = \infty ;$$

and if as x passes through any one of these values $\phi'(x)$ changes its sign, we have for that value of x a maximum or minimum value of $\phi(x)$, according as, when x increases, the change is from positive to negative or from negative to positive.

Example (1). Suppose $\phi(x) = x^3 - 9x^2 + 24x - 7,$
then
$$\phi'(x) = 3(x^2 - 6x + 8),$$
$$\phi''(x) = 6(x - 3).$$

If we put $\phi'(x) = 0$, we obtain $x = 2$, or $x = 4$;

when $x = 2$, $\phi''(x)$ is negative,
when $x = 4$, $\phi''(x)$ is positive.

Therefore when $x = 2$, $\phi(x)$ has a maximum value, and
when $x = 4$, $\phi(x)$ has a minimum value.

Example (2). Let $\phi(x) = e^x + e^{-x} + 2\cos x$;
therefore
$$\phi'(x) = e^x - e^{-x} - 2\sin x,$$
$$\phi''(x) = e^x + e^{-x} - 2\cos x,$$

$$\phi'''(x) = e^x - e^{-x} + 2\sin x,$$

$$\phi''''(x) = e^x + e^{-x} + 2\cos x.$$

If $x = 0$, we have $\phi'(x) = 0$, $\phi''(x) = 0$, $\phi'''(x) = 0$, and $\phi''''(x) = 4$. Hence, $\phi(x)$ is a minimum when $x = 0$.

It may be easily shewn that $x = 0$ is the only value of x for which $\phi'(x)$ vanishes; for

$$e^x = 1 + x + \frac{x^2}{\lfloor 2} + \frac{x^3}{\lfloor 3} + \dots,$$

$$e^{-x} = 1 - x + \frac{x^2}{\lfloor 2} - \frac{x^3}{\lfloor 3} + \dots,$$

$$2\sin x = 2\left\{x - \frac{x^3}{\lfloor 3} + \frac{x^5}{\lfloor 5} - \dots\right\};$$

therefore $\quad \phi'(x) = 4\left\{\frac{x^3}{\lfloor 3} + \frac{x^7}{\lfloor 7} + \frac{x^{11}}{\lfloor 11} + \dots\right\}.$

All the terms in $\phi'(x)$ being of the *same* sign, $\phi'(x)$ can never vanish except when $x = 0$.

Example (3). Suppose $\dfrac{du}{dx} = x(x-1)^2(x-3)^3$, for what values of x will u be a maximum or minimum? In this Example the method of Art. 214 is preferable. When x is negative $\dfrac{du}{dx}$ is positive; when x is positive and less than unity, $\dfrac{du}{dx}$ is negative. Hence $\dfrac{du}{dx}$ changes from positive to negative as x passes through the value 0, and $x = 0$ makes u a maximum. When $x = 1$, $\dfrac{du}{dx}$ vanishes; it does not how-ever change its sign, but continues negative until $x = 3$, and after that it is positive. Hence, when $x = 1$, u has neither a maximum nor minimum value, but has a minimum value when $x = 3$.

Suppose that in the Example last given we merely wish to ascertain if $x = 0$ gives a maximum or minimum value to u, and that we are required to proceed according to the method of Art. 211: we have

$$\frac{du}{dx} = x\,(x-1)^2\,(x-3)^3,$$

$$\frac{d^2u}{dx^2} = (x-1)^2\,(x-3)^3 + 2x\,(x-1)\,(x-3)^3 + 3x\,(x-1)^2\,(x-3)^2;$$

when $x = 0$ the first term in $\dfrac{d^2u}{dx^2}$ is negative, *and the other two terms vanish since they both have x as a factor.* Hence we need not have expressed them, but might have put

$$\frac{d^2u}{dx^2} = (x-1)^2\,(x-3)^3 + \text{terms vanishing when } x = 0.$$

This remark should be carefully noticed, because in Examples like the above we are saved the trouble of writing down superfluous terms.

Example (4). The following Example will introduce the reader to considerations by which the process for finding maxima and minima values may sometimes be abbreviated.

Through a given point P a straight line is drawn, meeting the axes Ox and Oy at A and B respectively: find the least length this straight line can have.

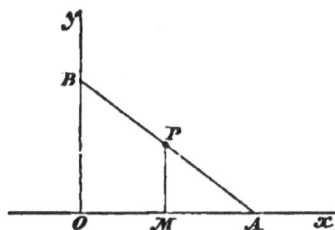

Let $OM = a$, $MP = b$, $PAO = \theta$.

Then
$$PA = \frac{b}{\sin\theta},$$

$$PB = \frac{a}{\cos\theta}.$$

Put $u = \dfrac{b}{\sin\theta} + \dfrac{a}{\cos\theta}$, and we have to find the least value of u.

Now
$$\frac{du}{d\theta} = -\frac{b\cos\theta}{\sin^2\theta} + \frac{a\sin\theta}{\cos^2\theta};$$

therefore $\dfrac{du}{d\theta}$ vanishes only when $\tan\theta = \sqrt[3]{\dfrac{b}{a}}$.

From the figure it appears that by making θ either as small as we please, or as nearly equal to a right angle as we please, the straight line AB may be made as great as we please. Also, as θ varies from 0 to $\frac{\pi}{2}$, there must be some value of θ which gives to the straight line AB the *least* length it can have, and this *least* length of AB will satisfy the definition of a *minimum* length. And as $\frac{du}{d\theta}$ for a value of θ between 0 and $\frac{\pi}{2}$ can never change its sign except when $\tan \theta = \sqrt[3]{\dfrac{b}{a}}$, this must be the value of θ that gives the least length we are seeking.

This value of θ gives for the least length the value

$$(a^{\frac{2}{3}} + b^{\frac{2}{3}})^{\frac{3}{2}}.$$

In this Example it is easy to see from the value of $\frac{du}{d\theta}$, that it *does* change sign from negative to positive when θ increases and passes through the value assigned ; but in more complicated questions it is often advisable to shew in the manner above exemplified, that a maximum or minimum *must* necessarily exist, and then we are saved the trouble of examining if the differential coefficient of the function changes sign when it vanishes.

215. If u be a function of x we have shewn that $\frac{du}{dx} = 0$ is the equation from which we are to find values of x which make u a maximum or a minimum. If then between two assigned values of x there exists no value which makes $\frac{du}{dx}$ vanish, we conclude that there is no maximum or minimum value of u between those assigned values of x; so that u either continually increases or continually decreases as x changes from the less to the greater of the assigned values. This principle has already been noticed in Art. 89, but its importance and its natural connexion with the subject of the present Chapter lead us to draw attention to it again.

For example, suppose

$$u = 2x - \tan^{-1} x - \log \{x + \sqrt{(1 + x^2)}\} ;$$

then $\qquad \dfrac{du}{dx} = 2 - \dfrac{1}{1 + x^2} - \dfrac{1}{\sqrt{(1 + x^2)}}$.

Hence $\dfrac{du}{dx}$ is positive and cannot vanish for any value of x lying between any assigned positive value and positive infinity. We conclude that u continually increases as x changes from zero to positive infinity.

216. *Maxima and minima values of an implicit function.*

Let $\phi(x, y) = 0$ be an equation connecting x and y; it is required to find the maxima or minima values of y. From the given equation we know that y must be *some* function of x, and if the equation admits of solution we can express y explicitly in terms of x, and then find the maxima or minima values of y by the foregoing Articles.

But instead of solving the given equation we may proceed thus: by Art. 177,

$$\frac{dy}{dx} = -\frac{\left(\dfrac{du}{dx}\right)}{\left(\dfrac{du}{dy}\right)} \; ;$$

where u stands for $\phi(x, y)$. But the values of x which make y a maximum or minimum must, by Art. 211, be found by solving the equation $\dfrac{dy}{dx} = 0$. Hence

$$\left(\frac{du}{dx}\right) = 0,$$

and this equation, combined with $u = 0$, will determine the values of x, which *may* make y a maximum or minimum. To determine whether such a value of x *does* make y a maximum or minimum, we must, by Art. 211, examine the value of $\dfrac{d^2y}{dx^2}$. By Art. 180, since $\left(\dfrac{du}{dx}\right) = 0$, we have

$$\frac{d^2y}{dx^2} = -\frac{\left(\dfrac{d^2u}{dx^2}\right)}{\left(\dfrac{du}{dy}\right)} .$$

Hence we have this rule : To find the maxima or minima values of y, which is an implicit function of x determined by $u = 0$, we must find values of x and y which satisfy $u = 0$ and $\left(\dfrac{du}{dx}\right) = 0$. If when these values are substituted in $\dfrac{\left(\dfrac{d^2u}{dx^2}\right)}{\left(\dfrac{du}{dy}\right)}$, the fraction is *positive*, we have obtained a maximum value of y; if the fraction be *negative*, we have a minimum value of y.

Example. If
$$x^3 - 3axy + y^3 = 0 \dots\dots\dots\dots\dots(1),$$
find the maxima or minima values of y.

Here
$$\frac{dy}{dx} = \frac{ay - x^2}{y^2 - ax};$$

therefore
$$ay - x^2 = 0 \dots\dots\dots\dots\dots\dots(2).$$

Combining (1) with (2), we have
$$x^6 - 2a^3 x^3 = 0 ;$$

therefore
$$x = 0,$$

or
$$x = a\sqrt[3]{2}.$$

The corresponding values of y are
$$y = 0,$$
$$y = a\sqrt[3]{4}.$$

If we substitute the values $x = a\sqrt[3]{2}$, $y = a\sqrt[3]{4}$, in $-\dfrac{\left(\dfrac{d^2u}{dx^2}\right)}{\left(\dfrac{du}{dy}\right)}$,

that is, in $-\dfrac{6x}{3(y^2 - ax)}$, we obtain $-\dfrac{2}{a}$. Hence there is a maximum value of y. The values $x = 0$, $y = 0$, which make the numerator of $\dfrac{dy}{dx}$ vanish, also make its denominator vanish; thus $\dfrac{dy}{dx}$ assumes an indeterminate form, and we must discover its real value. Forming the derived equations from the given equation, we have

$$(y^2 - ax) \frac{d^2y}{dx^2} + 2y \left(\frac{dy}{dx}\right)^2 - 2a \frac{dy}{dx} + 2x = 0,$$

$$(y^2 - ax) \frac{d^3y}{dx^3} + \left(6y \frac{dy}{dx} - 3a\right) \frac{d^2y}{dx^2} + 2 \left(\frac{dy}{dx}\right)^3 + 2 = 0.$$

When we put $x = 0$, $y = 0$, in these, the first equation gives $\frac{dy}{dx} = 0$, and the second equation gives $\frac{d^2y}{dx^2} = \frac{2}{3a}$. Hence, when $x = 0$, and $y = 0$, we have y a minimum.

217. If the values of x and y found from $u = 0$ and $\left(\frac{du}{dx}\right) = 0$, make $\frac{d^2y}{dx^2}$ vanish, then in order that they may make y a maximum or minimum, it will be necessary that $\frac{d^3y}{dx^3}$ should also vanish. This can be tested by making use of the value of $\frac{d^3y}{dx^3}$ given by Art. 184; and by obtaining a formula for $\frac{d^4y}{dx^4}$ similar to that for $\frac{d^3y}{dx^3}$ just referred to, we can ascertain whether $\frac{d^4y}{dx^4}$ is positive or negative for the specific values of x and y. On account however of the complexity of the general formulæ for $\frac{d^3y}{dx^3}$ and $\frac{d^4y}{dx^4}$, it is preferable to determine them in any example directly by the *method* of Art. 184, rather than to quote the results of that Article.

218. Suppose $u = \phi(x, y)$ and $\psi(x, y) = 0$; so that y is a function of x by the second equation, and therefore from the first equation u is a function of x; required the maxima and minima values of u. We may proceed theoretically thus: by solving the equation $\psi(x, y) = 0$, obtain y as a function of x; substitute this value of y in $\phi(x, y)$; then u becomes a function of x only, and its maxima and minima values can be found by previous rules. But we may avoid the difficulty of solving the equation $\psi(x, y) = 0$, thus.

By Art. 172, we have

$$\frac{du}{dx} = \left(\frac{du}{dx}\right) + \left(\frac{du}{dy}\right)\frac{dy}{dx}.$$

Also, putting v for $\psi(x, y)$, we have, by Art. 177,

$$\frac{dy}{dx} = -\frac{\left(\dfrac{dv}{dx}\right)}{\left(\dfrac{dv}{dy}\right)};$$

therefore

$$\frac{du}{dx} = \left(\frac{du}{dx}\right) - \frac{\left(\dfrac{du}{dy}\right)\left(\dfrac{dv}{dx}\right)}{\left(\dfrac{dv}{dy}\right)}.$$

Hence, the values of x and y that render u a maximum or minimum must be sought among those that satisfy simultaneously

$$\left(\frac{du}{dx}\right)\left(\frac{dv}{dy}\right) - \left(\frac{du}{dy}\right)\left(\frac{dv}{dx}\right) = 0,$$

and $\psi(x, y)$ or $v = 0$.

The value of $\dfrac{d^2u}{dx^2}$ must then be found by Art. 176, and we must examine whether the specific values of x and y render this positive or negative, in order to determine whether u is a minimum or a maximum.

Example. $\qquad u = x^2 + y^2,$

while $\qquad (x - a)^2 + (y - b)^2 - c^2 = 0, \quad \text{or} \quad v = 0.$

Here $\quad \left(\dfrac{du}{dx}\right) = 2x, \qquad\qquad \left(\dfrac{du}{dy}\right) = 2y,$

$$\left(\frac{dv}{dx}\right) = 2(x - a), \qquad \left(\frac{dv}{dy}\right) = 2(y - b).$$

Hence $\qquad x(y - b) - y(x - a) = 0\ ;$

therefore $\qquad ay = bx.$

Substitute the value of y in $v = 0$, and we have

$$x^2\left(1 + \frac{b^2}{a^2}\right) - 2x\left(a + \frac{b^2}{a}\right) + a^2 + b^2 = c^2 \; ;$$

therefore

$$x = a \pm \frac{ac}{\sqrt{(a^2 + b^2)}} \cdot$$

Upon examination it will be found, that if we take the upper sign in the value of x we obtain a maximum value for u, and if we take the lower sign, a minimum. This example is a solution of the geometrical question, "To find the points in the circumference of a given circle which are at a maximum or minimum distance from a given point."

219. The process for finding the maxima and minima values of an implicit function may be extended to the case in which one variable is connected with more than one other variable, the whole number of equations being one less than the whole number of variables. Suppose, for example, we have three equations,

$$F(x, y, z, u) = 0,$$
$$F_1(x, y, z, u) = 0,$$
$$F_2(x, y, z, u) = 0;$$

u being the variable of which we wish to find the maximum or minimum value.

From the given equations it follows that we may consider y, z, and u functions of the independent variable x. Hence

$$\left. \begin{aligned} \frac{dF}{dx} + \frac{dF}{dy}\frac{dy}{dx} + \frac{dF}{dz}\frac{dz}{dx} + \frac{dF}{du}\frac{du}{dx} &= 0 \\[4pt] \frac{dF_1}{dx} + \frac{dF_1}{dy}\frac{dy}{dx} + \frac{dF_1}{dz}\frac{dz}{dx} + \frac{dF_1}{du}\frac{du}{dx} &= 0 \\[4pt] \frac{dF_2}{dx} + \frac{dF_2}{dy}\frac{dy}{dx} + \frac{dF_2}{dz}\frac{dz}{dx} + \frac{dF_2}{du}\frac{du}{dx} &= 0 \end{aligned} \right\} \quad \ldots\ldots\ldots (1).$$

From these equations we can eliminate $\frac{dy}{dx}$ and $\frac{dz}{dx}$, and the value of $\frac{du}{dx}$ which we then obtain must be put equal to zero. Or, more simply, we may put $\dfrac{du}{dx} = 0$ in these equations, and then eliminate $\dfrac{dy}{dx}$ and $\dfrac{dz}{dx}$ from the resulting equations which are

$$\left.\begin{array}{l} \dfrac{dF}{dx} + \dfrac{dF}{dy}\dfrac{dy}{dx} + \dfrac{dF}{dz}\dfrac{dz}{dx} = 0 \\[2mm] \dfrac{dF_1}{dx} + \dfrac{dF_1}{dy}\dfrac{dy}{dx} + \dfrac{dF_1}{dz}\dfrac{dz}{dx} = 0 \\[2mm] \dfrac{dF_2}{dx} + \dfrac{dF_2}{dy}\dfrac{dy}{dx} + \dfrac{dF_2}{dz}\dfrac{dz}{dx} = 0 \end{array}\right\} \quad \dots\dots\dots\dots (2).$$

The equation obtained by eliminating $\dfrac{dy}{dx}$ and $\dfrac{dz}{dx}$, combined with the equations $F = 0$, $F_1 = 0$, $F_2 = 0$, will determine x, y, z and u.

By differentiating equations (1) again, we can obtain $\dfrac{d^2u}{dx^2}$, and by the sign which the values of x, y, z, u, already found, give to this quantity, we determine whether u is a maximum or minimum.

220. Suppose we have a function of n variables, the variables being connected by $n-1$ equations, and we require the maximum or minimum value of the function. For example, suppose three equations

$$F(x,\, y,\, z,\, u) = 0, \quad F_1(x,\, y,\, z,\, u) = 0, \quad F_2(x,\, y,\, z,\, u) = 0,$$

and that we wish to find the maximum or minimum of $f(x,\, y,\, z,\, u)$. In this case, to the equations (1) of the preceding Article, in which $\dfrac{du}{dx}$ must not be supposed zero, we must add

$$\dfrac{df}{dx} + \dfrac{df}{dy}\dfrac{dy}{dx} + \dfrac{df}{dz}\dfrac{dz}{dx} + \dfrac{df}{du}\dfrac{du}{dx} = 0.$$

From these four equations we must eliminate $\dfrac{dy}{dx}$, $\dfrac{dz}{dx}$, and $\dfrac{du}{dx}$. The resulting equation combined with the given equations $F = 0$, $F_1 = 0$, $F_2 = 0$, will determine x, y, z, and u. We should then form the second differential coefficient of $f(x, y, z, u)$ with respect to x. This will involve $\dfrac{d^2y}{dx^2}$, $\dfrac{d^2z}{dx^2}$, and $\dfrac{d^2u}{dx^2}$, which must be found by differentiating equations (1): by the sign of this second differential coefficient of $f(x, y, z, u)$ we shall settle whether the function is a maximum or a minimum.

221. In Art. 214 we obtained as the condition for $\phi(x)$ having a maximum or minimum value, that $\phi'(x)$ must change sign, and hence that $\phi'(x)$ must be zero or infinite. The cases in which $\phi'(x)$ is infinite occur but rarely, and in the Articles following Art. 214 we have always considered $\phi'(x)$ to *vanish* when $\phi(x)$ is a maximum or minimum. We shall here add one proposition which shews that according to the first view given of maxima and minima values (Arts. 209...213), a maximum or minimum *may* exist when the differential coefficient of the function considered becomes infinite.

Suppose that $\phi(x)$ is such a function of x that when $x = a$ we have some of the differential coefficients of $\phi(x)$ infinite, so that $\phi(a + h)$ cannot be expanded in powers of h by Taylor's Theorem.

Suppose that by some unexceptionable algebraical process we find

$$\phi(a + h) - \phi(a) = Ah^{\alpha} + Bh^{\beta} + Ch^{\gamma} + \ldots,$$

where α, β, γ, ..., are not necessarily positive integers. If any one of these exponents be a fraction in its lowest terms with an even denominator, then $\phi(a - h) - \phi(a)$ will be impossible, and the consideration of maxima and minima values becomes inapplicable. If none of the exponents be of this form, then $\phi(a - h) - \phi(a)$ will be a possible quantity. Now there may be cases in which, by taking h small enough, the sign of Ah^{α} determines the sign of $\phi(a + h) - \phi(a)$; for example, this happens if the number of terms in $\phi(a+h) - \phi(a)$ is finite, and the exponents α, β, γ, ..., all positive, and α the least. Let us suppose such a case, and let α be a proper fraction with an even numerator; then $\phi(a + h) - \phi(a)$ and $\phi(a - h) - \phi(a)$ are both positive if A be positive, and negative if A be negative, when h is taken small enough. Hence

$\phi(a)$ in the former case is a minimum value of $\phi(x)$ and in the latter a maximum value.

Also, since α is a *proper* fraction,

$$\frac{d\phi(a+h)}{dh} \text{ is infinite when } h = 0,$$

therefore $\phi'(x)$ is infinite when $x = a$.

Hence $\phi(x)$ *may* be a maximum or a minimum when $\phi'(x)$ is infinite.

Example. Suppose

$$\phi(x) = c + (x - a)^{\frac{2}{3}} + (x - a)^{\frac{4}{3}};$$

therefore

$$\phi(a + h) = c + h^{\frac{2}{3}} + h^{\frac{4}{3}},$$

$$\phi(a) = c,$$

$$\phi(a \pm h) - \phi(a) = h^{\frac{2}{3}} + h^{\frac{4}{3}}.$$

Hence $\phi(a + h)$ and $\phi(a - h)$ are both necessarily greater than $\phi(a)$. Hence $\phi(a)$ is a minimum value of $\phi(x)$, and it is obvious that $\phi'(x)$ is infinite when $x = a$.

222. *On certain cases of Geometrical Maxima and Minima.*

We occasionally meet in Geometry cases of maxima or minima values for which the ordinary analytical process appears to fail, though from geometrical considerations it is obvious that maxima or minima do exist. The following problem will introduce the difficulty which it is proposed to explain. "Find the maximum and minimum perpendicular from the focus on the tangent to an ellipse, the perpendicular being expressed in terms of the radius vector."

The equation which gives the perpendicular in terms of the radius vector is

$$p^2 = \frac{b^2 r}{2a - r};$$

therefore $p\dfrac{dp}{dr} = \dfrac{ab^2}{(2a - r)^2}$, which must $= 0$.

Now this can only be satisfied by $r = \pm \infty$, which values are not admissible, whereas we know from Geometry that p

has a maximum value $= a(1 + e)$, and a minimum value $= a(1 - e)$.

The reason we do not find these values by the above usual analytical process is this. In the ordinary theory of maxima and minima the function is considered to be expressed in terms of an *independent* variable which may assume all possible values. But in the example above r is not an *independent* variable; its values are limited to those found by ascribing *all possible values* to θ in the equation

$$r = \frac{a(1 - e^2)}{1 + e \cos \theta}.$$

Since r is thus a function of θ, we may consider p which is a function of r to be also a function of θ. Hence $\frac{dp}{d\theta} = \frac{dp}{dr}\frac{dr}{d\theta}$, and this may be made $= 0$ if we can make $\frac{dr}{d\theta} = 0$. This we can do, and thus p has a maximum or minimum value at the same time as r has.

Similar remarks apply to other examples. Thus generally, if $y = \phi(x)$, where x is not susceptible of all possible values, it may be impossible to make $\frac{dy}{dx} = 0$, and thus there may be, apparently, no maximum or minimum value of y. But in this case, if x can be expressed in terms of some variable θ which can assume all possible values, we must put $\frac{dx}{d\theta} = 0$, which makes $\frac{dy}{d\theta} = 0$, and thus we determine simultaneous maxima or minima values of x and y.

Example. To find the maximum and minimum length of the straight line drawn to a circle from a given external point.

Take the axis of x passing through the centre of the circle and the given external point, the former being the origin. Let $a =$ the radius of the circle, $c =$ the distance of the given point (A say) from the centre; and let x be the abscissa of a point P on the circumference; then $AP^2 = c^2 + a^2 - 2cx$.

The differential coefficient of this expression with respect to x is $-2c$, which cannot vanish. But if we put $x = a \cos \theta$, we have

$$AP^2 = c^2 + a^2 - 2ac \cos \theta,$$

$$\frac{d \cdot AP^2}{d\theta} = 2ac \sin \theta \, ;$$

and $\theta = 0$, $\theta = \pi$, give the minimum and maximum values respectively of AP^2.

In this Example the difficulty would not appear if we had so chosen our axes that x should not be a maximum simultaneously with AP. Calling b the ordinate of A, c the abscissa of A, and a the radius of the circle, we shall have

$$AP^2 = a^2 + b^2 + c^2 - 2b \sqrt{(a^2 - x^2)} - 2cx,$$

which has its minimum and maximum values, when

$$x = \pm \frac{ac}{\sqrt{(b^2 + c^2)}} \, .$$

Another solution of the problem is given in Art. 218.

The following is an analogous case. Find those conjugate diameters in an ellipse of which the sum is a maximum or minimum. Let r and r' be any two conjugate diameters, and $u = r + r'$, then u is to be a maximum or minimum, while $r^2 + r'^2 = a^2 + b^2 = c^2$, say;

thus

$$u = r + \sqrt{(c^2 - r^2)},$$

$$\frac{du}{dr} = 1 - \frac{r}{\sqrt{(c^2 - r^2)}} \, .$$

If $\dfrac{du}{dr}$ be put $= 0$, we get $r^2 = \dfrac{c^2}{2}$, and therefore $r'^2 = \dfrac{c^2}{2}$. This gives us the *equal conjugate diameters*, the sum of which we know to be a maximum. If we express r, and therefore r', in terms of some variable which can take all possible values, as for example ϕ the inclination of r to the axis major, we shall get an additional result. For $\dfrac{du}{d\phi} = \dfrac{du}{dr} \dfrac{dr}{d\phi}$, and therefore, if $\dfrac{dr}{d\phi} = 0$, we have also $\dfrac{du}{d\phi} = 0$. But $\dfrac{dr}{d\phi} = 0$ makes r a maximum or minimum, and thus we obtain the *two principal axes*, whose sum is a minimum. By a different method, we might have obtained at first the minimum value of $r + r'$.

For since $r^2 + r'^2 = a^2 + b^2$, and $rr' \sin \theta = ab$,

we have $(r + r')^2 = a^2 + b^2 + \dfrac{2ab}{\sin \theta}$,

where θ is the angle between r and r'. Differentiate with respect to θ, and we get $-\dfrac{2ab \cos \theta}{\sin^2 \theta} = 0$, therefore $\theta = \dfrac{\pi}{2}$; this gives the minimum value as before; $\dfrac{d\theta}{d\phi} = 0$ would give us a second result, which would be the maximum.

The foregoing Article has been derived from the third volume of the *Cambridge Mathematical Journal*, page 237. The following problem will furnish an exercise. Find the maximum or minimum length of the straight line drawn from the end of the minor axis of an ellipse to meet the curve. If x, y, be co-ordinates of the point where a straight line drawn from the end of the minor axis meets the curve, the length of the straight line can be expressed either as a function of x or of y; thus two solutions can be obtained and compared.

In the solution of some of the examples on maxima and minima the following results will be required: they may be established by means of the Integral Calculus.

The volume of a right cylinder is found by multiplying the area of its base by its altitude.

The convex surface of a right cylinder is found by multiplying the perimeter of its base by its altitude.

The volume of a right cone is one-third of the product of its base and altitude.

The convex surface of a right cone on a circular base is one-half the product of its slant side and the perimeter of its base.

If r be the radius of a sphere its volume is $\dfrac{4\pi r^3}{3}$ and its surface is $4\pi r^2$.

EXAMPLES OF MAXIMA AND MINIMA.

1. Shew that $x^5 - 5x^4 + 5x^3 - 1$ is a maximum when $x = 1$; a minimum when $x = 3$; neither when $x = 0$.

2. Shew that $x^3 - 3x^2 + 3x + 7$ is neither a maximum not a minimum when $x = 1$.

3. If $u = x^3 - 3x^2 + 6x + 7$, shew that it has neither a maximum nor a minimum value.

4. If $u = x^3 - 9x^2 + 15x - 3$, find its maximum and minimum value.

    A maximum when $x = 1$; a minimum when $x = 5$.

5. $u = (x - 1)^4 (x + 2)^3$.

    A maximum when $x = -\frac{5}{7}$; a minimum when $x = 1$; neither when $x = -2$.

6. $u = (1 + x^{\frac{3}{2}}) (7 - x)^2$.

    A maximum when $x = 1$; a minimum when $x = 0$, and when $x = 7$.

7. $u = 3x^5 - 125x^3 + 2160x$.

    A maximum when $x = -4$, and when $x = 3$; a minimum when $x = -3$, and when $x = 4$.

8. $u = \dfrac{1 - x + x^2}{1 + x - x^2}$.   A minimum when $x = \frac{1}{2}$.

9. $u = \dfrac{x^2 - 7x + 6}{x - 10}$.

    A maximum when $x = 4$; a minimum when $x = 16$.

10. If $\dfrac{du}{dx} = x^2 (x - 1)^3 (x - 2)^3 (x - 3)^4$, find when u is a maximum or minimum.

    A maximum when $x = 0$; a minimum when $x = 2$.

11. If $\dfrac{du}{dx} = (x - 1)(x - 2)^2 (x - 3)^3$, find when u is a maximum or minimum.

    A maximum when $x = 1$; a minimum when $x = 3$.

12. $u = x (a + x)^2 (a - x)^3$.

> A maximum when $x = \dfrac{a}{3}$, and when $x = -a$,
>
> and a minimum when $x = -\dfrac{a}{2}$.

13. $u = \dfrac{(a - x)^3}{a - 2x}$.

> A minimum when $x = \dfrac{a}{4}$.

14. $u = b + c (x - a)^{\frac{2}{3}}$.

> A minimum when $x = a$.

15. $u = \dfrac{a^2}{x} + \dfrac{b^2}{a - x}$.

> A minimum when $x = \dfrac{a^2}{a + b}$, and a maximum when
>
> $x = \dfrac{a^2}{a - b}$.

16. $u = \dfrac{3x^2 - a^2}{(a^2 + x^2)^3}$.

> A minimum when $x = 0$, and a maximum when $x = \pm a$.

17. $u = (mx + na)^{m+n} - (m + n)^{m+n} x^m a^n$.

> A minimum when $x = a$.

18. Shew that $\dfrac{x}{1 + x \tan x}$ is a maximum when $x = \cos x$.

19. Shew that $x^{\frac{1}{x}}$ is a maximum when $x = e$.

20. Shew that $\dfrac{\tan^3 x}{\tan 3x}$ is a maximum when $x = \dfrac{\pi}{8}$.

21. Shew that $\sin x (1 + \cos x)$ is a maximum when $x = \dfrac{\pi}{3}$.

22. If $xy (y - x) = 2a^3$, shew that y has a minimum value when $x = a$.

23. If $3a^2y^2 + xy^3 + 4ax^3 = 0$, shew that when $x = \dfrac{3a}{2}$, y has a maximum value, namely $-3a$, the value of $\dfrac{d^2y}{dx^2}$ being then $-\dfrac{8}{5a}$.

24. If $x^4 + 2ax^2y - ay^3 = 0$, shew that when $x = \pm a$, $y = -a$ and is a minimum. Also, when $y = -\dfrac{8a}{9}$, x is both a maximum and minimum, and is $= \pm \dfrac{4a\sqrt{6}}{9}$.

25. If $2x^5 + 3ay^4 - x^2y^3 = 0$, shew that $x = a \cdot 5^{\frac{4}{3}}$ makes y a minimum, and $= a \cdot 5^{\frac{4}{5}}$.

26. Find the maximum and minimum value of y, when
$$y^4 - 4c^3yx + x^4 = 0.$$
$x = c \sqrt[4]{3}$ makes $y = c \sqrt[4]{(27)}$ a maximum.
$x = -c \sqrt[4]{3}$ makes $y = -c \sqrt[4]{(27)}$ a minimum.

27. A person being in a boat 3 miles from the nearest point of the beach, wishes to reach in the shortest time a place 5 miles from that point along the shore: supposing he can walk 5 miles an hour, but row only at the rate of 4 miles an hour, required the place where he must land.

> One mile from the place to be reached.

28. The sides of a rectangle are a and b: shew that the greatest rectangle that can be drawn so as to have its sides passing through the corners of the given rectangle is a square, each side of which is $\dfrac{a+b}{\sqrt{2}}$.

29. If a rectangular piece of pasteboard, the sides of which are a and b, have a square cut out at each corner, find the side of the square that the remainder may form a box of maximum content.

> The side $= \dfrac{a + b - \sqrt{(a^2 - ab + b^2)}}{6}$.

30. A Norman window consists of a rectangle surmounted by a semicircle. Given the perimeter, required the height and breadth of the window when the quantity of light admitted is a maximum.

The radius of the semicircle must equal the height of the rectangle.

31. Shew that the altitude of the greatest equilateral triangle that can be circumscribed about a given triangle, is

$$\{a^2 + b^2 - 2ab \cos (\tfrac{1}{3}\pi + C)\}^{\frac{1}{2}}.$$

32. A straight line is drawn through the given point P, meeting the axes Ox and Oy at A and B respectively (see the figure on page 198); find the position of the straight line,

   (1)   When AB is a minimum.

   (2)   When $OA + OB$ is a minimum.

   (3)   When $OA \times OB$ is a minimum.

   (4)   When $OA + OB + AB$ is a minimum.

   (5)   When $OA \times OB \times AB$ is a minimum.

   (6)   When $OA^n + OB^n$ is a minimum.

Let θ denote the angle PAO, then we must have

   (1) $\tan \theta = \left(\dfrac{b}{a}\right)^{\frac{1}{3}},$

   (2) $\tan \theta = \left(\dfrac{b}{a}\right)^{\frac{1}{4}},$

   (3) $\tan \theta = \dfrac{b}{a},$

   (4) $\tan \theta = \dfrac{b + \sqrt{(2ab)}}{a + \sqrt{(2ab)}},$

   (5) $2a \tan^3 \theta - b \tan^2 \theta + a \tan \theta - 2b = 0,$

   (6) $\tan \theta = \left(\dfrac{b}{a}\right)^{\frac{1}{n+1}}.$

33. Having given an angle of a triangle and the opposite side, prove that the area will be a maximum when the given angle is equidistant from the other angles.

34. Having given an angle of a quadrilateral and the two opposite sides, prove that the area will be a maximum when the given angle is equidistant from the other angles.

It follows from the preceding Example that the two sides which contain the given angle must be equal in order to ensure a maximum area; for if they were not equal the area of the quadrilateral would be increased by changing these two sides into two equal sides.

35. Find the least ellipse which can be described about a given parallelogram, and shew that its area is to that of the parallelogram as π is to 2.

36. The least tangent to an ellipse intercepted by the axes is divided at the point of contact into two parts, which are equal to the semiaxes respectively.

37. Find the area and position of the greatest triangle that can be placed in a given parabolic segment, having the chord of the segment for its base.

38. Find the least triangle which can be described about a given ellipse, having a side parallel to the major axis and having the other sides equal.
    The height is three times the semi-minor axis.

39. Prove that of all circular sectors described with the same perimeter, the sector of greatest area is that in which the circular arc is double the radius.

40. A chord PSp is drawn through the focus S of an ellipse, and the points P, p, are joined with the other focus H: determine when the area PHp is a maximum.

Let e be the eccentricity of the ellipse and θ the angle between the chord PSp and the major axis of the ellipse. If $2e^2$ is greater than 1 the maximum is determined by $\cos^2 \theta = 2 - \dfrac{1}{e^2}$, and $\theta = \dfrac{\pi}{2}$ gives a minimum; if $2e^2$ is not greater than 1 the maximum is when $\theta = \dfrac{\pi}{2}$, and there is no minimum.

41. Find the length of the shortest normal chord in a parabola, and prove that it intersects the curve nearer the vertex than any other normal chord.

   If $4a$ be the latus rectum of the parabola the required length is $6a\sqrt{3}$.

42. Two ships are sailing uniformly with velocities u, v along straight lines inclined at an angle θ: shew that if a, b be their distances at one time from the point of intersection of the courses, the least distance of the ships is equal to $\dfrac{(av - bu)\sin\theta}{(u^2 + v^2 - 2uv\cos\theta)^{\frac{1}{2}}}$.

43. Of all the straight lines drawn from the vertex of a given ellipse to the circumference of the circumscribing circle, determine that for which the portion intercepted between the two curves is a maximum.

   If θ be the inclination of the straight line to the major axis of the ellipse, and e the eccentricity of the ellipse,
   $$2e^2\cos^2\theta = 3 - e^2 - \sqrt{\{(1 - e^2)(9 - e^2)\}}.$$

44. If an ellipse be described to touch a given semicircle and its diameter symmetrically, its area when a maximum will be $\dfrac{2\pi r^2}{3\sqrt{3}}$, r being the radius of the circle.

45. An ellipse is inscribed in an isosceles triangle, and has one of its axes coincident in direction with the straight line bisecting the vertical angle of the triangle: shew that this axis is two-thirds of the height of the triangle when the area of the ellipse is a maximum.

46. Find what sector must be taken out of a given circle, in order that the remainder may form the curved surface of a cone of maximum volume.

   The angle of the sector must be $\dfrac{2\pi(\sqrt{3} - \sqrt{2})}{\sqrt{3}}$.

47. Two focal chords are drawn in an ellipse at right angles, find when their sum is a maximum, and when a minimum.

[In the following problems the cones and cylinders are supposed to be *right* cones and cylinders on *circular* bases.]

48. Determine the greatest cylinder that can be inscribed in a given cone.

> If b be the height of the cone, and a the radius of its base, the volume of the cylinder is $\dfrac{4}{27}\, \pi a^2 b$.

49. Determine the cylinder of greatest convex surface that can be inscribed in the same cone.

$$\text{The surface} = \frac{\pi b a}{2}.$$

50. Determine the cylinder, so that its *whole* surface shall be a maximum.

> The radius of the cylinder $=\dfrac{ab}{2\,(b-a)}$; but by the nature of the problem this must be less than a; this leads to the condition that b must be greater than $2a$ in order to ensure a maximum. If b be not greater than $2a$ the whole surface of the cylinder *continually increases* as its radius increases, and there is no *maximum*.

51. Determine the greatest cylinder that can be inscribed in a given sphere.

> If r be the radius of the sphere the height of the cylinder is $\dfrac{2r}{\sqrt{3}}$.

52. Determine the cylinder inscribed in a given sphere which has the greatest convex surface.

$$\text{Height} = r\sqrt{2}.$$

53. Determine the cylinder so that its *whole* surface shall be a maximum.

$$\text{Height} = r\left\{2\left(1-\frac{1}{\sqrt{5}}\right)\right\}^{\frac{1}{2}}.$$

54. Determine the greatest cone that can be inscribed in a given sphere. $\qquad$ Height $= \frac{4}{3} r$.

55.   Determine the cone of the greatest convex surface that can be inscribed in a given sphere.

$$\text{Height} = \tfrac{4}{3} r.$$

56.   Determine the cone so that its *whole* surface shall be a maximum.

$$\text{Height} = \frac{r}{16} \left(23 - \sqrt{17} \right).$$

57.   Given the volume of a cylinder, find its height and radius when the sum of the areas of its convex surface and one end is a minimum.

The height is equal to the radius.

58.   Of all cones described about a given sphere, find that of minimum volume.

The sine of the semivertical angle must be $\tfrac{1}{3}$.

59.   A series of cones have their slant sides of the same length: find that which has the greatest volume.

The tangent of the semivertical angle $= \sqrt{2}$.

60.   Find the position of the chord which passes through a given point within a parabola, and cuts off from the parabola the least possible area.

61.   Find a point in an ellipse from which, if perpendiculars be drawn to two given conjugate diameters, the sum of their squares will be a maximum.

62.   Prove that $\phi \{ f(x) \}$ is necessarily either a maximum or minimum when $f(x)$ is a maximum. And so also when $f(x)$ is a minimum.

CHAPTER XIV.

EXPANSION OF A FUNCTION OF TWO INDEPENDENT VARIABLES.

223. LET $u = \phi(x, y)$ be a function of two independent variables, and suppose $\phi(x+h, y+k)$ is to be expanded in ascending powers of h and k. Put

$$h = ah', \quad k = ak',$$

then
$$\phi(x+h, y+k) = \phi(x+ah', y+ak');$$

the last expression may be considered a function of a, and denoted by $f(a)$. By Maclaurin's theorem,

$$f(a) = f(0) + f'(0) \cdot a + f''(0) \cdot \frac{a^2}{\lfloor 2} + \ldots\ldots;$$

we shall now shew how the differential coefficients of $f(a)$ may be conveniently expressed. Suppose

$$x + ah' = x', \quad y + ak' = y';$$

then $f(a)$ stands for $\phi(x', y')$ and since both x' and y' contain a, we have by Art. 169,

$$f'(a) = \frac{d\phi(x', y')}{dx'} \frac{dx'}{da} + \frac{d\phi(x', y')}{dy'} \frac{dy'}{da}$$

$$= h' \frac{d\phi(x', y')}{dx'} + k' \frac{d\phi(x', y')}{dy'}.$$

Also, by Art. 63,

$$\frac{d\phi(x', y')}{dx} = \frac{d\phi(x', y')}{dx'} \cdot \frac{dx'}{dx};$$

but
$$\frac{dx'}{dx} = 1;$$

therefore
$$\frac{d\phi(x', y')}{dx'} = \frac{d\phi(x', y')}{dx}.$$

Similarly
$$\frac{d\phi(x', y')}{dy'} = \frac{d\phi(x', y')}{dy};$$

hence
$$f'(a) = h'\frac{d\phi(x', y')}{dx} + k'\frac{d\phi(x', y')}{dy},$$

which, for shortness, may be written

$$f'(a) = h'\frac{df}{dx} + k'\frac{df}{dy}.$$

Similarly,

$$f''(a) = h'^2\frac{d^2f}{dx^2} + 2h'k'\frac{d^2f}{dx\,dy} + k'^2\frac{d^2f}{dy^2},$$

$$f'''(a) = h'^3\frac{d^3f}{dx^3} + 3h'^2k'\frac{d^3f}{dx^2dy} + 3h'k'^2\frac{d^3f}{dx\,dy^2} + k'^3\frac{d^3f}{dy^3}.$$

The law of the formation of the successive differential coefficients of $f(a)$ is thus obvious. When $a = 0$, $f(a)$ becomes u; hence we have

$$f(0) = u,$$

$$f'(0) = h'\frac{du}{dx} + k'\frac{du}{dy},$$

$$f''(0) = h'^2\frac{d^2u}{dx^2} + 2h'k'\frac{d^2u}{dx\,dy} + k'^2\frac{d^2u}{dy^2},$$

$$\dotsm\dotsm\dotsm$$

Restore h for ah', and k for ak'; then

$$\phi(x+h, y+k) = u + h\frac{du}{dx} + k\frac{du}{dy}$$

$$+ \frac{1}{\lfloor 2}\left\{h^2\frac{d^2u}{dx^2} + 2hk\frac{d^2u}{dx\,dy} + k^2\frac{d^2u}{dy^2}\right\}$$

$$+ \frac{1}{\lfloor 3}\left\{h^3\frac{d^3u}{dx^3} + 3h^2k\frac{d^3u}{dx^2dy} + 3hk^2\frac{d^3u}{dx\,dy^2} + k^3\frac{d^3u}{dy^3}\right\}$$

$$+ \dotsm\dotsm$$

224. If we wish the series for $\phi(x+h, y+k)$ to close after a finite number of terms, we can put the expansion for $f(a)$ under the form

$$f(a) = f(0) + f'(0) \cdot a + f''(0) \cdot \frac{a^2}{\lfloor 2} + \ldots + f^{n-1}(0) \cdot \frac{a^{n-1}}{\lfloor n-1}$$

$$+ f^n(\theta a) \cdot \frac{a^n}{\lfloor n};$$

and from this the required form for $\phi(x+h, y+k)$ can be obtained. For example, if $n = 3$,

$$\phi(x+h, y+k) = u + h\frac{du}{dx} + k\frac{du}{dy}$$

$$+ \frac{1}{\lfloor 2} \left\{ h^2\frac{d^2u}{dx^2} + 2hk\frac{d^2u}{dx\,dy} + k^2\frac{d^2u}{dy^2} \right\}$$

$$+ \frac{1}{\lfloor 3} \left\{ h^3\frac{d^3v}{dx^3} + 3h^2k\frac{d^3v}{dx^2\,dy} + 3hk^2\frac{d^3v}{dx\,dy^2} + k^3\frac{d^3v}{dy^3} \right\},$$

where v stands for $\phi(x+\theta h, y+\theta k)$.

225. In the formula established in Art. 223, put $x = 0$, and $y = 0$; then

$$\phi(h, k) = u_0 + h\frac{du_0}{dx} + k\frac{du_0}{dy}$$

$$+ \frac{1}{\lfloor 2} \left\{ h^2\frac{d^2u_0}{dx^2} + 2hk\frac{d^2u_0}{dx\,dy} + k^2\frac{d^2u_0}{dy^2} \right\}$$

$$+ \ldots\ldots\ldots\ldots$$

where u_0, $\dfrac{du_0}{dx}$, $\dfrac{du_0}{dy}$, $\dfrac{d^2u_0}{dy^2}$, $\ldots\ldots$ stand for the values of u, $\dfrac{du}{dx}$, $\dfrac{du}{dy}$, $\dfrac{d^2u}{dx^2}$, $\ldots\ldots$ when in these expressions we put $x = 0$, and $y = 0$. If we change h and k into x and y respectively in the above formula, we have

$$\phi(x, y) = u_0 + x\frac{du_0}{dx} + y\frac{du_0}{dy}$$

$$+ \frac{1}{\lfloor 2} \left\{ x^2\frac{d^2u_0}{dx^2} + 2xy\frac{d^2u_0}{dx\,dy} + y^2\frac{d^2u_0}{dy^2} \right\}$$

$$+ \ldots\ldots\ldots\ldots$$

x and y being each put equal to zero in u_0 and its differential coefficients after the differentiations have been performed.

In this manner the formula of Maclaurin is extended to the expansion of functions of two variables.

226. The expression for the nth differential coefficient of $f(a)$, in Art. 223, is

$$h'^n \frac{d^n f}{dx^n} + n h'^{n-1} k' \frac{d^n f}{dx^{n-1} dy} + \frac{n(n-1)}{\lfloor 2} h'^{n-2} k'^2 \frac{d^n f}{dx^{n-2} dy^2} \cdots + k'^n \frac{d^n f}{dy^n},$$

which, for abbreviation, may be written

$$\left(h' \frac{d}{dx} + k' \frac{d}{dy} \right)^n f$$

provided we interpret this expression thus: $\left(h' \dfrac{d}{dx} + k' \dfrac{d}{dy} \right)^n$ is to be expanded by the Binomial Theorem as if $h' \dfrac{d}{dx}$ were one term and $k' \dfrac{d}{dy}$ the other term : when the expansion is effected, every such term as $\left(h' \dfrac{d}{dx} \right)^{n-r} \left(k' \dfrac{d}{dy} \right)^r f$ which occurs is to be replaced by $h'^{n-r} k'^r \dfrac{d^n f}{dx^{n-r} dy^r}$. If we adopt this mode of abbreviation the result of Art. 223 may be written

$$\phi(x+h, y+k) = u + \left(h \frac{d}{dx} + k \frac{d}{dy} \right) u + \frac{1}{\lfloor 2} \left(h \frac{d}{dx} + k \frac{d}{dy} \right)^2 u$$

$$+ \ldots\ldots + \frac{1}{\lfloor n-1} \left(h \frac{d}{dx} + k \frac{d}{dy} \right)^{n-1} u + \frac{1}{\lfloor n} \left(h \frac{d}{dx} + k \frac{d}{dy} \right)^n v,$$

where $u = \phi(x, y)$, and $v = \phi(x + \theta h, y + \theta k)$.

By Art. 110 the last term of the expansion may, if we please, be replaced by

$$\frac{1}{\lfloor n-1} (1-\theta)^{n-1} \left(h \frac{d}{dx} + k \frac{d}{dy} \right)^n v.$$

The methods here given for the expansion of a function of two independent variables may be readily extended to the expansion of a function of more than two independent variables.

MISCELLANEOUS EXAMPLES.

1. Shew that if x and c are positive
$$2 \log \frac{x}{c+x} + \frac{c}{x} + \frac{c}{c+x}$$
decreases as x increases.

2. Shew that if x and c are positive
$$\left(\frac{x}{c+x}\right)^{c+2x}$$
increases as x increases.

3. If $u = (x-3)e^{2x} + 4xe^x + x + 3$ shew that $\dfrac{d^2u}{dx^2}$, $\dfrac{du}{dx}$, and u
    are positive for all positive values of x. See Ex. 10, p. 86.

4. Shew that for positive values of x the expression
$$\frac{e^{2x}(x-2) + e^x(x+2)}{(e^x-1)^3}$$
diminishes as x increases, and that its greatest value
is $\dfrac{1}{6}$.

5. Demonstrate the following approximate expression when
    x is small,
$$(1+x)^{\frac{1}{x}} = e \left\{ 1 - \frac{x}{2} + \frac{11x^2}{24} - \frac{7x^3}{16} \right\}.$$

6. Evaluate $\dfrac{(1+x)^{\frac{1}{x}} - e}{x}$ when $x = 0$.

    Result. $-\dfrac{e}{2}$.

7. Shew that when x is infinite
$$x \left(1 + \frac{1}{x}\right)^x - ex^2 \log \left(1 + \frac{1}{x}\right) = 0.$$

8. Find the value when x is infinite of
$$8x^2 \left(1 + \frac{1}{x}\right)^x - 8ex^3 \log \left(1 + \frac{1}{x}\right).$$

    Result. c.

9. Evaluate $\dfrac{\dfrac{\pi}{4} - \tan^{-1} x}{x^n - e^{\sin(\log x)}}$ when $x = 1$.

10. Evaluate $\dfrac{\log\left(\cot\dfrac{x}{2}\right)}{\cot x + \log x}$ when $x = 0$.

11. Evaluate $\dfrac{\sec^n x}{e^{\tan x}}$ when $x = \dfrac{\pi}{2}$.

12. Evaluate $\dfrac{\tan nx - \tan mx}{\sin(n^2 x - m^2 x)}$,

(1) when $x = 0$,   (2) when $n = m$.

13. In the equation $f(x+h) - f(x) = h f'(x + \theta h)$, shew that if $f''(x)$ is not zero the limiting value of θ as h is indefinitely diminished is $\dfrac{1}{2}$: also shew that if $f^r(x)$ is the first of the differential coefficients $f''(x), f'''(x),\ldots$ which is not zero, the limiting value of θ as h is indefinitely diminished is

$$\left(\frac{1}{r}\right)^{\frac{1}{r-1}}.$$

14. In the equation $f(x+h) - f(x) = h f'(x + \theta h)$ shew that if θ be the same for all values of h, it must equal $\dfrac{1}{2}$ and $f''(x)$ must be constant.

15. Change the independent variable from z to x in the equation

$$z^2 \frac{d^2y}{dz^2} + z \frac{dy}{dz} - 1 = (\log z)^2 \left\{ z^2 \frac{d^2y}{dz^2} + z \frac{dy}{dz} \right\},$$

where $z = e^{\sin x}$.

$\qquad\qquad$ Result. $\dfrac{d^2y}{dx^2} + \tan x \dfrac{dy}{dx} = 1$.

16. Transform the expression

$$\left\{ \left(\frac{du}{dx}\right)^2 + \left(\frac{du}{dy}\right)^2 + \left(\frac{du}{dz}\right)^2 \right\} \left\{ x\,\frac{du}{dx} + y\,\frac{du}{dy} + z\,\frac{du}{dz} \right\}^{-2}$$

into one in which r, θ, ϕ shall be the independent variables, having given

$$x = r \sin \theta \cos \phi, \quad y = r \sin \theta \sin \phi, \quad z = r \cos \theta.$$

17. If x, y and ξ, η be co-ordinates of the same point referred to two systems of rectangular co-ordinates, shew that

$$\frac{d^2\phi}{dx^2}\frac{d^2\phi}{dy^2} - \left(\frac{d^2\phi}{dx\,dy}\right)^2 = \frac{d^2\phi}{d\xi^2}\frac{d^2\phi}{d\eta^2} - \left(\frac{d^2\phi}{d\xi\,d\eta}\right)^2.$$

18. Shew that $x^2 + x \sin x + 4 \cos x$ is a minimum when $x = 0$.

19. CQ is the perpendicular from the centre C of an ellipse on the tangent at a point P: find the maximum value of PQ.

Result. $a - b$.

20. A straight line drawn from the extremity of the minor axis of an ellipse cuts the major axis at Q and the curve at P; from P the ordinate PN is drawn to the major axis: find when the area PQN is a maximum.

Result. $PN = \dfrac{b}{4}\,(\sqrt{17} - 1)$.

CHAPTER XV.

MAXIMA AND MINIMA VALUES OF A FUNCTION OF TWO INDEPENDENT VARIABLES.

227. DEFINITION. A function $\phi(x, y)$ of two independent variables is said to have a *maximum* value when $\phi(x+h, y+k)$ is *less* than $\phi(x, y)$ for all values of h and k. positive or negative, comprised between zero and certain finite limits however small. The function is said to have a *minimum* value when $\phi(x+h, y+k)$ is *greater* than $\phi(x, y)$ for all such values of h and k.

228. *To investigate the conditions that a function of two independent variables may have a maximum or minimum value.*

Let
$$u = \phi(x, y),$$
$$v = \phi(x+\theta h, y+\theta k);$$
then, by Art. 226,
$$\phi(x+h, y+k) = u + h\frac{du}{dx} + k\frac{du}{dy} + R,$$
where
$$R = \frac{1}{\lfloor 2} \left\{ h^2\frac{d^2v}{dx^2} + 2hk\frac{d^2v}{dx\,dy} + k^2\frac{d^2v}{dy^2} \right\}.$$

Now, if $h\dfrac{du}{dx} + k\dfrac{du}{dy}$ be not zero, by taking h and k sufficiently small, we can always make R less than $h\dfrac{du}{dx} + k\dfrac{du}{dy}$, and hence the sign of $\phi(x+h, y+k) - \phi(x, y)$ will depend on that of $h\dfrac{du}{dx} + k\dfrac{du}{dy}$, and will therefore change by changing that of h and k; it is impossible then that $\phi(x, y)$ can have a maximum or minimum value unless
$$h\frac{du}{dx} + k\frac{du}{dy} = 0.$$

Since the quantities h and k are *independent*, we must have

$$\frac{du}{dx} = 0, \quad \frac{du}{dy} = 0.$$

Find values of x and y from these equations, say $x = a$, $y = b$; let the values of $\dfrac{d^2u}{dx^2}$, $\dfrac{d^2u}{dx\,dy}$, $\dfrac{d^2u}{dy^2}$, when these values are assigned to x and y, be denoted by A, B, C, respectively. We have then by Art. 226,

$$\phi\,(a + h,\ b + k) - \phi\,(a,\ b) = \frac{1}{\lfloor 2} \{Ah^2 + 2Bhk + Ck^2\} + R_1,$$

where $R_1 = \dfrac{1}{\lfloor 3} \left\{ h^3 \dfrac{d^3v}{dx^3} + 3h^2k \dfrac{d^3v}{dx^2\,dy} + 3hk^2 \dfrac{d^3v}{dx\,dy^2} + k^3 \dfrac{d^3v}{dy^3} \right\}$,

x being made $= a$, and $y = b$, after the differentiations have been performed.

If A, B, and C do not all vanish, the sign of

$$\phi\,(a + h,\ b + k) - \phi\,(a,\ b)$$

will, when h and k are taken small enough, depend on that of $Ah^2 + 2Bhk + Ck^2$, or of $\dfrac{k^2}{A} \left\{ \left(A\dfrac{h}{k} + B \right)^2 + AC - B^2 \right\}$.

If $AC - B^2$ be negative, it will be possible, by ascribing a suitable value to $\dfrac{h}{k}$, to make the last expression vanish and change its sign; and then $\phi\,(a,\ b)$ is neither a maximum nor minimum value of $\phi\,(x,\ y)$. Hence *generally* we must have $AC - B^2$ *positive* as a condition for the existence of a maximum or minimum. In this case A and C will have the same sign, and $Ah^2 + 2Bhk + Ck^2$ will have the same sign as A or C; and if that sign be *positive*, $\phi\,(a,\ b)$ is a *minimum* value of $\phi\,(x,\ y)$, if *negative*, $\phi\,(a,\ b)$ is a *maximum* value.

We say that *generally* $AC - B^2$ must be positive; because, in fact, there *may be* a maximum or minimum value when $AC - B^2 = 0$, as we shall now proceed to shew.

229.   *To investigate the additional conditions for the existence of a maximum or minimum when $AC - B^2 = 0$.*

If $AC - B^2 = 0$, then

$$Ah^2 + 2Bhk + Ck^2 = \frac{k^2}{A} \left(A\frac{h}{k} + B \right)^2;$$

hence $\phi(a+h, b+k) - \phi(a, b)$ is always of the same sign as A, when h and k are taken small enough, *except when $\dfrac{h}{k}$ is equal to $-\dfrac{B}{A}$*; and then the sign is as yet unknown and further investigation is required. Let P, Q, S, T stand for the values of

$$\frac{d^3u}{dx^3}, \quad \frac{d^3u}{dx^2\,dy}, \quad \frac{d^3u}{dx\,dy^2}, \quad \frac{d^3u}{dy^3}$$

respectively, when $x = a$ and $y = b$; and let

$$R_2 = \frac{1}{\lfloor 4} \left\{ h^4 \frac{d^4v}{dx^4} + 4h^3k \frac{d^4v}{dx^3\,dy} + \ldots + k^4 \frac{d^4v}{dy^4} \right\},$$

x being made $= a$ and $y = b$ after the differentiations.

Suppose $\dfrac{h}{k}$ is equal to $-\dfrac{B}{A}$, then $Ah^2 + 2Bhk + Ck^2$ vanishes, and

$$\phi(a+h, b+k) - \phi(a, b) = \frac{1}{\lfloor 3}\{Ph^3 + 3Qh^2k + 3Shk^2 + Tk^3\} + R_2.$$

Hence if h and k be taken small enough the sign of

$$\phi(a+h, b+k) - \phi(a, b)$$

will be the same as the sign of

$$Ph^3 + 3Qh^2k + 3Shk^2 + Tk^3,$$

and will therefore change by changing the sign of h and k; it is impossible then that $\phi(a, b)$ can be a maximum or minimum value unless

$$Ph^3 + 3Qh^2k + 3Shk^2 + Tk^3$$

vanishes when $\dfrac{h}{k}$ is equal to $-\dfrac{B}{A}$.

Suppose this condition to be satisfied, then the sign of

$$\phi(a+h, b+k) - \phi(a, b),$$

when $\dfrac{h}{k}$ is equal to $-\dfrac{B}{A}$, is the same as the sign of R_2; and when $\dfrac{h}{k}$ is *not* equal to $-\dfrac{B}{A}$, and h and k are taken small enough, the sign of $\phi(a+h,\ b+k) - \phi(a,\ b)$ is the same as the sign of A. But in order that $\phi(a,\ b)$ may be a maximum or minimum value the sign of $\phi(a+h,\ b+k) - \phi(a,\ b)$ must be invariable when h and k are taken small enough. Hence we have the condition that the sign of R_2 when $\dfrac{h}{k}$ is equal to $-\dfrac{B}{A}$, and h and k are taken small enough, must be the same as the sign of A.

If these two additional conditions are satisfied $\phi(a,\ b)$ is a maximum value if A be negative, and a minimum value if A be positive.

230. If $A = 0$, $B = 0$, and $C = 0$, we must proceed thus:

$$\phi(a+h,\ b+k) - \phi(a,\ b) = \frac{1}{\underline{|3}}\{Ph^3 + 3Qh^2k + 3Shk^2 + Tk^3\} + R_2,$$

where P, Q, S, T, stand for the values of $\dfrac{d^3u}{dx^3}$, $\dfrac{d^3u}{dx^2dy}$, ...

when $x = a$ and $y = b$, and

$$R_2 = \frac{1}{\underline{|4}}\left\{h^4\frac{d^4v}{dx^4} + 4h^3k\frac{d^4v}{dx^3dy} + \dots + k^4\frac{d^4v}{dy^4}\right\},$$

x being made $= a$, and $y = b$, after the differentiations.

Hence, that $\phi(a,\ b)$ may be a maximum or minimum, it is necessary that P, Q, S, T, should all vanish. Also, R_2 must be of invariable sign; but the conditions to ensure this are too complicated to find investigation here.

231. The following is another method of investigating the conditions that a function of two independent variables may admit of a maximum or minimum value.

Let $u = \phi(x,\ y)$, where x and y are independent: required the maxima and minima values of u.

If y, instead of being independent of x, were equal to some function of x, say $\psi(x)$, then u would be a function of one variable x. We should then have

$$\frac{du}{dx} = \left(\frac{du}{dx}\right) + \left(\frac{du}{dy}\right)\psi'(x),$$

$$\frac{d^2u}{dx^2} = \left(\frac{d^2u}{dx^2}\right) + 2\left(\frac{d^2u}{dx\,dy}\right)\psi'(x) + \left(\frac{d^2u}{dy^2}\right)\{\psi'(x)\}^2 + \left(\frac{du}{dy}\right)\psi''(x).$$

In order that u may be a maximum or minimum, we must have, by Art. 211,

$$\frac{du}{dx} = 0,$$

therefore $\qquad \left(\frac{du}{dx}\right) + \left(\frac{du}{dy}\right) \psi'(x) = 0.$

Hence, since y is really independent of x, this equation must hold whatever be the function $\psi'(x)$; ·

therefore $\qquad \left(\frac{du}{dx}\right) = 0, \qquad \left(\frac{du}{dy}\right) = 0.$

In order that u may be a *maximum*, the values of x and y derived from the last equations must make $\dfrac{d^2u}{dx^2}$ negative, whatever $\psi'(x)$ may be; hence, denoting by A, B, C, the values which $\left(\dfrac{d^2u}{dx^2}\right)$, $\left(\dfrac{d^2u}{dx\,dy}\right)$, and $\left(\dfrac{d^2u}{dy^2}\right)$, respectively assume for the values of x and y under consideration, we require that

$$A + 2B\psi'(x) + C\left\{\psi'(x)\right\}^2$$

should be always negative, whatever $\psi'(x)$ may be. Hence as in Art. 228, A must be *negative*, and *generally* $AC - B^2$ must be positive. Similarly, that u may be a *minimum* we must have A *positive*, and *generally* $AC - B^2$ positive.

The preceding method may be rendered more symmetrical by supposing both x and y functions of a third variable t. Putting for shortness Dx for $\dfrac{dx}{dt}$, and Dy for $\dfrac{dy}{dt}$, we have

$$\frac{du}{dt} = \left(\frac{du}{dx}\right) Dx + \left(\frac{du}{dy}\right) Dy,$$

$$\frac{d^2u}{dt^2} = \left(\frac{d^2u}{dx^2}\right)(Dx)^2 + 2\left(\frac{d^2u}{dx\,dy}\right) Dx\,Dy + \left(\frac{d^2u}{dy^2}\right)(Dy)^2$$
$$+ \left(\frac{du}{dx}\right)\frac{dDx}{dt} + \left(\frac{du}{dy}\right)\frac{dDy}{dt}.$$

Hence we must have

$$\left(\frac{du}{dx}\right) = 0, \qquad \left(\frac{du}{dy}\right) = 0.$$

Also for values of x and y found from these equations,

$$\left(\frac{d^2u}{dx^2}\right)(Dx)^2 + 2\left(\frac{d^2u}{dx\,dy}\right)Dx\,Dy + \left(\frac{d^2u}{dy^2}\right)(Dy)^2$$

must preserve an invariable sign, whatever be the signs and values of Dx and Dy. From this we deduce the same results as in the preceding Article.

232. There is no theoretical difficulty in finding the maximum or minimum value of an *implicit* function of two independent variables, nor in finding the maximum or minimum value of a variable which is connected with any number of other variables by equations, when the whole number of equations is *two* less than the whole number of variables. For example, suppose we have two equations

$$f_1(x, y, z, u) = 0, \quad f_2(x, y, z, u) = 0\ldots\ldots\ldots(1),$$

involving four variables x, y, z, u, and we wish to find the maximum or minimum value of u. We may eliminate one of the three variables x, y, z between the two equations; suppose we eliminate z; then we obtain one equation connecting x, y, and u; from this we find u in terms of x and y, and proceed in the ordinary way to investigate the maximum or minimum value of u. Or if we wish to avoid the elimination we may adopt the following method: consider x and y as the independent variables and differentiate the given equations (1); thus

$$\left.\begin{array}{l} \dfrac{df_1}{dx} + \dfrac{df_1}{dz}\dfrac{dz}{dx} + \dfrac{df_1}{du}\dfrac{du}{dx} = 0 \\[2ex] \dfrac{df_1}{dy} + \dfrac{df_1}{dz}\dfrac{dz}{dy} + \dfrac{df_1}{du}\dfrac{du}{dy} = 0 \\[2ex] \dfrac{df_2}{dx} + \dfrac{df_2}{dz}\dfrac{dz}{dx} + \dfrac{df_2}{du}\dfrac{du}{dx} = 0 \\[2ex] \dfrac{df_2}{dy} + \dfrac{df_2}{dz}\dfrac{dz}{dy} + \dfrac{df_2}{du}\dfrac{du}{dy} = 0 \end{array}\right\}\ldots\ldots\ldots\ldots(2).$$

From these equations we can eliminate $\dfrac{dz}{dx}$ and $\dfrac{dz}{dy}$, and find $\dfrac{du}{dx}$ and $\dfrac{du}{dy}$; then for a maximum or minimum value of u

the values of $\dfrac{du}{dx}$ and $\dfrac{du}{dy}$ must be zero. Thus, more simply, we may put $\dfrac{du}{dx} = 0$ and $\dfrac{du}{dy} = 0$ in equations (2), and then eliminate $\dfrac{dz}{dx}$ and $\dfrac{dz}{dy}$; the two resulting equations combined with (1) will determine the values of x, y, z and u, which *may* correspond to a maximum or minimum value of u. And by differentiating equations (2) with respect to x and y we can find $\dfrac{d^2u}{dx^2}$, $\dfrac{d^2u}{dxdy}$, and $\dfrac{d^2u}{dy^2}$, and so settle whether u is really a maximum or minimum.

Practically the solution of problems of this class is facilitated by the method of *indeterminate multipliers*, which is explained in the following Chapter.

233. The student will find it advantageous to illustrate this Chapter by means of the Geometry of Three Dimensions. If $z = \phi(x, y)$ be the equation to a surface, to find the maxima and minima values of z amounts to finding those points on the surface which are at a greater or a less distance from the plane of (x, y) than adjacent points. The conditions $\dfrac{dz}{dx} = 0$, and $\dfrac{dz}{dy} = 0$, make the *tangent plane* at any one of the points in question parallel to the plane of (x, y). The interpretation of the case in which $B^2 - AC = 0$ will be seen from what is stated in Art. 235.

The method given in Art. 231 admits of clear geometrical illustration. If, for example, there be a point on the given surface which is at a *maximum* distance from the plane of (x, y), then in passing from that point to an adjacent point, *along any curve whatever lying on the surface*, we must approach nearer to the plane of (x, y). Now, by combining the equation $z = \phi(x, y)$ with $y = \psi(x)$, we obtain a curve lying on the given surface, and by giving every variety of form to $\psi(x)$ we may obtain as many curves as we please. Hence we see that if we put $y = \psi(x)$, and leave the form of the function $\psi(x)$ arbitrary, we do not really break the restriction that x and y are to be independent.

234. A function u of two variables *may* have a maximum or minimum value for values of x and y which render $\dfrac{du}{dx}$ and $\dfrac{du}{dy}$ *indeterminate* or *infinite*. Such exceptional cases must be examined specially, as there is no general theory applicable to them. For example, suppose

$$u = (x^2 + y^2)^{\frac{1}{3}},$$

$$\frac{du}{dx} = \frac{2x}{3\,(x^2 + y^2)^{\frac{2}{3}}}, \quad \frac{du}{dy} = \frac{2y}{3\,(x^2 + y^2)^{\frac{2}{3}}}.$$

Here, when x and y vanish $\dfrac{du}{dx}$ and $\dfrac{du}{dy}$ become indeterminate. If we put $y = ax$, we have

$$\frac{du}{dx} = \frac{2}{3x^{\frac{1}{3}}(1 + a^2)^{\frac{2}{3}}}, \quad \frac{du}{dy} = \frac{2a}{3x^{\frac{1}{3}}(1 + a^2)^{\frac{2}{3}}}.$$

Hence $\dfrac{du}{dx}$ and $\dfrac{du}{dy}$ are infinite when $x = 0$, and $y = 0$. But u is really a *minimum* then, for it vanishes only when x and y vanish and is never negative.

235. *On a case of maxima or minima values of a function of two independent variables.*

If u denote a function of two independent variables x and y, the values of x and y that make u a maximum or minimum are found from the two equations

$$\frac{du}{dx} = 0, \quad \frac{du}{dy} = 0.$$

If these equations are satisfied by a *single* relation between x and y, we cannot determine a *finite* number of values of x and y, that render u a maximum or minimum. This case we propose to examine.

Suppose $\quad u = \phi\,(x,\,y)$ ....(1),

$$\frac{du}{dx} = U.\,M, \quad \frac{du}{dy} = V.\,M \;............\;.........(2),$$

where U, V, M are functions of x and y.

If $\qquad M = 0$ ....(3),

both $\dfrac{du}{dx}$ and $\dfrac{du}{dy}$ vanish.

From equations (2) we deduce

$$\frac{d^2u}{dx^2} = U \cdot \frac{dM}{dx} + M \cdot \frac{dU}{dx} = U \cdot \frac{dM}{dx} \text{ when (3) is satisfied,}$$

$$\frac{d^2u}{dy^2} = V \cdot \frac{dM}{dy} + M \cdot \frac{dV}{dy} = V \cdot \frac{dM}{dy} \text{ when (3) is satisfied,}$$

$$\frac{d^2u}{dx\,dy} = V \cdot \frac{dM}{dx} + M \cdot \frac{dV}{dx} = V \cdot \frac{dM}{dx} \text{ when (3) is satisfied,}$$

$$\frac{d^2u}{dy\,dx} = U \cdot \frac{dM}{dy} + M \cdot \frac{dU}{dy} = U \cdot \frac{dM}{dy} \text{ when (3) is satisfied.}$$

But $\dfrac{d^2u}{dx\,dy} = \dfrac{d^2u}{dy\,dx}$ always; hence, when (3) is satisfied,

$$\left(\frac{d^2u}{dx\,dy} \right)^2 = U \cdot V \cdot \frac{dM}{dx} \cdot \frac{dM}{dy}.$$

If then A, B, C denote the values of $\dfrac{d^2u}{dx^2}$, $\dfrac{d^2u}{dx\,dy}$, and $\dfrac{d^2u}{dy^2}$ when (3) is satisfied, we have

$$A C = B^2 \qquad ....(4).$$

Now suppose that from $M = 0$, we find y in terms of x, say $y = \psi(x)$, and substitute in u; we thus make u a function of x only. On this hypothesis

$$\frac{du}{dx} = \left(\frac{du}{dx} \right) + \left(\frac{du}{dy} \right) \frac{dy}{dx}$$

$$= U \cdot M + V \cdot M \frac{dy}{dx}, \text{ by (2),}$$

$$= 0, \text{ since } M = 0 \text{ by hypothesis.}$$

Hence, this substitution of $\psi(x)$ for y has reduced u to a *constant*, since $\dfrac{du}{dx}$ vanishes without our assigning any particular value to x.

Let us now return to equations (1) and (2). Change in $\phi(x, y)$ the variables x and y to $x+h$ and $y+k$ respectively. Calling u' the new value of u, we get

$$u' = u + h\frac{du}{dx} + k\frac{du}{dy} + \frac{h^2}{\lfloor 2}\left\{\frac{d^2u}{dx^2} + \frac{2k}{h}\frac{d^2u}{dx\,dy} + \frac{k^2}{h^2}\frac{d^2u}{dy^2}\right\} + R.$$

Let us now assign to x and y any values consistent with (3), leaving however the ratio of k to h quite arbitrary, and examine whether u' becomes less or greater than u when k and h are sufficiently diminished. The coefficient of $\frac{h^2}{2}$ in the above value of u', is

$$\frac{d^2u}{dx^2} + \frac{2k}{h}\frac{d^2u}{dx\,dy} + \frac{k^2}{h^2}\frac{d^2u}{dy^2} \text{ or } A + \frac{2k}{h}B + \frac{k^2}{h^2}C.$$

Now by (4) this

$$= A\left(1 + \frac{kB}{hA}\right)^2,$$

and is therefore necessarily positive if A be positive, and necessarily negative if A be negative, *whatever be the ratio of k to h, except for that particular value of the ratio which makes the expression vanish.* Hence the conclusion will be this: if we assign to x and y values consistent with $M=0$, then when h and k are sufficiently diminished, u' is certainly less than u if $\frac{d^2u}{dx^2}$ be negative, and certainly greater than u if $\frac{d^2u}{dx^2}$ be positive, *excepting only when k has to h one particular ratio.* This latter case would require further examination, had we not *already shewn that by a certain supposition u is reduced to a constant,* so that when k has to h the *one* particular ratio, u' is ultimately neither greater than u nor less than u, but equal to it.

The whole theory may be illustrated geometrically; for example, if

$$z^2 = a^2 - x^2 - y^2 + (x\cos\alpha + y\sin\alpha)^2 \ldots\ldots\ldots(1),$$

find maxima or minima values of z;

$$z\,\frac{dz}{dx} = -x + (x\cos\alpha + y\sin\alpha)\cos\alpha$$

$$= (y\cos\alpha - x\sin\alpha)\sin\alpha,$$

$$z\,\frac{dz}{dy} = -(y\cos\alpha - x\sin\alpha)\cos\alpha\;;$$

therefore, when $\qquad y\cos\alpha - x\sin\alpha = 0$ ....(2),

$\dfrac{dz}{dx}$ and $\dfrac{dz}{dy}$ both vanish.

Under these circumstances z becomes $= \pm\,a$.

Now equation (1) represents a cylinder having its axis parallel to the plane of (x, y). Equation (2) represents a plane which passes through the axis of the cylinder, and which cuts the surface in two parallel straight lines. Along the upper straight line we have $z = a$. All points in this straight line are at the same distance from the plane of (x, y), and *at a greater distance than any points not in this straight line*. This straight line is in fact a *ridge* in the surface.

Another example may be seen in the equation

$$z^2 = 2a\,\sqrt{(x^2 + y^2)} - (x^2 + y^2).$$

This surface is that formed by the revolution of a circle about a tangent line which is the axis of z. The highest point of the circle will by revolution generate a circle, all the points of which are at the same distance from the plane of (x, y), and at a greater distance than any adjacent points of the surface.

EXAMPLES.

1. Let $\qquad u = x^2 + xy + y^2 + \dfrac{a^3}{x} + \dfrac{a^3}{y},$

$$\frac{du}{dx} = 2x + y - \frac{a^3}{x^2}, \qquad \frac{du}{dy} = 2y + x - \frac{a^3}{y^2};$$

therefore $\qquad 2x + y - \dfrac{a^3}{x^2} = 0, \qquad 2y + x - \dfrac{a^3}{y^2} = 0;$

therefore $\qquad (2x + y)\,x^2 = a^3 = (2y + x)\,y^2\;;$

therefore $\qquad 2\,(x^3 - y^3) = xy\,(y - x)\,;$

therefore $\qquad 2\,(x - y)\,(x^2 + xy + y^2) = xy\,(y - x)\,:$

either then $\qquad\qquad x = y,$

or $\qquad\qquad 2x^2 + 3xy + 2y^2 = 0.$

The latter leads to an impossible result; the former gives

$$x = y = \frac{a}{\sqrt[3]{3}}\,.$$

Also $\qquad\qquad \dfrac{d^2u}{dx^2} = 2 + \dfrac{2a^3}{x^3}\,,$

$$\frac{d^2u}{dx\,dy} = 1,$$

$$\frac{d^2u}{dy^2} = 2 + \frac{2a^3}{y^3}\,;$$

therefore $\dfrac{d^2u}{dx^2}\dfrac{d^2u}{dy^2} - \left(\dfrac{d^2u}{dx\,dy}\right)^2$ is positive when x and y have the assigned values, and $\dfrac{d^2u}{dx^2}$ is positive; hence u is then a minimum.

2. Let $\quad u = \cos x \cos \alpha + \sin x \sin \alpha \cos (y - \beta),$

$$\frac{du}{dx} = -\sin x \cos \alpha + \cos x \sin \alpha \cos (y - \beta),$$

$$\frac{du}{dy} = -\sin \alpha \sin x \sin (y - \beta).$$

Hence $\dfrac{du}{dy}$ vanishes when $y = \beta$, and then $\dfrac{du}{dx}$ becomes $\sin (\alpha - x)$, and vanishes when $x = \alpha.$

Also $\quad \dfrac{d^2u}{dx^2} = -\cos x \cos \alpha - \sin x \sin \alpha \cos (y - \beta),$

$$\frac{d^2u}{dx\,dy} = -\cos x \sin \alpha \sin (y - \beta),$$

$$\frac{d^2u}{dy^2} = -\sin \alpha \sin x \cos (y - \beta).$$

The first expression becomes -1, the second becomes 0, and the third becomes $-\sin^2 a$, when the assigned values of x and y are substituted. Hence $\dfrac{d^2u}{dx^2}\dfrac{d^2u}{dy^2} - \left(\dfrac{d^2u}{dx\,dy}\right)^2$ is positive, and u is a maximum.

3. Suppose $u = e^{-x^2-y^2}(ax^2 + by^2)$,

$$\frac{du}{dx} = 2x\,(a - ax^2 - by^2)\,e^{-x^2-y^2},$$

$$\frac{du}{dy} = 2y\,(b - ax^2 - by^2)\,e^{-x^2-y^2}.$$

Here $\dfrac{du}{dx} = 0$, and $\dfrac{du}{dy} = 0$, give as one pair of values $x = 0$, $y = 0$. And these values make

$$\frac{d^2u}{dx^2} = 2a, \quad \frac{d^2u}{dx\,dy} = 0, \quad \frac{d^2u}{dy^2} = 2b;$$

therefore u has then a minimum value.

Another pair of values is given by

$$x = 0,$$

and $\qquad\qquad b - ax^2 - by^2 = 0,$

that is, $\qquad\qquad x = 0$, and $y = \pm 1$.

With these values we have

$$\frac{d^2u}{dx^2} = 2\,(a - b)\,e^{-1}, \quad \frac{d^2u}{dx\,dy} = 0, \quad \frac{d^2u}{dy^2} = -4be^{-1}.$$

Hence, if a is less than b, we have a maximum value of u, and if a is greater than b, we have neither a maximum nor a minimum.

There is only one other solution, namely, that found by combining

$$y = 0, \text{ and } a - ax^2 - by^2 = 0\,;$$

therefore $\qquad\qquad y = 0$, and $x = \pm 1$.

Here we should find that if a is less than b, there is neither a maximum nor a minimum, and if a is greater than b, there is a maximum value of u.

If in this example $a = b$, we arrive at the anomalous case considered in Art. 235.

4. Let $u = \sin x + \sin y + \cos(x + y)$,

$$\frac{du}{dx} = \cos x - \sin(x + y),$$

$$\frac{du}{dy} = \cos y - \sin(x + y).$$

If $\dfrac{du}{dx}$ and $\dfrac{du}{dy}$ vanish, we must have

$$\cos x = \cos y = \sin(x + y).$$

These equations admit of numerous solutions. For example,

if $\qquad\qquad \cos x = \cos y,$

we have $\qquad\qquad x = y,$ as one solution.

Hence we have $\cos x = \sin 2x$

$$= 2 \sin x \cos x;$$

therefore, either $\cos x = 0$, or $\sin x = \tfrac{1}{2}$.

If we take the first, and put $x = y = \dfrac{\pi}{2}$, we have neither a maximum nor a minimum; if we put

$$x = y = \frac{3\pi}{2},$$

we obtain a minimum.

If we take $\sin x = \tfrac{1}{2}$, and put

$$x = y = \frac{\pi}{6},$$

we obtain a maximum value for u.

5.   To find a point such that the sum of the straight lines joining it with the angular points of a given triangle shall be a minimum.

Let ABC be the given triangle; let $BC = a$, $CA = b$, $AB = c$. Take any point P and draw PM perpendicular to AB; let $AM = x$, $PM = y$. Also let $AP = u$, $BP = v$, $CP = w$; the angle $APM = \theta$, $BPM = \phi$, $CPM = \psi$.

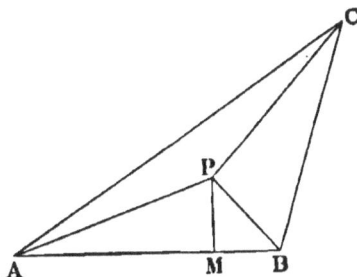

Then
$$u^2 = x^2 + y^2,$$
$$v^2 = (c - x)^2 + y^2,$$
$$w^2 = (b \cos A - x)^2 + (b \sin A - y)^2.$$

For a minimum value of $u + v + w$ we must have

$$\frac{du}{dx} + \frac{dv}{dx} + \frac{dw}{dx} = 0 \quad\dots\dots\dots\dots\dots (1),$$

and

$$\frac{du}{dy} + \frac{dv}{dy} + \frac{dw}{dy} = 0 \quad\dots\dots\dots\dots\dots (2).$$

Now
$$\frac{du}{dx} = \frac{x}{u} = \sin\theta,$$

$$\frac{dv}{dx} = -\frac{c - x}{v} = -\sin\phi,$$

$$\frac{dw}{dx} = -\frac{b \cos A - x}{w} = -\sin\psi,$$

$$\frac{du}{dy} = \frac{y}{u} = \cos\theta,$$

$$\frac{dv}{dy} = \frac{y}{v} = \cos\phi,$$

$$\frac{dw}{dy} = -\frac{b \sin A - y}{w} = \cos\psi.$$

Hence, from (1) and (2),
$$\sin\theta = \sin\phi + \sin\psi,$$
$$\cos\theta = -\cos\phi - \cos\psi.$$

Square and add; thus

$$1 = 2 + 2 \cos (\psi - \phi),$$

therefore $\cos (\psi - \phi) = -\tfrac{1}{2} = \cos 120^\circ.$

Thus the angle CPB must be 120°. Similarly it may be shewn that APB and APC must each be 120°. Hence we have the following result: describe on the sides of the given triangle segments of circles each containing an angle of 120°, and their common point of intersection is the point required.

It is obvious that there must be a point for which the proposed sum is a minimum, and therefore we need not examine the criteria depending on the second differential coefficients.

If the given triangle has an angle equal to 120°, then that angular point is the point required; if it has an angle greater than 120°, the method fails to give the solution. It may however be shewn that when the triangle has an angle greater than 120°, the vertex of the obtuse angle is the point required.

For suppose the point P *inside* the triangle and very near to the angle B of the triangle; let $PB = r$, $PBA = a$, $PBC = \gamma$; then

$$u = \sqrt{(c^2 - 2cr \cos a + r^2)}, \quad v = r,$$

$$w = \sqrt{(a^2 - 2ar \cos \gamma + r^2)}.$$

Thus neglecting squares and higher powers of r we have approximately

$$u + v + w = a + c + r - r (\cos a + \cos \gamma)$$

$$= a + c + r - 2r \cos \frac{a + \gamma}{2} \cos \frac{a - \gamma}{2}.$$

Now $2 \cos \dfrac{a + \gamma}{2}$ is less than unity if B is greater than

120°, and thus $a + c + r - 2r \cos \dfrac{a + \gamma}{2} \cos \dfrac{a - \gamma}{2}$ is *greater*

than $a + c$. And it is obvious that if P be *outside* the triangle the sum of its distances from A, B, and C is greater

T. D. C.   R than $a + c$. Therefore in passing from B to any adjacent point either inside or outside the triangle the sum of the distances is *increased;* and therefore at the point B the sum is a minimum.

The values of $\dfrac{dv}{dx}$ and $\dfrac{dv}{dy}$ take the form $\dfrac{0}{0}$ at the point B; and this is the reason that the solution failed to indicate the point B. We have already remarked in Art. 234 that a maximum or minimum value may exist corresponding to such indeterminate values of the differential coefficients.

6. Find the maximum and minimum value of

$$\frac{(hx + ky - a)\,(hx + ky - b)}{1 + x^2 + y^2}.$$

Let u denote the expression, and let v denote

$$1 + x^2 + y^2\,;$$

then $u = v^{-1}\,(hx + ky - a)\,(hx + ky - b)\,;$

$$\frac{du}{dx} = \frac{h\,(2hx + 2ky - a - b)}{v} - \frac{2x\,(hx + ky - a)\,(hx + ky - b)}{v^2},$$

$$\frac{du}{dy} = \frac{k\,(2hx + 2ky - a - b)}{v} - \frac{2y\,(hx + ky - a)\,(hx + ky - b)}{v^2}.$$

Put $\dfrac{du}{dx} = 0$, and $\dfrac{du}{dy} = 0$; thus we deduce

$$\frac{x}{h} = \frac{y}{k} = r \text{ suppose.}$$

Substitute rh for x and rk for y in $\dfrac{du}{dx} = 0$ or $\dfrac{du}{dy} = 0$; we shall obtain after reduction the following quadratic equation in r:

$$r^2\,(h^2 + k^2)\,(a + b) + 2r\,(h^2 + k^2 - ab) - (a + b) = 0\,;$$

thus the values of r are possible, and one is positive and the other is negative.

If we differentiate the values of $\dfrac{du}{dx}$ and $\dfrac{du}{dy}$, and after differentiation use the relations which arise from $\dfrac{du}{dx} = 0$ and $\dfrac{du}{dy} = 0$, we shall find

$$\frac{d^2u}{dx^2} = -\frac{h\,(2ky - a - b)}{xv} = -\frac{2k^2r - a - b}{rv},$$

$$\frac{d^2u}{dy^2} = -\frac{k\,(2hx - a - b)}{yv} = -\frac{2h^2r - a - b}{rv},$$

$$\frac{d^2u}{dx\,dy} = \frac{2hk}{v}.$$

Hence the sign of $AC - B^2$ is the same as the sign of

$$\frac{(2k^2r - a - b)\,(2h^2r - a - b)}{r^2} - 4h^2k^2,$$

and is therefore the same as the sign of

$$(a + b)^2 - 2r\,(h^2 + k^2)\,(a + b).$$

Now it may be shewn that if $a + b$ be not zero and a be not equal to b, the sign of the last expression is *positive* for both the values which r can have. For suppose $a + b$ *positive*; then we have to shew that $\dfrac{a + b}{2\,(h^2 + k^2)} - r$ is positive, that is, we have to shew that $\dfrac{a + b}{2\,(h^2 + k^2)}$ is greater than the positive root of the quadratic in r. Substitute the positive quantity $\dfrac{a + b}{2\,(h^2 + k^2)}$ for r in the expression which forms the left-hand member of the quadratic; we shall obtain a *positive* result if a and b are unequal; this shews that $\dfrac{a + b}{2\,(h^2 + k^2)}$ *is* greater than the positive root of the quadratic (*Algebra*, Art. 339). Similarly we may establish the result if $a + b$ is negative.

Hence the necessary conditions for a maximum or minimum are fulfilled.

Since $AC - B^2$ is positive A and C have the same sign, and that sign is the same as the sign of $A + C$, and therefore the same as the sign of

$$\frac{a + b - (h^2 + k^2)\, r}{r}.$$

If $a + b$ is positive this expression is positive or negative according as r is positive or negative; if $a + b$ is negative it is positive or negative according as r is negative or positive. Thus we can discriminate between the maximum and minimum value of u.

Two particular cases which have been excepted above remain to be noticed.

I. Suppose $a = b$. Here we shall have

$$\frac{du}{dx} = 2v^{-2}\,(hx + ky - a)\,\{hv - x\,(hx + ky - a)\},$$

$$\frac{du}{dy} = 2v^{-2}\,(hx + ky - a)\,\{kv - y\,(hx + ky - a)\}.$$

If we suppose $hx + ky - a = 0$ we arrive at the case discussed in Art. 235, in which there is not strictly a maximum or minimum. If we take the other factors in $\frac{du}{dx}$ and $\frac{du}{dy}$ and put

$$hv - x\,(hx + ky - a) = 0 \text{ and } kv - y\,(hx + ky - a) = 0,$$

we shall obtain

$$x = -\frac{h}{a}, \quad y = -\frac{k}{a};$$

these values will be found to make u a maximum.

The quadratic equation for r, when $a = b$, has for its roots

$$r = \frac{a}{h^2 + k^2} \text{ or } r = -\frac{1}{a};$$

the former value leads to values of x and y which satisfy $hx + ky - a = 0$; the latter leads to the values

$$x = -\frac{h}{a}, \quad y = -\frac{k}{a}.$$

II. Suppose $a + b = 0$. The original investigation becomes inapplicable; it may be shewn that the only values of

x and y which make $\dfrac{du}{dx}$ and $\dfrac{du}{dy}$ vanish are $x = 0$, $y = 0$; and these give a minimum value to u.

7. Find the maximum value of $x^3 y^2 (6 - x - y)$.

Result. Maximum when $x = 3$, $y = 2$.

8. If $u = (2ax - x^2)(2by - y^2)$, find its maximum or minimum value.

Result. $x = a$, $y = b$, make u a maximum.

9. If $u = x^4 + y^4 - 2x^2 + 4xy - 2y^2$, shew that when $x = 0$, and $y = 0$, u is neither a maximum nor minimum; when $x = \pm \sqrt{2}$, and $y = \mp \sqrt{2}$, u is a minimum.

10. If $u = y^4 - 8y^3 + 18y^2 - 8y + x^3 - 3x^2 - 3x$, then $3 + 4\sqrt{2}$ is a maximum value of u and $-6 - 4\sqrt{2}$ is a minimum value of u.

11. If $u = x^2 + xy + y^2 - ax - by$, then $\frac{1}{3}(ab - a^2 - b^2)$ is a minimum value of u.

12. Divide a number n into three parts, x, y, and z, such that $\dfrac{xy}{2} + \dfrac{xz}{3} + \dfrac{yz}{4}$ shall be a maximum or minimum, and determine which it is.

Result. $\dfrac{x}{21} = \dfrac{y}{20} = \dfrac{z}{6}$ a maximum.

13. If $u = x^3 + y^3 + 3axy$, then a^3 is a maximum value of u.

14. Find the maximum or minimum of $x(x^2 + y^2) - 3axy$.

15. Find the maximum or minimum of $\dfrac{1 + x^2 + y^2}{1 - ax - by}$.

Result. $\dfrac{x}{a} = \dfrac{y}{b} = \dfrac{1 \pm \sqrt{(1 + a^2 + b^2)}}{a^2 + b^2}$;

with the upper sign there is a maximum, with the lower a minimum.

16. If $u = \sqrt{\{(c - x)(c - y)(x + y - c)\}}$, shew that it is a maximum when $x = y = \dfrac{2c}{3}$.

17. Shew that $\dfrac{a + bx + cy}{\sqrt{(1 + x^2 + y^2)}}$ is a maximum when $x = \dfrac{b}{a}$, $y = \dfrac{c}{a}$.

18. Shew that $xe^{y + x \sin y}$ has neither a maximum nor a minimum.

19. Find the minimum value of $x + y + z$, subject to the condition

$$\frac{a}{x} + \frac{b}{y} + \frac{c}{z} = 1.$$

Result. When $\dfrac{x}{\sqrt{a}} = \dfrac{y}{\sqrt{b}} = \dfrac{z}{\sqrt{c}} = \sqrt{a} + \sqrt{b} + \sqrt{c}$.

20. Find the minimum value of $x^p y^q z^r$ subject to the same condition as in the preceding Example.

Result. When $\dfrac{px}{a} = \dfrac{qy}{b} = \dfrac{rz}{c} = p + q + r$.

21. Having given the three sides of a triangle, find a point within it, such that, if perpendiculars be drawn from it to the sides, their continued product shall be a maximum. Shew that straight lines joining this point with the corners of the given triangle will divide it into three equal triangles.

22. Find the maximum value of xyz subject to the condition

$$\frac{x^2}{a^2} + \frac{y^2}{b^2} + \frac{z^2}{c^2} = 1.$$

Result. $\dfrac{abc}{3\sqrt{3}}$.

23. Determine a point within a triangle, such that the sum of the squares on the distances from the three sides is a minimum.

Result. If p, q, r, be the perpendiculars on the sides a, b, c, respectively, then

$$\frac{p}{a} = \frac{q}{b} = \frac{r}{c} = \frac{2 \text{ area of triangle}}{a^2 + b^2 + c^2}.$$

24. Determine a point within a triangle such that the sum of the squares on the distances from the three angles is a minimum.

   Result. The centre of gravity of the triangle.

25. Through a point within a triangle three straight lines are drawn parallel to the sides dividing the triangle into three parallelograms and three triangles: shew that the sum of these triangles is least when the straight lines are drawn through the centre of gravity of the triangle.

26. A triangular space is to be diminished by fencing off the corners, each fence being circular and having the nearest corner as centre: shew how to leave the greatest possible central space with a given length of fence.

   Result. The radii of the circular fences are equal.

27. Given the sum of the three edges of a rectangular parallelepiped, find its form that its surface may be a maximum.

28. In a given sphere inscribe a rectangular parallelepiped whose volume is a maximum. Also one whose surface is a maximum.

   Result. A cube.

29. Of all triangles of the same perimeter find that which will generate the greatest double cone by revolving about a side.

   Result. The fixed side must be two-thirds of each of the other sides of the triangle.

30. A rectangular parallelepiped is so constructed that a plane which passes through three of its corners, but through no edge, contains a point whose distances from the three faces adjacent to one of the other corners are given. Shew that the shortest diagonal which such a parallelepiped can have, is $(a^{\frac{2}{3}} + b^{\frac{2}{3}} + c^{\frac{2}{3}})^{\frac{3}{2}}$, where a, b, c are the given distances.

CHAPTER XVI.

MAXIMA AND MINIMA VALUES OF A FUNCTION OF SEVERAL VARIABLES.

236. LET $u = \phi(x, y, z)$ be a function of three independent variables, of which we require the maxima and minima values. By an investigation similar to that in Art. 224,

$$\phi(x + h, y + k, z + l) - \phi(x, y, z)$$

$$= h\frac{du}{dx} + k\frac{du}{dy} + l\frac{du}{dz}$$

$$+ \frac{h^2}{2}\frac{d^2u}{dx^2} + \frac{k^2}{2}\frac{d^2u}{dy^2} + \frac{l^2}{2}\frac{d^2u}{dz^2} + kl\frac{d^2u}{dy\,dz} + hl\frac{d^2u}{dx\,dz} + hk\frac{d^2u}{dx\,dy}$$

$$+ R;$$

where R is a function involving powers and products of h, k, l of the third degree, which may be expressed for abbreviation by

$$\frac{1}{\underline{3}}\left\{h\frac{d}{dx} + k\frac{d}{dy} + l\frac{d}{dz}\right\}^3 v,$$

v denoting $\qquad \phi(x + \theta h, \ y + \theta k, \ z + \theta l).$

If we make h, k, l small enough, the sign of

$$\phi(x + h, \ y + k, \ z + l) - \phi(x, y, z)$$

will in general depend upon that of the terms involving only the first powers of h, k, l; hence, to ensure a maximum or minimum, we must have

$$h\frac{du}{dx} + k\frac{du}{dy} + l\frac{du}{dz} = 0,$$

and therefore, since h, k, l are independent,

$$\frac{du}{dx} = 0, \quad \frac{du}{dy} = 0, \quad \frac{du}{dz} = 0.$$

Let values of x, y, z be found from these equations, and when these values are substituted in $\dfrac{d^2u}{dx^2}$, $\dfrac{d^2u}{dy^2}$, ..., let

$$\frac{d^2u}{dx^2} = A, \quad \frac{d^2u}{dy^2} = B, \quad \frac{d^2u}{dz^2} = C,$$

$$\frac{d^2u}{dy\,dz} = A', \quad \frac{d^2u}{dx\,dz} = B', \quad \frac{d^2u}{dx\,dy} = C'.$$

The sign of

$$\phi\,(x+h,\ y+k,\ z+l) - \phi\,(x,\ y,\ z)$$

can, with the values of x, y, z just found, be made to depend on that of

$$Ah^2 + Bk^2 + Cl^2 + 2A'kl + 2B'hl + 2C'hk \,\ldots\ldots\ldots(1).$$

Hence, that u may have a maximum or minimum value, the expression (1) must retain the same sign, whatever be the signs and values of h, k, l comprised between zero and fixed finite limits. If we put

$$h = sl, \quad k = tl,$$

it follows that

$$As^2 + Bt^2 + C + 2A't + 2B's + 2C'st \,\ldots\ldots\ldots\ldots(2),$$

must be of invariable sign, whatever be the signs and values of s and t.   Multiply (2) by A, and rearrange the terms; then

$$(As + B' + C't)^2 + (AB - C'^2)\,t^2 + 2\,(AA' - B'C')\,t + AC - B'^2$$
$$\ldots\ldots\ldots\ldots\ldots (3),$$

must retain an invariable sign.

Hence, $(AB - C'^2)\,t^2 + 2\,(AA' - B'C')\,t + AC - B'^2$ must be incapable of becoming negative; therefore

$$AB - C'^2 \text{ must be positive, and} \ldots\ldots\ldots\ldots (4),$$

$(AA' - B'C')^2$ less than $(AB - C'^2)\,(AC - B'^2)$ ...... (5);

(4) and (5) are the conditions that must be satisfied in order that u may be a maximum or minimum.   Conversely, if they

are satisfied, u is a maximum or minimum; for then (3) is necessarily positive, therefore (2) has always the same sign as A, and u is a maximum if A be negative, and a minimum if A be positive.

Hence the necessary and sufficient conditions for the existence of a maximum or minimum value of a function u of three independent variables, are, that the values of x, y, z drawn from

$$\frac{du}{dx} = 0, \quad \frac{du}{dy} = 0, \quad \frac{du}{dz} = 0,$$

should make $\quad \dfrac{d^2u}{dx^2}\dfrac{d^2u}{dy^2} - \left(\dfrac{d^2u}{dx\,dy}\right)^2$ positive,

and $\quad \left(\dfrac{d^2u}{dx^2}\dfrac{d^2u}{dy\,dz} - \dfrac{d^2u}{dx\,dy}\dfrac{d^2u}{dx\,dz}\right)^2$ less than

$$\left\{\frac{d^2u}{dx^2}\frac{d^2u}{dy^2} - \left(\frac{d^2u}{dx\,dy}\right)^2\right\}\left\{\frac{d^2u}{dx^2}\frac{d^2u}{dz^2} - \left(\frac{d^2u}{dx\,dz}\right)^2\right\}.$$

It follows of course from these conditions, that

$$\frac{d^2u}{dx^2}\frac{d^2u}{dz^2} - \left(\frac{d^2u}{dx\,dz}\right)^2 \text{ must be positive,}$$

and thus $\dfrac{d^2u}{dx^2}$, $\dfrac{d^2u}{dy^2}$, $\dfrac{d^2u}{dz^2}$ must all have the same sign, and u is a maximum if that sign be negative, and a minimum if it be positive.

From the conditions (4) and (5), we should conjecture by the principle of symmetry, that $BC - A'^2$ will also be positive if (4) and (5) hold. This is easily verified, for from (5) we find that

$$A\left\{ABC + 2A'B'C' - AA'^2 - BB'^2 - CC'^2\right\}$$

is positive, and therefore, since A and B have the same sign, by (4)

$$B\left\{ABC + 2A'B'C' - AA'^2 - BB'^2 - CC'^2\right\}$$

is positive, and therefore

$$(BB' - A'C')^2 \text{ is less than } (BC - A'^2)(BA - C'^2),$$

from which it follows that $BC - A'^2$ is positive.

237. Example. Let $u = \dfrac{xyz}{(a+x)\,(x+y)\,(y+z)\,(z+b)}$;

$$\frac{du}{dx} = \frac{yz\,(ay - x^2)}{(a+x)^2\,(x+y)^2\,(y+z)\,(z+b)} = \frac{u\,(ay - x^2)}{x\,(a+x)\,(x+y)}.$$

Similarly, $\quad \dfrac{du}{dy} = \dfrac{u\,(xz - y^2)}{y\,(x+y)\,(y+z)}$,

$$\frac{du}{dz} = \frac{u\,(by - z^2)}{z\,(y+z)\,(z+b)}.$$

Hence, if $ay - x^2 = 0$, $xz - y^2 = 0$, and $by - z^2 = 0$, u *may* be a maximum or minimum : these equations give

$$\frac{x}{a} = \frac{y}{x} = \frac{z}{y} = \frac{b}{z} ;$$

therefore each of these fractions $= \sqrt[4]{\left(\dfrac{x}{a} \cdot \dfrac{y}{x} \cdot \dfrac{z}{y} \cdot \dfrac{b}{z} \right)}$ or $\sqrt[4]{\dfrac{b}{a}}$.

Call this r ; then

$$x = ar, \ y = xr = ar^2, \ z = yr = ar^3.$$

Proceeding to the second differential coefficients of u, we have

$$\frac{d^2u}{dx^2} = - \frac{2xu}{x\,(a+x)\,(x+y)} + \&c.,$$

the terms included in the &c. being such as vanish when the specific values are assigned to x, y, z.

Hence $\quad A = - \dfrac{2u}{a^2 r\,(1+r)^2} = - \dfrac{2}{a^3 r\,(1+r)^6}.$

Similarly B, C, ... can be found, and we shall finally arrive at the result that u is a maximum.

238. Suppose it required to determine the maxima and minima values of a function $\phi\,(x, y, z, \ldots)$ of m variables, these variables being connected by n equations, of which the general form is

$$F_r\,(x, y, z, \ldots) = 0 \ \ldots\ldots\ldots\ldots\ldots(1).$$

The m variables involved in ϕ are of course not all independent, since by means of the given equations n of them may be expressed in terms of the remaining $m - n$. The simplest *theoretical* method of investigating the maxima and minima values of ϕ would be to express by means of the given equations the values of n of the variables in terms of the rest, and to substitute these values in ϕ; thus ϕ would become a function of $m - n$ independent variables, and we might proceed to ascertain its maxima and minima values in the manner already given for functions of one, two, or three independent variables. But this method would be often impracticable on account of the difficulty of solving the given equations, and the following method is therefore adopted.

Suppose $x, y, z...$ all functions of some new variable t, of which consequently ϕ becomes a function. Put for shortness

$$\frac{dx}{dt} = Dx, \quad \frac{dy}{dt} = Dy, \quad \frac{dz}{dt} = Dz,...$$

then
$$\frac{d\phi}{dt} = \frac{d\phi}{dx} Dx + \frac{d\phi}{dy} Dy + \frac{d\phi}{dz} Dz + \ldots\ldots\ldots(2).$$

From the n given equations (1) we deduce

$$\left.\begin{array}{l} \dfrac{dF_1}{dx} Dx + \dfrac{dF_1}{dy} Dy + \dfrac{dF_1}{dz} Dz + ... = 0 \\[2mm] \dfrac{dF_2}{dx} Dx + \dfrac{dF_2}{dy} Dy + \dfrac{dF_2}{dz} Dz + ... = 0 \\[2mm] \cdots\cdots\cdots\cdots\cdots\cdots\cdots\cdots \\[2mm] \dfrac{dF_n}{dx} Dx + \dfrac{dF_n}{dy} Dy + \dfrac{dF_n}{dz} Dz + ... = 0 \end{array}\right\} \ldots\ldots(3).$$

By solving the linear equations (3) we can express n of the quantities $Dx, Dy, Dz...$ in terms of the remaining $m - n$. Substitute these values in (2), then only $m - n$ of the quantities $Dx, Dy, Dz ...$ remain, and we have a result which may be written

$$\frac{d\phi}{dt} = X . Dx + Y . Dy + Z . Dz ... + Q . Dq \ldots\ldots(4),$$

where $X, Y, Z, ...$ do not involve any of the quantities $Dx, Dy, Dz, ...$ Since, consistently with the given equations, we may consider the $m - n$ quantities $Dx, Dy, Dz, ...$

to be quite arbitrary, it follows, in the same manner as in Art. 232, that if ϕ is to be a maximum or minimum, we must have

$$X = 0, \quad Y = 0, \quad Z = 0, \quad \ldots\ldots Q = 0 \ldots\ldots\ldots (5).$$

From these $m - n$ equations, combined with the n given equations, we can find the values of the variables for which ϕ *may* be a maximum or minimum. To determine whether ϕ *is* a maximum or minimum we must express $\dfrac{d^2\phi}{dt^2}$. From (4), *with the use of* (5), we have

$$\frac{d^2\phi}{dt^2} = \frac{dX}{dx}(Dx)^2 + \frac{dY}{dx} Dx\, Dy + \frac{dZ}{dx} Dx\, Dz + \ldots$$

$$+ \frac{dX}{dy} Dy\, Dx + \frac{dY}{dy}(Dy)^2 + \ldots$$

$$+ \ldots\ldots\ldots\ldots\ldots\ldots\ldots\ldots\ldots\ldots$$

We should then examine whether the above expression retains an invariable sign, when the specific values of the variables x, y, z, ... are used, whatever be the arbitrary values assigned to Dx, Dy, Dz, ... If it does, then ϕ is a maximum if that sign be negative, and a minimum if it be positive.

239. The practical solution of any example according to the above theory is facilitated by making use of *indeterminate multipliers*. Multiply the first of equations (3) by λ_1, the second by λ_2, ... the nth by λ_n, the values of λ_1, λ_2, ... λ_n being at present undetermined. Add the results to (2), then we may write

$$\frac{d\phi}{dt} = \left\{ \frac{d\phi}{dx} + \lambda_1 \frac{dF_1}{dx} + \lambda_2 \frac{dF_2}{dx} + \lambda_3 \frac{dF_3}{dx} + \ldots \right\} Dx$$

$$+ \left\{ \frac{d\phi}{dy} + \lambda_1 \frac{dF_1}{dy} + \lambda_2 \frac{dF_2}{dy} + \lambda_3 \frac{dF_3}{dy} + \ldots \right\} Dy$$

$$+ \left\{ \frac{d\phi}{dz} + \lambda_1 \frac{dF_1}{dz} + \lambda_2 \frac{dF_2}{dz} + \lambda_3 \frac{dF_3}{dz} + \ldots \right\} Dz$$

$$+ \ldots\ldots\ldots\ldots\ldots\ldots\ldots\ldots \ldots\ldots\ldots (6).$$

If we equate the coefficients of n of the quantities Dx, $Dy,...$ to zero, we shall arrive at n equations for determining $\lambda_1, \lambda_2, ... \lambda_n$. Substitute these values of $\lambda_1, \lambda_2, ... \lambda_n$, in the remaining terms of (6), and $\dfrac{d\phi}{dt}$ takes the form given in (4); we must therefore equate to zero the coefficients of the remaining $m - n$ of the quantities Dx, $Dy, ...$ Hence we have the rule: "Equate to zero the coefficients of *every one* of the quantities Dx, $Dy, ...$ in (6); the m equations thus found, together with the n given equations, will enable us to eliminate the n quantities $\lambda_1, \lambda_2 ... \lambda_n$, and to find the values of the quantities $x, y, z...$"

240. The concluding part of the theory in Art. 238, in which we are directed to examine the sign of $\dfrac{d^2\phi}{dt^2}$, frequently becomes in practice excessively complicated. In fact the examples of this method are generally such as allow us to predict that a maximum or minimum *must* exist, and to dispense with the second part of the investigation.

EXAMPLES.

1. Find the maximum or minimum value of
$$x^2 + y^2 + z^2,$$
subject to the conditions
$$\left. \begin{array}{l} ax + by + cz - 1 = 0, \\ a'x + b'y + c'z - 1 = 0, \end{array} \right\} \quad ....(1).$$

Putting ϕ for $x^2 + y^2 + z^2$, we have
$$\frac{d\phi}{dt} = 2x\,Dx + 2y\,Dy + 2z\,Dz.$$

Also from equations (1),
$$\left. \begin{array}{l} a\,Dx + b\,Dy + c\,Dz = 0, \\ a'\,Dx + b'\,Dy + c'\,Dz = 0, \end{array} \right\} \quad .... (2).$$

Hence, multiplying equations (2) by λ_1 and λ_2 respectively, we may put

$$\frac{d\phi}{dt} = (2x + \lambda_1 a + \lambda_2 a') \, Dx + (2y + \lambda_1 b + \lambda_2 b') \, Dy$$

$$+ (2z + \lambda_1 c + \lambda_2 c') \, Dz.$$

Therefore
$$\begin{rcases} 2x + \lambda_1 a + \lambda_2 a' = 0, \\ 2y + \lambda_1 b + \lambda_2 b' = 0, \\ 2z + \lambda_1 c + \lambda_2 c' = 0, \end{rcases} \dots\dots\dots (3).$$

Multiply equations (3) by a, b, c, respectively and add; then we have, by (1),

$$2 + \lambda_1 (a^2 + b^2 + c^2) + \lambda_2 (aa' + bb' + cc') = 0 \dots\dots (4).$$

Similarly,

$$2 + \lambda_2 (a'^2 + b'^2 + c'^2) + \lambda_1 (aa' + bb' + cc') = 0 \dots\dots (5).$$

Equations (4) and (5) determine λ_1 and λ_2, and then by (3) we find x, y, z. Also multiplying (3) by x, y, z, respectively and adding, we have

$$2\phi + \lambda_1 + \lambda_2 = 0,$$

which finds ϕ. This is the solution of the following question in Geometry of Three Dimensions: "In the line of inter-section of two given planes to find the nearest point to the origin of co-ordinates." From the nature of the question it is evident there must be a minimum value of ϕ.

2. Determine the greatest quadrilateral which can be formed with the four given sides a, β, γ, δ, taken in this order.

Let x denote the angle between a and β, y the angle between γ and δ. The area of the figure is $\frac{1}{2}(a\beta \sin x + \gamma\delta \sin y)$, therefore we may put

$$\phi(x, y) = a\beta \sin x + \gamma\delta \sin y \dots\dots\dots (1).$$

If we draw a diagonal of the figure from the intersection of β and γ to the intersection of a and δ, we have from the two different values which can be found for the length of this diagonal, $a^2 + \beta^2 - 2a\beta \cos x = \gamma^2 + \delta^2 - 2\gamma\delta \cos y$.

Thus $\quad a^2 + \beta^2 - 2a\beta \cos x - \gamma^2 - \delta^2 + 2\gamma\delta \cos y = 0 \dots\dots (2).$

From (1) and (2),

$$\frac{d\phi}{dt} = \alpha\beta \cos x\, Dx + \gamma\delta \cos y\, Dy \dots\dots\dots(3),$$

$$0 = \alpha\beta \sin x\, Dx - \gamma\delta \sin y\, Dy \dots\dots\dots(4),$$

therefore $\quad \dfrac{d\phi}{dt} = \alpha\beta \left\{ \cos x + \dfrac{\sin x \cos y}{\sin y} \right\} Dx \dots \dots\dots(5).$

Hence, since the coefficient of Dx must vanish,

$$\sin (x + y) = 0.$$

Therefore $x + y$ must be zero, or some multiple of π; the only solution applicable to the present question is

$$x + y = \pi \dots\dots\dots\dots\dots\dots(6).$$

Hence $\cos y = -\cos x$: substituting this value of $\cos y$ in equation (2), we have

$$\cos x = \frac{\alpha^2 + \beta^2 - \gamma^2 - \delta^2}{2(\alpha\beta + \gamma\delta)}.$$

Since by (5) $\quad \dfrac{d\phi}{dt} = \dfrac{\alpha\beta \sin(x+y)}{\sin y} Dx,$

we have, neglecting such terms as vanish, by (6),

$$\frac{d^2\phi}{dt^2} = \frac{\alpha\beta \cos(x+y)}{\sin y} Dx\,(Dx + Dy),$$

which, by means of (4) and (6), becomes

$$-\frac{\alpha\beta}{\sin y}\left(1 + \frac{\alpha\beta}{\gamma\delta}\right)(Dx)^2.$$

Hence, since $\dfrac{d^2\phi}{dt^2}$ is *negative*, we have found a maximum value of ϕ, namely, when the sum of two opposite angles of the figure is equal to two right angles.

Thus the quadrilateral must be capable of being inscribed in a circle.

It may now be shewn that when all the sides of a rectilineal figure are given the area is greatest when the figure can be inscribed in a circle. For let PQ, QR, RS, ST represent any four consecutive sides. Then, by what we have just seen, P, Q, R, S must lie on the circumference of a circle: for otherwise the area could be increased, by leaving the rest of the figure unchanged, and shifting PQ, QR, RS until the points P, Q, R, S did lie on the circumference of a circle. Similarly Q, R, S, T must lie on the circumference of a circle. And *this circle is the same as the former circle*, for it is the circle described round the triangle QRS. In this manner we shew that when the area is greatest the figure must have all its angular points on the circumference of a circle.

Suppose an indefinitely large number of consecutive sides of the figure to become indefinitely small: then the corresponding portion of the boundary of the greatest area becomes an arc of the circle of which the remaining sides are chords. Hence we obtain the following general result: if an area is to be bounded by given straight rods and strings, the area is greatest when the strings are all arcs of the same circle, and the straight rods all chords of that circle.

The following problem is analogous to that which we have been considering. Required to determine the greatest area which can be inclosed by a quadrilateral *three* of whose sides are given.

Let a, b, c denote the lengths of the three given sides, taken in order of contiguity. Let θ denote the angle between the sides b and c, and ϕ the angle between the side a and that diagonal which passes through the angle between a and b. Then the area of the figure is

$$\frac{1}{2} bc \sin \theta + \frac{1}{2} a \sqrt{(b^2 + c^2 - 2bc \cos \theta)} \sin \phi.$$

This is a function of the two independent variables θ and ϕ; but we can obtain the result which we require without going through the usual process for finding the maximum value of a function of two independent variables. For we see that to ensure the greatest area ϕ must be a right angle. In a similar manner we might shew that the angle between the side c and that diagonal which passes through the angle between b and c must also be a right angle. Hence the quadrilateral figure must be capable of being inscribed in a circle of which the side not given must be the diameter.

T. D. C.   S

It may now be shewn that when all the sides of a recti-lineal figure are given except one, the area is greatest when the figure can be inscribed in a circle of which the side not given is the diameter.

For let QR represent the side not given, and PQ an adjacent side. Then the whole figure must be capable of being inscribed in a circle : for otherwise the area could be increased without changing the length of any side. And the angle QPR must be a right angle: for otherwise we might leave PQ and PR unchanged, and by changing QR replace the triangle PQR by a larger triangle. And since QPR is a right angle, QR is a diameter of the circle surrounding the figure.

3.   Find the maximum and minimum value of u^2 when

$$u^2 = a^2x^2 + b^2y^2 + c^2z^2 \quad\text{.....................} (1),$$

while

$$x^2 + y^2 + z^2 = 1 \quad\text{.......................} (2),$$

and

$$lx + my + nz = 0 \quad\text{....... ....} (3).$$

From (1), (2), and (3), we deduce

$$0 = a^2x\,Dx + b^2y\,Dy + c^2z\,Dz \quad\text{........... ....} (4),$$

$$0 = x\,Dx + y\,Dy + z\,Dz \quad\text{.....................} (5),$$

$$0 = l\,Dx + m\,Dy + n\,Dz \quad\text{.....................} (6).$$

Multiply (5) by λ_1 and (6) by λ_2 and add to (4); then equate to zero the coefficients of Dx, Dy, Dz; thus

$$a^2x + \lambda_1 x + \lambda_2 l = 0 \quad\text{.....................} (7),$$

$$b^2y + \lambda_1 y + \lambda_2 m = 0 \quad\text{.......... ....} (8),$$

$$c^2z + \lambda_1 z + \lambda_2 n = 0 \quad\text{.....................} (9).$$

Multiply (7) by x, (8) by y, and (9) by z, and add; then by (2) and (3),

$$a^2x^2 + b^2y^2 + c^2z^2 + \lambda_1 = 0.$$

Hence

$$\lambda_1 = -u^2.$$

Therefore, from (7), (8), and (9),

$$x = \frac{\lambda_2 l}{u^2 - a^2},$$

$$y = \frac{\lambda_2 m}{u^2 - b^2},$$

$$z = \frac{\lambda_2 n}{u^2 - c^2},$$

and thus, from equation (3),

$$\frac{l^2}{u^2 - a^2} + \frac{m^2}{u^2 - b^2} + \frac{n^2}{u^2 - c^2} = 0.$$

This equation is a quadratic in u^2, from which two values of u^2 can be determined, one of which will be a maximum and the other a minimum. It is obvious that a maximum and a minimum value of u^2 must exist, for x, y, z, cannot all vanish simultaneously, and no one of them can be greater than unity; hence u^2 must lie between the limits 0 and $a^2 + b^2 + c^2$.

4. Find the values of x, y, z, when x^4yz^2 is a maximum or minimum, subject to the condition

$$a^2x^2 + 2by^3 + z^4 = c^4.$$

We have, putting u for x^4yz^2,

$$4x^3yz^2 Dx + x^4z^2 Dy + 2x^4yz Dz = 0,$$

or $\qquad u\left\{\dfrac{4Dx}{x} + \dfrac{Dy}{y} + \dfrac{2Dz}{z}\right\} = 0.$

Also $\qquad a^2x Dx + 3by^2 Dy + 2z^3 Dz = 0.$

Therefore $\qquad \dfrac{4}{x} + \lambda a^2 x = 0,$

$$\frac{1}{y} + 3\lambda by^2 = 0,$$

$$\frac{1}{z} + \lambda z^3 = 0.$$

Multiply the first of these equations by x, the second by $\frac{2y}{3}$, and the third by z, and add; then

$$\frac{17}{3} + \lambda \{a^2x^2 + 2by^3 + z^4\} = 0;$$

therefore $\qquad \lambda = -\dfrac{17}{3c^4}.$

Hence $\qquad a^2x^2 = \dfrac{12c^4}{17}, \qquad by^3 = \dfrac{c^4}{17}, \qquad z^4 = \dfrac{3c^4}{17}.$

5. To find the maximum and minimum value of r^2 when

$$r^2 = (x - a)^2 + (y - \beta)^2 + (z - \gamma)^2,$$

the variables and constants being connected by the equations

$$\frac{x^2}{a^2} + \frac{y^2}{b^2} + \frac{z^2}{c^2} = 1 \dots\dots\dots\dots\dots(1),$$

$$lx + my + nz = p \dots\dots\dots\dots\dots(2),$$

$$la + m\beta + n\gamma = p \dots\dots\dots\dots\dots(3),$$

$$\frac{a}{a^2l} = \frac{\beta}{b^2m} = \frac{\gamma}{c^2n} \dots\dots\dots\dots\dots(4).$$

[The student who is acquainted with Geometry of Three Dimensions will see that (1) is the equation to an ellipsoid, and (2) is the equation to a plane; a, β, γ are the co-ordinates of the centre of the curve of intersection of the plane and the ellipsoid, and r is the radius vector drawn from the centre of this curve to any point of the curve.]

Since r^2 is to be a maximum or minimum, we have

$$(x - a) Dx + (y - \beta) Dy + (z - \gamma) Dz = 0 \dots\dots(5);$$

also from (1) and (2)

$$\frac{xDx}{a^2} + \frac{yDy}{b^2} + \frac{zDz}{c^2} = 0 \dots\dots\dots\dots(6),$$

$$lDx + mDy + nDz = 0 \dots\dots\dots\dots(7).$$

Multiply (6) by λ, and (7) by μ, and add to (5); then equate to zero the coefficients of Dx, Dy, and Dz; thus,

$$x - a + \frac{\lambda x}{a^2} + \mu l = 0 \quad \ldots\ldots\ldots\ldots\ldots (8),$$

$$y - \beta + \frac{\lambda y}{b^2} + \mu m = 0 \quad \ldots\ldots\ldots\ldots\ldots (9),$$

$$z - \gamma + \frac{\lambda z}{c^2} + \mu n = 0 \quad \ldots\ldots\ldots\ldots\ldots (10).$$

Multiply (8), (9), and (10) by $x - a$, $y - \beta$, and $z - \gamma$ respectively, and add; thus by (2) and (3)

$$r^2 + \lambda \left\{ \frac{x(x-a)}{a^2} + \frac{y(y-\beta)}{b^2} + \frac{z(z-\gamma)}{c^2} \right\} = 0,$$

that is,
$$r^2 + \lambda \left\{ 1 - \frac{xa}{a^2} - \frac{y\beta}{b^2} - \frac{z\gamma}{c^2} \right\} = 0 \ldots (11).$$

Now by (4)

$$\frac{a}{a^2 l} = \frac{\beta}{b^2 m} = \frac{\gamma}{c^2 n} = \frac{al + \beta m + \gamma n}{a^2 l^2 + b^2 m^2 + c^2 n^2} = \frac{p}{a^2 l^2 + b^2 m^2 + c^2 n^2},$$

thus (11) becomes with the help of (2)

$$r^2 = \lambda \left\{ \frac{p^2}{a^2 l^2 + b^2 m^2 + c^2 n^2} - 1 \right\} = \lambda k \text{ say.}$$

Thus (8), (9), and (10) may be written

$$\left. \begin{aligned} x \left(1 + \frac{r^2}{ka^2} \right) &= a - \mu l, \\ y \left(1 + \frac{r^2}{kb^2} \right) &= \beta - \mu m, \\ z \left(1 + \frac{r^2}{kc^2} \right) &= \gamma - \mu n, \end{aligned} \right\} \quad \ldots\ldots\ldots\ldots\ldots (12).$$

By substituting the values of x, y, and z from these in (2), we obtain

$$\frac{lka^2(a - \mu l)}{ka^2 + r^2} + \frac{mkb^2(\beta - \mu m)}{kb^2 + r^2} + \frac{nkc^2(\gamma - \mu n)}{kc^2 + r^2} = p;$$

also
$$la \quad + \quad m\beta \quad + \quad n\gamma \quad = p.$$

By subtraction

$$\frac{a^2l^2\left(k\mu+\dfrac{r^2\alpha}{a^2l}\right)}{ka^2+r^2}+\frac{b^2m^2\left(k\mu+\dfrac{r^2\beta}{b^2m}\right)}{kb^2+r^2}+\frac{c^2n^2\left(k\mu+\dfrac{r^2\gamma}{c^2n}\right)}{kc^2+r^2}=0.$$

Now $k\mu+\dfrac{r^2\alpha}{a^2l}$, $k\mu+\dfrac{r^2\beta}{b^2m}$, and $k\mu+\dfrac{r^2\gamma}{c^2n}$ are of equal value by (4); and this value cannot be zero, because then by (12) we should obtain the inadmissible results $x=\alpha$, $y=\beta$, $z=\gamma$. Hence dividing out we have

$$\frac{a^2l^2}{ka^2+r^2}+\frac{b^2m^2}{kb^2+r^2}+\frac{c^2n^2}{kc^2+r^2}=0.$$

This quadratic will give two values of r^2, one will be the maximum value of r^2 and the other the minimum value.

$\Big[$The *product* of the values of r^2 will be

$$\frac{k^2a^2b^2c^2\,(l^2+m^2+n^2)}{a^2l^2+b^2m^2+c^2n^2}\;;$$

and π times the square root of this product is the *area* of the curve of intersection of the ellipsoid and plane; hence taking the *positive* value of the square root we have for the area

$$\frac{\pi abc\,(a^2l^2+b^2m^2+c^2n^2-p^2)\,\sqrt{(l^2+m^2+n^2)}}{(a^2l^2+b^2m^2+c^2n^2)^{\frac{3}{2}}}\Big].$$

6. Find the maximum or minimum value of u when $u=x^2y^3z^4$, and $2x+3y+4z=a$.

$\qquad$ *Result.* $\left(\dfrac{a}{9}\right)^9$ is a maximum value.

7. Find the minimum value of u from the equation

$$u=x^2+y^2+z^2+\;\ldots..,$$

the variables being connected by the equation

$$ax+by+cz+\ldots\ldots=k.$$

$\qquad$ *Result.* $u=\dfrac{k^2}{a^2+b^2+c^2+\ldots\ldots}.$

8. Find the minimum value of

$$x^2 + y^2 + z^2 + x - 2z - xy.$$

   Result. $x = -\frac{2}{3}$, $y = -\frac{1}{3}$, $z = 1$.

9. Find the minimum value of $x^4 + y^4 + z^4$, where $xyz = c^3$.

10. If x, y, z are the angles of a triangle, find the values of x, y, z which make $\sin^m x \sin^n y \sin^p z$ a maximum.

11. Find the maximum or minimum value of $x^p y^q z^r$ subject to the condition $lx + my + nz = a$. Hence find the parallelepiped of maximum volume which has for its three edges the axes of x, y, z, and has the intersection of its opposite edges in a given plane.

12. If $ax^2 + bxy + cy^2 = f$, and $r^2 = x^2 + y^2$, shew that the maximum and minimum values of r^2 are given by the equation

$$(b^2 - 4ac)\, r^4 + 4f(a + c)\, r^2 - 4f^2 = 0.$$

13. Find the maximum value of

$$(ax + by + cz)\, e^{-a^2 x^2 - \beta^2 y^2 - \gamma^2 z^2}.$$

   Result. $x = \dfrac{\mu a}{a^2}$, $\qquad y = \dfrac{\mu b}{\beta^2}$, $\qquad z = \dfrac{\mu c}{\gamma^2}$;

   where $\qquad \dfrac{1}{\mu} = \sqrt{2} \sqrt{\left(\dfrac{a^2}{a^2} + \dfrac{b^2}{\beta^2} + \dfrac{c^2}{\gamma^2} \right)}.$

14. A given volume V of metal is to be formed into a rectangular vessel; the sides of the vessel are to be of a given thickness a, and there is to be no lid. Determine the shape of the vessel so that it may have a maximum capacity.

   Result. If x, y, and z, are the external length, breadth, and depth;

$$x = y = a + \left(\frac{V - a^3}{3a} \right)^{\frac{1}{3}}; \qquad z = \frac{x}{2}.$$

15.   If $r^2 = x^2 + y^2 + z^2$, where

$$ax^2 + by^2 + cz^2 + 2a'yz + 2b'zx + 2c'xy = 1,$$

and $$lx + my + nz = 0,$$

find the maximum and minimum values of r^2.

Result.   They are determined by the equation

$$l^2\left(b - \frac{1}{r^2}\right)\left(c - \frac{1}{r^2}\right) + m^2\left(c - \frac{1}{r^2}\right)\left(a - \frac{1}{r^2}\right) + n^2\left(a - \frac{1}{r^2}\right)\left(b - \frac{1}{r^2}\right)$$

$$- 2mna'\left(a - \frac{1}{r^2}\right) - 2nlb'\left(b - \frac{1}{r^2}\right) - 2lmc'\left(c - \frac{1}{r^2}\right)$$

$$+ 2mnl'c' + 2nlc'a' + 2lma'b' - l^2a'^2 - m^2b'^2 - n^2c'^2 = 0.$$

241. WE may make use of differentiation in order to eliminate from an equation involving variables and constants one or more of the constants. For example, let

$$(y - b)^2 + (x - a)^2 - c^2 = 0 \quad \ldots\ldots\ldots\ldots\ldots(1).$$

Differentiate three times, giving

$$(y - b)\frac{dy}{dx} + x - a = 0 \ldots\ldots\ldots\ldots\ldots\ldots(2),$$

$$(y - b)\frac{d^2y}{dx^2} + \left(\frac{dy}{dx}\right)^2 + 1 = 0 \ldots\ldots\ldots\ldots(3),$$

$$(y - b)\frac{d^3y}{dx^3} + 3\frac{dy}{dx}\frac{d^2y}{dx^2} = 0 \ldots\ldots\ldots\ldots(4).$$

From these four equations we may deduce an equation free from the three constants : we have

$$\frac{dy}{dx} = -\frac{x - a}{y - b},$$

$$\frac{d^2y}{dx^2} = -\frac{(x - a)^2 + (y - b)^2}{(y - b)^3} = -\frac{c^2}{(y - b)^3},$$

$$\frac{d^3y}{dx^3} = -\frac{3\frac{dy}{dx}\frac{d^2y}{dx^2}}{y - b} = -\frac{3c^2(x - a)}{(y - b)^5}.$$

Hence $\quad \left\{1 + \left(\frac{dy}{dx}\right)^2\right\}\frac{d^3y}{dx^3} - 3\frac{dy}{dx}\left(\frac{d^2y}{dx^2}\right)^2 = 0 \ldots\ldots\ldots\ldots(5).$

242. In general, if we have an equation between x and y and n arbitrary constants, and we differentiate m times successively, we have $m+1$ equations between which we can eliminate m constants, and this will give a result involving $\dfrac{d^m y}{dx^m}$ and inferior differential coefficients of y. There will also be $n-m$ constants in the resulting equation; and as we can choose at pleasure the m constants we eliminate, we can form as many resulting equations containing $n-m$ constants, as the number of combinations that can be formed out of n things taken m at a time; that is,

$$\frac{n\,(n-1)\,\ldots\,(n-m+1)}{\underline{|m}}.$$

Each of these resulting equations is called *a differential equation of the mth order*, $\dfrac{d^m y}{dx^m}$ being the highest differential coefficient of y which occurs in it.

When the original equation is differentiated n times successively, we have $n+1$ equations, between which *all* the constants can be eliminated, giving us a differential equation of the nth order.

243. If we recur to the example in Art. 241, we have for one of the three differential equations of the first order,

$$(y-b)\,\frac{dy}{dx} + x - a = 0.$$

If we find a from this equation in terms of x, y, b, and $\dfrac{dy}{dx}$, and substitute in the given equation, we obtain another differential equation of the first order. If we find b in terms of x, y, a, and $\dfrac{dy}{dx}$, and substitute in the given equation, we obtain the remaining differential equation of the first order.

The three differential equations of the second order which can be obtained by combining equations (1), (2), and (3) of Art. 241, are

$$b = y + \frac{1 + \left(\dfrac{dy}{dx}\right)^2}{\dfrac{d^2y}{dx^2}} ,$$

$$a = x - \frac{\dfrac{dy}{dx}\left\{1 + \left(\dfrac{dy}{dx}\right)^2\right\}}{\dfrac{d^2y}{dx^2}} ,$$

$$c^2 = \frac{\left\{1 + \left(\dfrac{dy}{dx}\right)^2\right\}^3}{\left(\dfrac{d^2y}{dx^2}\right)^2} .$$

It will be found on trial, that if we take *any one* of the differential equations of the first order, and differentiate twice, we shall obtain the same result if we eliminate the two constants involved in these three equations, as we have already found in equation (5) of Art. 241. Also, if we take *any one* of the differential equations of the second order, differentiate once, and eliminate the constant involved in these two equations, we shall still arrive at the equation (5) of Art. 241.

244. The process by which, as in the preceding Article, we may deduce differential equations by differentiation and elimination of constants, has not in itself much interest or value. But the method of passing from the differential equations to the primitive equation from which they were deduced, forms a most important branch of mathematics. In fact all investigations in physical science lead to *differential equations*, which must be solved before we can be said to understand the subject we are considering. We do not enter here on the solution of differential equations, but it is usual, in treatises on the Differential Calculus to devote some space to the consideration of the formation of such equations by elimination, as this process throws light on the methods to be adopted for their solution.

245. Not only constants may be eliminated, but functions. Suppose, for example,

$$y = \sin x \, ;$$

then

$$\frac{dy}{dx} = \cos x$$

$$= \sqrt{(1 - y^2)} \, ;$$

therefore

$$y^2 + \left(\frac{dy}{dx}\right)^2 - 1 = 0.$$

Hence the function $\sin x$ has been eliminated.

Again, let

$$y = \tan (x + y) \, ;$$

therefore

$$\frac{dy}{dx} = \{1 + \tan^2 (x + y)\} \left\{1 + \frac{dy}{dx}\right\}$$

$$= (1 + y^2) \left(1 + \frac{dy}{dx}\right).$$

Hence $\tan (x + y)$ has been eliminated.

In these examples *given* functions have been eliminated: we proceed to cases in which *unknown* functions are eliminated.

246. Suppose $z = \phi \left(\dfrac{x}{y}\right)$, where ϕ denotes some function the form of which is not given, and which is therefore called an *arbitrary function*. The variables x and y are supposed independent.

Put $\dfrac{x}{y} = t$; then

$$z = \phi (t),$$

$$\frac{dz}{dx} = \phi' (t) \frac{dt}{dx} = \frac{1}{y} \, \phi' (t),$$

$$\frac{dz}{dy} = \phi' (t) \frac{dt}{dy} = -\frac{x}{y^2} \, \phi' (t) \, ;$$

therefore

$$x \frac{dz}{dx} + y \frac{dz}{dy} = 0.$$

Hence this last equation is true *whatever* be the form of the function ϕ; for example, if $z = \log\left(\dfrac{x}{y}\right)$, or $z = \sin\dfrac{x}{y}$, or $z = \left(\dfrac{x}{y}\right)^m$, in each case we have that equation subsisting.

247. Suppose $u = \phi(v)$, where u and v are known functions of x, y, and z, but the *form* of ϕ is not given. The variables x and y are supposed independent. If we differentiate both members of the equation with respect to x and y successively, we have

$$\frac{du}{dx} + \frac{du}{dz}\frac{dz}{dx} = \phi'(v)\left\{\frac{dv}{dx} + \frac{dv}{dz}\frac{dz}{dx}\right\},$$

$$\frac{du}{dy} + \frac{du}{dz}\frac{dz}{dy} = \phi'(v)\left\{\frac{dv}{dy} + \frac{dv\cdot dz}{dz\,dy}\right\}.$$

Therefore, whatever be the form of ϕ,

$$\left(\frac{du}{dx} + \frac{du}{dz}\frac{dz}{dx}\right)\left(\frac{dv}{dy} + \frac{dv}{dz}\frac{dz}{dy}\right) = \left(\frac{du}{dy} + \frac{du}{dz}\frac{dz}{dy}\right)\left(\frac{dv}{dx} + \frac{dv}{dz}\frac{dz}{dx}\right).$$

In other words we have eliminated the arbitrary function ϕ.

248. Suppose

$$\alpha_1 = f_1(x,\ y,\ z),$$
$$\alpha_2 = f_2(x,\ y,\ z),$$

two known functions of x, y, z, which enter into an equation.

$$F\{x,\ y,\ z,\ \phi_1(\alpha_1),\ \phi_2(\alpha_2)\} = 0 \dots\dots\dots\dots(1),$$

ϕ_1 and ϕ_2 being *arbitrary functions*. If we form the equations

$$\frac{dF}{dx} = 0, \qquad \frac{dF}{dy} = 0 \dots\dots\dots\dots\dots\dots(2),$$

$$\frac{d^2F}{dx^2} = 0, \qquad \frac{d^2F}{dx\,dy} = 0, \qquad \frac{d^2F}{dy^2} = 0 \dots\dots\dots(3),$$

we introduce the unknown functions

$$\phi_1'(\alpha_1), \quad \phi_2'(\alpha_2), \quad \phi_1''(\alpha_1), \quad \phi_2''(\alpha_2),$$

and these, with $\phi_1(\alpha_1)$, and $\phi_2(\alpha_2)$, form *six* quantities to be eliminated between the *six* equations (1), (2), (3). This cannot *generally* be effected. Proceeding to the equations

$$\frac{d^3F}{dx^3} = 0, \quad \frac{d^3F}{dx^2\,dy} = 0, \quad \frac{d^3F}{dx\,dy^2} = 0, \quad \frac{d^3F}{dy^3} = 0 \dots (4_j,$$

we shall introduce only *two* new unknown functions, namely $\phi_1'''(\alpha_1)$ and $\phi_2'''(\alpha_2)$. Hence we can obtain by elimination an equation between z and its partial differential coefficients with respect to y and x of the third order inclusive, which will be free from the functions $\phi_1(\alpha_1)$ and $\phi_2(c_2)$ and their derived functions. Since we have ten equations and eight quantities to be eliminated, two resulting equations can generally be obtained.

249. We say that generally, in the case supposed in the preceding Article, we *cannot eliminate* the arbitrary functions by proceeding as far as the second derived equations. Cases however occur, in which, owing to the forms of α_1 and α_2, this elimination *can* be effected ; for example, suppose

$$z = \phi_1(x + ay) + \phi_2(x - ay).$$

Here $$\frac{dz}{dx} = \phi_1'(x + ay) + \phi_2'(x - ay),$$

$$\frac{dz}{dy} = a\phi_1'(x + ay) - a\phi_2'(x - ay),$$

$$\frac{d^2z}{dx^2} = \phi_1''(x + ay) + \phi_2''(x - ay),$$

$$\frac{d^2z}{dy^2} = a^2\phi_1''(x + ay) + a^2\phi_2''(x - ay) ;$$

therefore $$\frac{d^2z}{dy^2} = a^2 \frac{d^2z}{dx^2}.$$

250. Suppose we have an equation between three variables of the form

$$F\{x, y, z, \phi_1(\alpha_1), \phi_2(\alpha_2), \dots\dots \phi_n(\alpha_n)\} = 0,$$

involving n *arbitrary functions* $\phi_1, \phi_2, \dots\dots \phi_n$ of the n *known functions* $\alpha_1, \alpha_2, \dots\dots \alpha_n$ respectively.

If we proceed in the manner of Art. 248, and deduce from this equation all its derived equations up to those of the mth order inclusive, we shall obtain

$$1 + 2 + 3 + 4 + \ldots\ldots (m + 1)$$

equations, that is $\dfrac{(m + 1)\,(m + 2)}{2}$ equations.

The number of unknown functions will be $(m + 1)\,n$, and therefore, that we may be able to eliminate the arbitrary functions, we must have generally

$$\dfrac{(m + 1)\,(m + 2)}{2} \text{ greater than } (m + 1)\,n,$$

therefore $\qquad \dfrac{m + 2}{2}$ greater than n ;

therefore $\qquad m = 2n - 1$ at least.

If $m = 2n - 1$, the number of equations will be $n\,(2n + 1)$, and the number of functions to be eliminated, $2n^2$; hence, there will be generally n resulting equations.

251. Suppose however that the known functions $\alpha_1,\ \alpha_2, \ldots \alpha_n$, are all the *same* function ; we shall find that it will be sufficient to proceed to the derived equations of the nth order inclusive, in order to be able to eliminate the arbitrary functions. For let

$$F\{x,\ y,\ z,\ \phi_1(\alpha),\ \phi_2(\alpha),\ \ldots\ldots \phi_n(\alpha)\} = 0 ;$$

differentiate with respect to x and y ; thus

$$\frac{dF}{dx} + \frac{dF}{dz}\frac{dz}{dx} + \frac{dF}{d\alpha}\left(\frac{d\alpha}{dx} + \frac{d\alpha}{dz}\frac{dz}{dx}\right) = 0,$$

$$\frac{dF}{dy} + \frac{dF}{dz}\frac{dz}{dy} + \frac{dF}{d\alpha}\left(\frac{d\alpha}{dy} + \frac{d\alpha}{dz}\frac{dz}{dy}\right) = 0.$$

Eliminate $\dfrac{dF}{d\alpha}$; thus

$$\frac{\dfrac{dF}{dx} + \dfrac{dF}{dz}\dfrac{dz}{dx}}{\dfrac{dF}{dy} + \dfrac{dF}{dz}\dfrac{dz}{dy}} = \frac{\dfrac{d\alpha}{dx} + \dfrac{d\alpha}{dz}\dfrac{dz}{dx}}{\dfrac{d\alpha}{dy} + \dfrac{d\alpha}{dz}\dfrac{dz}{dy}}.$$

This result involves only the same arbitrary functions as the original equation, namely,

$$\phi_1(a), \ \phi_2(a), \ldots \ldots \phi_n(a) \ ;$$

it also involves $\dfrac{dz}{dx}$ and $\dfrac{dz}{dy}$; we may denote it by

$$F\left\{x, \ y, \ z, \ \frac{dz}{dx}, \ \frac{dz}{dy}, \ \phi_1(a), \ \phi_2(a), \ \ldots \ldots \phi_n(a)\right\} = 0.$$

Differentiate this equation with respect to x and y as before; thus we obtain another result which involves only the same arbitrary functions as the original equation. By continuing the process until we introduce the differential coefficients of z of the nth order, we find that we have on the whole $n+1$ equations, from which the n arbitrary functions may be eliminated.

252. Suppose we have two equations of the form

$$F\{x, \ y, \ z, \ a, \ \phi_1(a), \ \phi_2(a), \ \ldots \ldots \phi_n(a)\} = 0,$$
$$f\{x, \ y, \ z, \ a, \ \phi_1(a), \ \phi_2(a), \ \ldots \ldots \phi_n(a)\} = 0,$$

where a is an *unknown* function of x, y, and z, and ϕ_1, $\phi_2, \ldots \phi_n$ denote arbitrary functions; and let it be required to eliminate a and the arbitrary functions of a. In this case also we shall find that it will be sufficient to proceed to the derived equations of the nth order inclusive.

As in the preceding Article we differentiate the first equation and thus obtain

$$\frac{\dfrac{dF}{dx} + \dfrac{dF}{dz}\dfrac{dz}{dx}}{\dfrac{dF}{dy} + \dfrac{dF}{dz}\dfrac{dz}{dy}} = \frac{\dfrac{da}{dx} + \dfrac{da}{dz}\dfrac{dz}{dx}}{\dfrac{da}{dy} + \dfrac{da}{dz}\dfrac{dz}{dy}} \ldots \ldots \ldots \ldots \ldots (1).$$

But as a is not a *known* function the right-hand member of (1) is not a known function. But from the second of the given equations we obtain in the same manner

$$\frac{\dfrac{da}{dx} + \dfrac{da}{dz}\dfrac{dz}{dx}}{\dfrac{da}{dy} + \dfrac{da}{dz}\dfrac{dz}{dy}} = \frac{\dfrac{df}{dx} + \dfrac{df}{dz}\dfrac{dz}{dx}}{\dfrac{df}{dy} + \dfrac{df}{dz}\dfrac{dz}{dy}} \ldots \ldots \ldots \ldots (2) ;$$

so that we can replace the right-hand member of (1) by the right-hand member of (2). Hence, as in the preceding Article, we obtain a result which we may write

$$F_1\left\{x,\, y,\, z,\, \frac{dz}{dx},\, \frac{dz}{dy},\, a,\, \phi_1(a),\, \phi_2(a),\, \ldots\ldots\, \phi_n(a)\right\} = 0.$$

Differentiate this again and make use of (1) or of (2); thus we obtain another result involving only the same arbitrary quantities. By continuing the process until we introduce the differential coefficients of z of the nth order, we find that we have on the whole $n + 2$ equations from which we may eliminate a and the n arbitrary functions of a.

253. As an example of the preceding, suppose only *one* arbitrary function $\phi(a)$. The given equations become

$$f\{x,\, y,\, z,\, a,\, \phi(a)\} = 0,$$
$$F\{x,\, y,\, z,\, a,\, \phi(a)\} = 0.$$

Differentiate each with respect to x and y. We thus have six equations, from which we may eliminate

$$a,\, \frac{da}{dx},\, \frac{da}{dy},\, \phi(a),\, \text{and}\, \phi'(a),$$

leaving one equation between

$$x,\, y,\, z,\, \frac{dz}{dx},\, \text{and}\, \frac{dz}{dy}.$$

254. The conclusions obtained in Arts. 251, 252 are due to Dr Salmon; see his *Geometry of Three Dimensions*, Chapter XII. It had been usual to overestimate the number of derived equations which are necessary in order to effect the elimination in Art. 252. Suppose, for example, there are two arbitrary functions so that

$$F\{x,\, y,\, z,\, a,\, \phi_1(a),\, \phi_2(a)\} = 0,$$
$$f\{x,\, y,\, z,\, a,\, \phi_1(a),\, \phi_2(a)\} = 0;$$

then it might appear that by forming the derived equations up to the second order inclusive, as in Art. 248, we should obtain *twelve* equations, but have *twelve* quantities to eliminate, namely

$$a, \frac{da}{dx}, \frac{d\alpha}{dy}, \frac{d^2\alpha}{dx^2}, \frac{d^2\alpha}{dx\,dy}, \frac{d^2\alpha}{dy^2},$$

$$\phi_1(\alpha), \ \phi_1'(\alpha), \ \phi_1''(\alpha), \ \phi_2(\alpha), \ \phi_2'(\alpha), \ \phi_2''(\alpha).$$

But the fact is that by adopting the method of Art. 252, we have $\phi_1'(\alpha)$ and $\phi_2'(\alpha)$ occurring in such a way that they *disappear together* in our elimination of $\frac{dF}{d\alpha}$ and $\frac{df}{d\alpha}$. Hence it happens that we are able to effect the required elimination without proceeding beyond the derived equations of the second order.

255.   In particular cases the elimination may be effected without proceeding to so many differentiations as the general theory indicates.   Suppose, for example, that we have three arbitrary functions, we should generally have to form the derived equations of the *third* order by Art. 252.   But if the three arbitrary functions are so related, that

$$\phi_2(\alpha) = \phi_1'(\alpha),$$
$$\phi_3(\alpha) = \phi_1''(\alpha),$$

the given equations take the form

$$F\{x, y, z, \alpha, \phi_1(\alpha), \ \phi_1'(\alpha), \phi_1''(\alpha)\} = 0,$$
$$f\{x, y, z, \alpha, \phi_1(\alpha), \ \phi_1'(\alpha), \phi_1''(\alpha)\} = 0 ;$$

and by proceeding as far as the *second* derived equations, we obtain *twelve* equations and *eleven* quantities to be eliminated, namely

$$a, \frac{d\alpha}{dx}, \frac{da}{dy}, \frac{d^2\alpha}{dx^2}, \frac{d^2\alpha}{dy^2}, \frac{d^2\alpha}{dx\,dy},$$

$$\phi_1(\alpha), \ \phi_1'(\alpha), \ \phi_1''(\alpha), \ \phi_1'''(\alpha), \ \phi_1''''(\alpha).$$

Thus we can deduce *one* resulting equation involving x, y, z, and partial differential coefficients of z up to those of the second order inclusive.

256. We will give one case in which more than three variables are involved. Suppose

$$F\{u,\ x,\ y,\ z,\ \phi\ (a,\ \beta)\} = 0 \ldots\ldots\ldots\ldots\ldots (1),$$

in which $\phi\ (a,\ \beta)$ designates an arbitrary function of the two quantities a and β, which are themselves both known functions of u, x, y, and z. If we differentiate (1) with respect to each of the independent variables x, y, z, we obtain three equations

$$\frac{dF}{dx} = 0, \qquad \frac{dF}{dy} = 0, \qquad \frac{dF}{dz} = 0 \ldots\ldots\ldots\ldots (2).$$

In these equations, besides the arbitrary function ϕ, we have its two derived functions $\dfrac{d\phi}{da}$ and $\dfrac{d\phi}{d\beta}$. Hence, between the four equations (1) and (2), we shall be able to eliminate the three arbitrary functions, and arrive at an equation involving u, x, y, z, $\dfrac{du}{dx}$, $\dfrac{du}{dy}$, and $\dfrac{du}{dz}$.

EXAMPLES.

1. Eliminate the constant from

$$xy - c = (x + y)\ (c - 1).$$

$$Result. \quad (x^2 + x + 1)\frac{dy}{dx} + y^2 + y + 1 = 0.$$

2. Eliminate e^x and $\cos x$ from

$$y - e^x \cos x = 0.$$

$$Result. \quad \frac{d^2y}{dx^2} - 2\frac{dy}{dx} + 2y = 0.$$

3. If $x^2 - 2ay - a^2 - b = 0$, shew that

$$x\frac{d^2y}{dx^2} - \frac{dy}{dx} = 0.$$

4. If $y = ae^{mx} \sin nx$, shew that

$$\frac{d^2y}{dx^2} - 2m\frac{dy}{dx} + (m^2 + n^2)\ y = 0.$$

T 2

5.   If $y = a \sin x + b \cos x$, then

$$\frac{d^2y}{dx^2} + y = 0.$$

6.   Eliminate the exponentials from

$$xy = ae^x + be^{-x}.$$

                       Result. $x\dfrac{d^2y}{dx^2} + 2\dfrac{dy}{dx} - xy = 0.$

7.   Eliminate the constants from

$$y^2 + bx^2 = a.$$

             Result. $xy\dfrac{d^2y}{dx^2} + x\left(\dfrac{dy}{dx}\right)^2 - y\dfrac{dy}{dx} = 0.$

8.   Eliminate the constants and exponentials from

$$ae^y + be^{-y} = fe^x + ge^{-x}.$$

      Result. $\left\{\dfrac{d^3y}{dx^3} + \left(\dfrac{dy}{dx}\right)^3 - \dfrac{dy}{dx}\right\}\left\{\left(\dfrac{dy}{dx}\right)^2 - 1\right\} = 3\dfrac{dy}{dx}\left(\dfrac{d^2y}{dx^2}\right)^2.$

9.   If $(x+y)(c + \log x) = xe^{\frac{y}{x}}$, then

$$xy\frac{dy}{dx} - y^2 = (x+y)^2 e^{-\frac{y}{x}}.$$

10.   Eliminate a and b from

$$y = \frac{a}{\sqrt{x}} \cos\left(\frac{\sqrt{7}}{2} \log x + b\right).$$

               Result. $x^2\dfrac{d^2y}{dx^2} + 2x\dfrac{dy}{dx} + 2y = 0.$

11.   Eliminate the constants from the equation

$$1 = ax^2 + 2bxy + cy^2.$$

            Result. $\dfrac{d^3y}{dx^3}\left(y - x\dfrac{dy}{dx}\right) + 3x\left(\dfrac{d^2y}{dx^2}\right)^2 = 0.$

12.   If $\dfrac{1}{z} - \dfrac{1}{x} = f\left(\dfrac{1}{y} - \dfrac{1}{x}\right)$, shew that

$$x^2\frac{dz}{dx} + y^2\frac{dz}{dy} = z^2.$$

13. If $\log z = \phi (ay + bx) + \psi (ay - bx)$, then

$$a^2 \left\{ z \frac{d^2z}{dx^2} - \left(\frac{dz}{dx} \right)^2 \right\} = b^2 \left\{ z \frac{d^2z}{dy^2} - \left(\frac{dz}{dy} \right)^2 \right\}.$$

14. If $z = e^{\frac{z}{x+y}} \phi (x+y)$, then $\dfrac{dz}{dx} - \dfrac{dz}{dy} = \dfrac{z}{x+y}$.

15. If $z = \phi (e^x \sin y)$, then $\sin y \dfrac{dz}{dy} = \cos y \dfrac{dz}{dx}$.

16. If $\dfrac{dz}{dx} + f(z) \dfrac{dz}{dy} = 0$, then

$$\frac{d^2z}{dx^2} \left(\frac{dz}{dy} \right)^2 - 2 \frac{d^2z}{dx\,dy} \frac{dz}{dx} \frac{dz}{dy} + \frac{d^2z}{dy^2} \left(\frac{dz}{dx} \right)^2 = 0.$$

17. If $z = f\left(\dfrac{y - nz}{x - mz} \right)$, then

$$(x - mz) \frac{dz}{dx} + (y - nz) \frac{dz}{dy} = 0.$$

18. Eliminate the arbitrary functions from

$$z = x\phi (ax + by) + y\psi (ax + by).$$

Result. $a^2 \dfrac{d^2z}{dy^2} - 2ab \dfrac{d^2z}{dx\,dy} + b^2 \dfrac{d^2z}{dx^2} = 0.$

19. Eliminate the arbitrary and exponential functions from

$$u = e^{nx} F (x+y) + e^{-nx} f (x-y).$$

Result. $\dfrac{d^2u}{dx^2} = n^2 u + 2n \dfrac{du}{dy} + \dfrac{d^2u}{dy^2}.$

20. Eliminate the circular and logarithmic functions from

(1) $y = \sin \log x$, (2) $y = \log \sin x$.

Results. (1) $x^2 \dfrac{d^2y}{dx^2} + x \dfrac{dy}{dx} + y = 0$, (2) $\dfrac{d^2y}{dx^2} + \left(\dfrac{dy}{dx} \right)^2 + 1 = 0$

21. If $z = \dfrac{y^2}{2} + \phi\left(\dfrac{1}{x} + \log y\right)$, then

$$y\,\frac{dz}{dy} + x^2\,\frac{dz}{dx} = y^2.$$

22. Eliminate the functions from $y = xf(z) + \phi(z)$.

Result. The same as in Example 16.

23. If $z + mx + ny = f\{(x-a)^2 + (y-b)^2 + (z-c)^2\}$, then

$$\{y - b - n(z-c)\}\,\frac{dz}{dx} - \{x - a - m(z-c)\}\,\frac{dz}{dy} = n(x-a) - m(y-b).$$

24. If $z = x^2(ax + by) + \phi(y^2 + x^2) + \psi(y^2 - x^2)$,

then $\dfrac{1}{x^2}\dfrac{d^2z}{dx^2} - \dfrac{1}{y^2}\dfrac{d^2z}{dy^2} - \dfrac{1}{x^3}\dfrac{dz}{dx} + \dfrac{1}{y^3}\dfrac{dz}{dy} = \dfrac{3a}{x} + \dfrac{bx^2}{y^3}$.

25. If $z = \phi\{x + f(y)\}$, then

$$\frac{d^2z}{dx\,dy}\frac{dz}{dx} - \frac{dz}{dy}\frac{d^2z}{dx^2} = 0.$$

26. Eliminate the arbitrary functions from

$$z = f\left(\frac{x}{y}\right) \cdot \phi\left(\frac{x^2 - y^2}{x^2 + y^2}\right) \cdot \chi(xy).$$

Result. $\left(x^2\dfrac{d^2z}{dx^2} - y^2\dfrac{d^2z}{dy^2}\right)z + \left(z - x\dfrac{dz}{dx} - y\dfrac{dz}{dy}\right)\left(x\dfrac{dz}{dx} - y\dfrac{dz}{dy}\right) = 0.$

27. If $u + y + z = x^2 f\{x(u-y),\ x(y-z)\}$, then

$$x\,\frac{du}{dx} + (u+z)\,\frac{du}{dy} + (u+y)\,\frac{du}{dz} = y + z.$$

28. If $u = \phi\left\{F(y^2 - xz),\ f\left(\dfrac{3zy}{x} - \dfrac{2y^3}{x^2} - t\right)\right\}$, then

$$x\,\frac{du}{dy} + 2y\,\frac{du}{dz} + 3z\,\frac{du}{dt} = 0.$$

29. If $u = xyz \cdot F\{f_1(x^2 + y^2 + z^2), \ f_2(xy + xz + yz)\}$, then

$$(y - z)\frac{du}{dx} + (z - x)\frac{du}{dy} + (x - y)\frac{du}{dz}$$

$$= u\left(\frac{y - z}{x} + \frac{z - x}{y} + \frac{x - y}{z}\right).$$

30. Eliminate z from the equations

$$\frac{d^2x}{dz^2} = \phi(x, y), \qquad\qquad \frac{d^2y}{dz^2} = \psi(x, y).$$

Result. $2\phi(x, y) = \dfrac{d}{dx} \dfrac{\psi(x, y) - \phi(x, y)\dfrac{dy}{dx}}{\dfrac{d^2y}{dx^2}}$.

31. Eliminate the arbitrary functions from

$$z = x^n f\left(\frac{y}{x}\right) + \frac{1}{y^n}\, \phi\left(\frac{x}{y}\right).$$

Result. $x^2 \dfrac{d^2z}{dx^2} + 2xy\dfrac{d^2z}{dx\,dy} + y^2\dfrac{d^2z}{dy^2} + x\dfrac{dz}{dx} + y\dfrac{dz}{dy} = n^2 z$.

32. Shew how to eliminate the n arbitrary functions from

$$z = \phi_1\left(\frac{y}{x}\right) + x\phi_2\left(\frac{y}{x}\right) \ldots\ldots + x^{n-1}\phi_n\left(\frac{y}{x}\right).$$

CHAPTER XVIII.

TANGENT AND NORMAL TO A PLANE CURVE.

257. DEFINITION. Let P, Q, be two points on a curve, and suppose a straight line drawn through them; the limiting position of this straight line, as Q moves along the curve and approaches indefinitely near to P, is called *the tangent to the curve at the point P.*

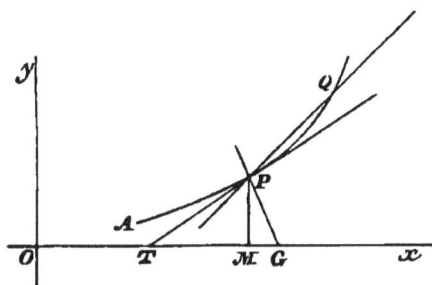

To find the equation to the tangent at a given point of a curve.

Let x, y, be the co-ordinates of the given point P,

$x + \Delta x$, $y + \Delta y$, the co-ordinates of another point Q on the curve.

Then x', y', being current co-ordinates, we have for the equation to the straight line PQ,

$$y' - y = \frac{y + \Delta y - y}{x + \Delta x - x}(x' - x),$$

that is,

$$y' - y = \frac{\Delta y}{\Delta x}(x' - x).$$

Now let Q approach indefinitely near to P; the limit of

$\frac{\Delta y}{\Delta x}$ is $\frac{dy}{dx}$, and the equation to the tangent at P is

$$y' - y = \frac{dy}{dx}(x' - x).$$

258. DEFINITION. The normal to a curve at any point is a straight line drawn through that point at right angles to the tangent at that point.

To find the equation to the normal at any point of a curve. Since the equation to the tangent at the point (x, y) is

$$y' - y = \frac{dy}{dx}(x' - x),$$

the equation to the normal at the same point is

$$y' - y = -\frac{1}{\frac{dy}{dx}}(x' - x),$$

supposing the axes rectangular.

259. Let the tangent and normal at the point P meet the axis of x at the points T and G respectively; draw the ordinate PM; then

MT is called the *subtangent*,

MG is called the *subnormal*.

Now $\qquad \dfrac{MP}{MT} = $ the tangent of PTx

$$= \frac{dy}{dx};$$

therefore $\qquad MT = \dfrac{y}{\frac{dy}{dx}} = y\,\dfrac{dx}{dy}.$

Also $\dfrac{MG}{MP} = $ tangent of $GPM = $ tangent of PTx

$$= \frac{dy}{dx};$$

therefore $\qquad MG = y\,\dfrac{dy}{dx}.$

In these expressions for the subnormal and subtangent, it is to be observed that the subtangent is measured from M towards the *left*, and the subnormal is measured from M towards the *right*. If in any curve $y \dfrac{dy}{dx}$ is a *negative* quantity, it indicates that G lies to the *left* of M, and, as in that case $y \dfrac{dx}{dy}$ is also negative, T lies to the *right* of M.

260. In the equation to the tangent put $y' = 0$, then

$$x' = x - y \frac{dx}{dy};$$

this therefore is the value of OT.

Similarly, if we put $x' = 0$, we find

$$y' = y - x \frac{dy}{dx},$$

which gives the ordinate of the point where the tangent meets the axis of y.

261. The length of the perpendicular from the origin on the tangent is, by the usual formulæ of analytical geometry,

$$\frac{x \dfrac{dy}{dx} - y}{\pm \sqrt{\left\{ 1 + \left(\dfrac{dy}{dx} \right)^2 \right\}}}.$$

262. If the equation to a curve be given in the form $\phi(x, y) = 0$, we have, by Art. 177,

$$\frac{dy}{dx} = - \frac{\left(\dfrac{d\phi}{dx} \right)}{\left(\dfrac{d\phi}{dy} \right)}.$$

Thus the equation to the tangent becomes

$$(y' - y) \left(\frac{d\phi}{dy} \right) + (x' - x) \left(\frac{d\phi}{dx} \right) = 0,$$

and the equation to the normal becomes

$$(y' - y) \left(\frac{d\phi}{dx} \right) - (x' - x) \left(\frac{d\phi}{dy} \right) = 0.$$

The length of the perpendicular on the tangent from the origin is, neglecting the sign,

$$\frac{x\left(\frac{d\phi}{dx}\right) + y\left(\frac{d\phi}{dy}\right)}{\sqrt{\left\{\left(\frac{d\phi}{dx}\right)^2 + \left(\frac{d\phi}{dy}\right)^2\right\}}}.$$

263. It is sometimes convenient to determine a curve by the two equations

$$y = \psi\,(t), \qquad x = \chi\,(t),$$

so that x and y are both functions of a variable t, by eliminating which between the given equations, a result of the usual form $y = f(x)$ may be obtained. With this supposition, we have

$$\frac{dy}{dx} = \frac{\frac{dy}{dt}}{\frac{dx}{dt}}.$$

Hence the equation to the tangent becomes

$$(y' - y)\frac{dx}{dt} = (x' - x)\frac{dy}{dt},$$

and the equation to the normal becomes

$$(y' - y)\frac{dy}{dt} = -(x' - x)\frac{dx}{dt}.$$

In the figure we have supposed the axes rectangular; if they are oblique no change is made either in the investigation of the equation to the tangent or in the result. But the equation to the normal is

$$y' - y = -\frac{1 + \cos\omega\,\frac{dy}{dx}}{\cos\omega + \frac{dy}{dx}}\,(x' - x),$$

where ω is the angle of inclination of the axes.

264. Example (1). The general equation to a curve of the second order is

$$Ay^2 + 2Bxy + Cx^2 + 2Dy + 2Ex + F = 0.$$

Hence, by Art. 262, the equation to the tangent at the point (x, y) is

$$(y' - y)(Ay + Bx + D) + (x' - x)(Cx + By + E) = 0,$$

which reduces by means of the given equation to

$$y'(Ay + Bx + D) + x'(Cx + By + E) + Dy + Ex + F = 0.$$

Example (2).   Suppose the equation to the curve to be

$$y = ae^{\frac{x}{c}},$$

therefore

$$\frac{dy}{dx} = \frac{a}{c} e^{\frac{x}{c}} = \frac{y}{c} ;$$

and the equation to the tangent becomes

$$y' - y = \frac{y}{c}(x' - x).$$

The subtangent $MT = \dfrac{y}{\dfrac{dy}{dx}} = c$, and is therefore constant in this curve which is called the *logarithmic curve.*

Example (3).   The equation to the logarithmic spiral is

$$\tan^{-1}\frac{y}{x} = k \log \sqrt{(x^2 + y^2)}.$$

Hence

$$\frac{x\dfrac{dy}{dx} - y}{x^2 + y^2} = \frac{k\left(y\dfrac{dy}{dx} + x\right)}{x^2 + y^2},$$

therefore

$$\frac{dy}{dx} = \frac{kx + y}{x - ky};$$

and the equation to the tangent is

$$y' - y = \frac{kx + y}{x - ky}(x' - x).$$

Example (4). Suppose that the equation $\phi(x, y) = 0$, or $u = 0$, can be put in the form

$$v_n + v_{n-1} + v_{n-2} + \ldots\ldots + v_1 + v_0 = 0,$$

where v_n, v_{n-1}, $\ldots\ldots$ are *homogeneous* functions of the degree n, $n-1$, $\ldots\ldots$ respectively; hence

$$\frac{du}{dx} = \frac{dv_n}{dx} + \frac{dv_{n-1}}{dx} + \ldots\ldots\ldots,$$

$$\frac{du}{dy} = \frac{dv_n}{dy} + \frac{dv_{n-1}}{dy} + \ldots\ldots\ldots,$$

and the equation to the tangent is

$$(y' - y)\left(\frac{dv_n}{dy} + \frac{dv_{n-1}}{dy} + \ldots\right) + (x' - x)\left(\frac{dv_n}{dx} + \frac{dv_{n-1}}{dx} + \ldots\right) = 0.$$

But by the property of homogeneous functions (see Example 3 at the end of Chapter VIII.)

$$y\frac{dv_n}{dy} + x\frac{dv_n}{dx} = nv_n,$$

$$y\frac{dv_{n-1}}{dy} + x\frac{dv_{n-1}}{dx} = (n-1)v_{n-1}.$$

$$\ldots\ldots\ldots\ldots\ldots\ldots$$

Hence the equation to the tangent becomes

$$y'\left(\frac{dv_n}{dy} + \frac{dv_{n-1}}{dy} + \ldots\ldots\right) + x'\left(\frac{dv_n}{dx} + \frac{dv_{n-1}}{dx} + \ldots\ldots\right)$$

$$= nv_n + (n-1)v_{n-1} + (n-2)v_{n-2} + \ldots\ldots,$$

or, since $\quad v_n + v_{n-1} + v_{n-2} \ldots + v_1 + v_0 = 0,$

$$y'\left(\frac{dv_n}{dy} + \frac{dv_{n-1}}{dy} + \ldots\ldots\right) + x'\left(\frac{dv_n}{dx} + \frac{dv_{n-1}}{dx} + \ldots\ldots\right)$$

$$+ v_{n-1} + 2v_{n-2} + \ldots + (n-1)v_1 + nv_0 = 0.$$

Example (5). Determine a point in a given curve so that the area of the triangle formed by the tangent at that point and the co-ordinate axes may be a maximum or a minimum.

By Art. 260, the area varies as the product of

$$x - y\frac{dx}{dy}, \quad \text{and} \quad y - x\frac{dy}{dx};$$

put

$$u = \frac{\left(y - x\frac{dy}{dx}\right)^2}{\frac{dy}{dx}},$$

then we require the maximum or minimum value of u.

It will be found that

$$\frac{du}{dx} = -\frac{\left(y - x\frac{dy}{dx}\right)\left(x\frac{dy}{dx} + y\right)\frac{d^2y}{dx^2}}{\left(\frac{dy}{dx}\right)^2}.$$

Now, as we shall see in Chapter XXI., where $\frac{d^2y}{dx^2} = 0$, the curve has in general a singular point called a *point of inflection*. Where $y - x\frac{dy}{dx} = 0$, the tangent passes through the origin and *the area in question vanishes*. It will be often obvious when any particular curve is considered, that neither of these exceptional cases can hold. We have then $x\frac{dy}{dx} + y = 0$ as the condition which determines the point required.

When $x\frac{dy}{dx} + y = 0$, we have, by Art. 260,

$$x' = 2x, \quad \text{and} \quad y' = 2y.$$

Hence in general when the area is a maximum or a minimum the portion of the tangent between the axes is *bisected at the point of contact*. It will in general be obvious from the figure in the case of any particular curve whether the area is a maximum or minimum.

265. If the equation to a curve be given in the form

$$F(x, y) - c = 0,$$

the equation to the tangent at the point (x, y), will be

$$(y' - y) \frac{dF}{dy} + (x' - x) \frac{dF}{dx} = 0 \ldots\ldots\ldots\ldots(1);$$

and the equation to the normal

$$(y' - y) \frac{dF}{dx} - (x' - x) \frac{dF}{dy} = 0 \ldots\ldots\ldots\ldots(2).$$

If we consider x', y', as constant, equation (1) combined with $F(x, y) = c$, will give the co-ordinates of the points where the tangents drawn from the point (x', y') meet the curve represented by $F(x, y) = c$; and equation (2) combined with $F(x, y) = c$ will give the co-ordinates of the points where the normals drawn from the point (x', y') meet the curve represented by $F(x, y) = c$.

Since the equations (1) and (2) are independent of c, they will represent the geometrical loci of the points where the curves which we obtain by ascribing different values to c in the equation $F(x, y) = c$, are met by their tangents or their normals respectively, which pass through the point (x', y'). Thus, if we want to draw *tangents* from the point (x', y') to any one of the curves $F(x, y) = c$, we must construct the curve

$$(x' - x) \left(\frac{dF}{dx}\right) + (y' - y)\left(\frac{dF}{dy}\right) = 0;$$

and determine where it intersects the particular curve $F(x, y) = c$ which we are considering; join the point or points of intersection with the point (x', y') and we have the required tangent or tangents. Similarly, we may draw *normals* from (x', y') to any one of the curves $F(x, y) = c$.

EXAMPLES.

1. In the curve $y(x - 1)(x - 2) = x - 3$, shew that the tangent is parallel to the axis of x at the points for which $x = 3 \pm \sqrt{2}$.

2. In the curve $y^3 = (x-a)^2 (x-c)$, shew that the tangent is parallel to the axis of x at the point for which
$$x = \frac{2c+a}{3}.$$

3. In the curve $x^2 y^2 = a^3 (x+y)$, the tangent at the origin is inclined at an angle of $135°$ to the axis of x.

4. In the curve $x^2 (x+y) = a^2 (x-y)$, the equation to the tangent at the origin is $y = x$.

5. In the curve $x^{\frac{2}{3}} + y^{\frac{2}{3}} = a^{\frac{2}{3}}$ find the length of the perpendicular from the origin on the tangent at (x, y); also find the length of that part of the tangent which is intercepted between the two axes.

$\qquad\qquad$ *Results.* (1) $\sqrt[3]{(axy)}$; $\qquad$ (2) a.

6. If x_1, y_1, be the parts of the axes of x and y intercepted by the tangent at the point (x, y) to the curve
$$\left(\frac{x}{a}\right)^{\frac{2}{3}} + \left(\frac{y}{b}\right)^{\frac{2}{3}} = 1, \text{ then } \frac{x_1^2}{a^3} + \frac{y_1^2}{b^2} = 1.$$

7. Shew that all the curves represented by the equation
$$\left(\frac{x}{a}\right)^n + \left(\frac{y}{b}\right)^n = 2,$$

different values being assigned to n, touch each other at the point (a, b).

8. In the curve $y^n = a^{n-1}x$, express the equation to the tangent in its simplest form ; and determine the value of n when the area included between the tangent and the co-ordinate axes is constant.

9. If the normal to the curve $x^{\frac{2}{3}} + y^{\frac{2}{3}} = a^{\frac{2}{3}}$, make an angle ϕ with the axis of x, shew that its equation is
$$y \cos \phi - x \sin \phi = a \cos 2\phi.$$

10. Find at what angle the curve $y^2 = 2ax$ cuts the curve $x^3 - 3axy + y^3 = 0$.

Results. The curves meet at the origin; here the first curve has the axis of y for its tangent, and the second curve has both the axes for tangents. The curves also meet at the point $x = a\sqrt[3]{2},\ y = a\sqrt[3]{4}$; and here they meet at an angle whose cotangent is $\sqrt[3]{4}$.

11. Tangents are drawn to the ellipse $\dfrac{x^2}{a^2} + \dfrac{y^2}{b^2} = 1$, and the circle $x^2 + y^2 - a^2 = 0$, at the points where a common ordinate cuts them: shew that if ϕ be the greatest inclination of these tangents

$$\tan \phi = \frac{a - b}{2\sqrt{(ab)}}.$$

12. If tangents be drawn from a point (h, k) to the curve whose equation is $\left(\dfrac{x}{a}\right)^3 + \left(\dfrac{y}{b}\right)^3 = 1$, an ellipse whose semiaxes are $a\left(\dfrac{a}{h}\right)^{\frac{1}{2}}$, and $b\left(\dfrac{b}{k}\right)^{\frac{1}{2}}$ will pass through the points of contact.

13. Shew that all the points of the curve $y^2 = 4a\left(x + a\sin\dfrac{x}{a}\right)$ at which the tangent is parallel to the axis of x lie on a certain parabola.

14. The normal to a parabola at any point P is produced to meet the directrix at Q, and the tangent at P meets the directrix at R: find (1) when QR is a minimum, (2) when the triangle PQR is a minimum.

Results. (1) $x = \dfrac{a}{3}$, (2) $x = \dfrac{a}{5}$; where $y^2 = 4ax$ is the equation to the parabola.

CHAPTER XIX.

ASYMPTOTES.

266. Suppose one or more of the branches of a curve to extend to an infinite distance from the origin, and that at successive points of such a branch we draw tangents. Then two different cases may exist with respect to the *directions* of these tangents; they either, as we pass from point to point along the curve, approach some definite limit or they do not. And with respect to the *position* of these tangents, two cases are possible; the intercepts cut from the axes of co-ordinates either tend to a finite limit or they do not. If the direction has a limit, and one or both of the intercepts a limit, there exists a straight line towards which the successive tangents continually approach. Such a straight line is called an asymptote to the curve; hence we have the definition which follows.

267. Definition. An asymptote to a curve is the limiting position of the tangent when the point of contact moves to an infinite distance from the origin.

To find whether a proposed curve has an asymptote, we must first ascertain if $\dfrac{dy}{dx}$ has a limiting value as we proceed to an infinite distance from the origin. If it *has not* there is generally no asymptote. If $\dfrac{dy}{dx}$ *has* a limiting value, we must then ascertain if the intercept on the axis of x, which by Art. 260 is $x - y\dfrac{dx}{dy}$, has a limiting value. Suppose it has, and let it be denoted by c while μ denotes the limit of $\dfrac{dy}{dx}$, then $y = \mu(x - c)$ is the equation to an asymptote.

268. If $\frac{dy}{dx}$ increases without limit, *and at the same time* $x - y\frac{dx}{dy}$ *has a finite limit*, we have an asymptote parallel to the axis of y.

Also we *may* have an asymptote when the limit of $x - y\frac{dx}{dy}$ is infinite, namely in the case where the limit of $\frac{dy}{dx}$ is zero, and the limit of $y - x\frac{dy}{dx}$, which is the intercept on the axis of y, is finite. The asymptote is then parallel to the axis of x.

269. We will now take some simple examples.

(1) The equation to the parabola is $y^2 = 4ax$; so that we have $y = \pm 2\sqrt{ax}$; therefore $\frac{dy}{dx} = \pm\sqrt{\frac{a}{x}}$; hence, when x increases indefinitely the limit of $\frac{dy}{dx}$ is zero; but $y - x\frac{dy}{dx} = \pm(2\sqrt{ax} - \sqrt{ax}) = \pm\sqrt{ax}$, which has no finite limit. Therefore there is no asymptote.

(2) The equation to the hyperbola is $y^2 = \frac{b^2}{a^2}(x^2 - a^2)$; so that we have $y = \pm\frac{b}{a}\sqrt{(x^2 - a^2)}$; therefore $\frac{dy}{dx} = \pm\frac{bx}{a\sqrt{(x^2 - a^2)}}$, and $x - y\frac{dx}{dy} = x - \frac{x^2 - a^2}{x} = \frac{a^2}{x}$. Hence the limit of $\frac{dy}{dx}$ when x is infinite is $\pm\frac{b}{a}$, and the limit of $x - y\frac{dx}{dy}$ is 0. Therefore $y = \frac{b}{a}x$ is the equation to one asymptote; and $y = -\frac{b}{a}x$ is the equation to another asymptote.

(3) Suppose $y = \frac{a^3}{(x-b)^2} + c$ to be the equation to a curve, then we have $\frac{dy}{dx} = -\frac{2a^3}{(x-b)^3}$, and $x - y\frac{dx}{dy} = x + \frac{x-b}{2} + \frac{c(x-b)^3}{2a^3}$.

As x approaches b, y and $\dfrac{dy}{dx}$ increase without limit. The limit of $x - y \dfrac{dx}{dy}$ is b, and, by Art. 268, there is an asymptote parallel to the axis of y, having for its equation $x = b$.

270.   An asymptote may also be defined as a *straight line, the distance of which from a point in a curve diminishes without limit as the point in the curve moves to an infinite distance from the origin.*

Suppose $$y = \mu x + \beta$$
the equation to a straight line, and
$$y = \mu x + \beta + v$$
the equation to a curve, then if v diminish without limit as x and y increase without limit, the straight line will be an asymptote to the curve.   For if x, y, be the co-ordinates of a point in the curve, the perpendicular distance of that point from the straight line is
$$\frac{y - \mu x - \beta}{\sqrt{(1 + \mu^2)}} \text{ or } \frac{v}{\sqrt{(1 + \mu^2)}},$$
and this diminishes without limit when x and y increase without limit.

271.   That the two definitions of an asymptote lead in general to the same results may be seen by considering different examples, or by the following proof.   Let $y = \mu x + \beta + v$ be the equation to a curve, where μ and β are constants, and v diminishes without limit as x and y increase without limit. From the given equation
$$\frac{y}{x} = \mu + \frac{\beta + v}{x}.$$

Hence μ is the limit of $\dfrac{y}{x}$ when x and y increase without limit.   But, by Art. 148,
$$\text{the limit of } \frac{y}{x} = \text{the limit of } \frac{\frac{dy}{dx}}{1} \text{ or } \frac{dy}{dx}.$$

Also β is the limit of $y - \mu x$; but $\mu =$ the limit of $\dfrac{dy}{dx}$;

therefore *in general* $\beta =$ the limit of $y - \dfrac{dy}{dx} x$. Hence the equation to the tangent to the curve at the point (x, y), which is

$$y' - y = \frac{dy}{dx}(x' - x),$$

becomes, when x and y are indefinitely increased,

$$y' = \mu x' + \beta;$$

that is, the equation to the asymptote found according to the first definition is the same as the equation found according to the second definition.

272. We say in the last Article that *in general* the limit of $y - \mu x =$ the limit of $y - \dfrac{dy}{dx} x$. Suppose, for example, that the equation to a curve is

$$y = Ax + B + \frac{a}{x};$$

therefore $\qquad \dfrac{y}{x} = A + \dfrac{B}{x} + \dfrac{a}{x^2}.$

Hence $\mu =$ the limit of $\dfrac{y}{x} = A$, and

$$y - \mu x = B + \frac{a}{x}.$$

Also $\qquad \dfrac{dy}{dx} = A - \dfrac{a}{x^2},$

therefore $\qquad y - \dfrac{dy}{dx} x = B + \dfrac{2a}{x}.$

Here $y - \dfrac{dy}{dx} x$ and $y - \mu x$ have the same limit, namely B.

But suppose $y = Ax + B + \dfrac{a + \sin x}{x}$.

Here, as before, $\qquad \mu = A.$

Also
$$y - \mu x = B + \frac{a + \sin x}{x}.$$

And
$$\frac{dy}{dx} = A + \frac{\cos x}{x} - \frac{a + \sin x}{x^2},$$

therefore
$$y - \frac{dy}{dx}x = B - \cos x + \frac{2(a + \sin x)}{x}.$$

Here we cannot assert that $y - \mu x$ and $y - \frac{dy}{dx}x$ have the same limit: the limit of the former is B, but the latter cannot be said to have a limit, on account of the term $\cos x$, which does not tend to any limit as x increases indefinitely. In this case the curve

$$y = Ax + B + \frac{a + \sin x}{x}$$

has an asymptote according to the definition of Art. 270, namely, $y = Ax + B$, but not according to the definition of Art. 267.

The demonstration in Art. 270 might, of course, start with the equation $x = \mu y + \beta + v$; so that, should the asymptote be parallel to the axis of y, by taking the second form we avoid having μ infinite.

273. We have hitherto confined ourselves to *rectilinear* asymptotes; we now extend the definition to curvilinear asymptotes.

DEFINITION. When the difference of the ordinates of two curves corresponding to a common abscissa diminishes without limit, or the difference of the abscissæ corresponding to a common ordinate diminishes without limit, as we pass from point to point along either curve, each curve is said to be an asymptote to the other.

Hence, if the equation to a curve can be put in the form

$$y = A_0 x^n + A_1 x^{n-1} + \dots + A_{r-1}x + A_n + \frac{B_1}{x} + \frac{B_2}{x^2} + \frac{B_3}{x^3} + \dots,$$

then $y = A_0 x^n + A_1 x^{n-1} + \dots + A_{n-1}x + A_n$

is the equation to a curve which is an asymptote to the former. So also is

$$y = A_0 x^n + A_1 x^{n-1} + \ldots + A_{n-1} x + A_n + \frac{B_1}{x},$$

and $y = A_0 x^n + A_1 x^{n-1} + \ldots + A_{n-1} x + A_n + \frac{B_1}{x} + \frac{B_2}{x^2},$

and so on.

Example. Find asymptotes to the curve

$$x^3 - xy^2 + ay^2 = 0.$$

Here $y^2 = \frac{x^3}{x - a}$; therefore $y = \pm \sqrt{\left(\frac{x^3}{x - a}\right)}.$

As x approaches the value a, both y and $\frac{dy}{dx}$ increase without limit, and $x = a$ is the equation to a rectilinear asymptote.

Putting y in the form $\pm x \left(1 - \frac{a}{x}\right)^{-\frac{1}{2}}$, and expanding by the Binomial Theorem, we have

$$y = \pm x \left\{ 1 + \frac{a}{2x} + \frac{3a^2}{8x^2} + \frac{5a^3}{16x^3} + \ldots \right\} \ldots\ldots\ldots\ldots (1).$$

Hence $y = \pm \left(x + \frac{a}{2}\right)$ are the equations to two rectilinear asymptotes. We may obtain as many curvilinear asymptotes as we please by making use of the series in (1). For example,

$$y = \pm \left(x + \frac{a}{2} + \frac{3a^2}{8x}\right)$$

are the equations to two asymptotic curves of the second order. The student will remember that by Art. 114 we may use the Binomial Theorem in the above Example as a *true arithmetical expansion* when $\frac{a}{x}$ is *less* than unity, which will certainly be the case when x is increased indefinitely.

274. The following method will furnish the rectilinear asymptotes with great readiness in many cases. Suppose the equation to a curve, $F(x, y) = 0$, to be such that $F(x, y)$ is the sum of different *homogeneous* functions of x and y, so that the equation may be put in the form

$$x^n \phi \left(\frac{y}{x} \right) + x^p \psi \left(\frac{y}{x} \right) + x^q \chi \left(\frac{y}{x} \right) + \ldots = 0 \ldots\ldots\ldots (1),$$

where n, p, q, are arranged in descending order of magnitude. For example, every rational integral algebraical equation between x and y can be put in this form. From (1) we have

$$\phi \left(\frac{y}{x} \right) + \frac{1}{x^{n-p}} \psi \left(\frac{y}{x} \right) + \frac{1}{x^{n-q}} \chi \left(\frac{y}{x} \right) + \ldots = 0 \ldots\ldots\ldots(2).$$

Now in finding an asymptote we must first by Art. 271 ascertain the limit of $\frac{y}{x}$ when x and y are infinite. If we call that limit μ, and suppose it to be finite, we have from (2)

$$\phi (\mu) = 0.$$

Let μ_1 be a value of μ obtained from this equation; we have next to find the limit of $y - \mu_1 x$. Put $y - \mu_1 x = \beta$, then from (2)

$$\phi \left(\mu_1 + \frac{\beta}{x} \right) + \frac{1}{x^{n-p}} \psi \left(\mu_1 + \frac{\beta}{x} \right) + \ldots = 0 \ldots\ldots\ldots(3).$$

But, by Art. 92,

$$\phi \left(\mu_1 + \frac{\beta}{x} \right) = \phi (\mu_1) + \frac{\beta}{x} \phi' \left(\mu_1 + \frac{\theta \beta}{x} \right)$$

$$= \frac{\beta}{x} \phi' \left(\mu_1 + \frac{\theta \beta}{x} \right),$$

since $\phi (\mu_1) = 0.$

Thus (3) becomes

$$\beta \phi' \left(\mu_1 + \frac{\theta \beta}{x} \right) + \frac{1}{x^{n-p-1}} \psi \left(\mu_1 + \frac{\beta}{x} \right) + \ldots = 0 \ldots\ldots\ldots(4).$$

In equation (4) let x be supposed to increase indefinitely, then we shall have different results depending on the value of p.

If p be *greater* than $n-1$ the value of β is infinite, and there is no asymptote for the root μ_1 of the equation

$$\phi(\mu) = 0.$$

If p be *equal* to $n-1$ and $\phi'(\mu_1)$ be not zero, the limit of β is $-\dfrac{\psi(\mu_1)}{\phi'(\mu_1)}$; and the equation to an asymptote is

$$y = \mu_1 x - \frac{\psi(\mu_1)}{\phi'(\mu_1)}.$$

If p be *less* than $n-1$ and $\phi'(\mu_1)$ be not zero, the limit of β is 0 and the equation to an asymptote is

$$y = \mu_1 x.$$

In the last case the equations

$$y = \mu x, \quad \phi(\mu) = 0,$$

give for determining the asymptotes

$$\phi\left(\frac{y}{x}\right) = 0, \quad \text{or} \quad x^n \phi\left(\frac{y}{x}\right) = 0;$$

hence when the equation to a curve can be exhibited in such a form that the sum of a number of homogeneous functions is zero, and the degree n of the highest of these functions exceeds by more than unity the degree of any of the others, all the asymptotes *in general* pass through the origin and may be found by equating to zero the homogeneous function of the nth degree. We say *in general* because there is the limitation that $\phi'(\mu_1)$ is not to be zero; that is, by the theory of equations $\phi(\mu) = 0$ must not have *equal roots*.

275. We will now consider the case in which $\phi'(\mu_1)$ is zero.

First suppose p *equal* to $n-1$.

If $\psi(\mu_1)$ is not zero β becomes infinite, and there is no asymptote for the root μ_1 of the equation $\phi(\mu) = 0$. But if $\psi(\mu_1) = 0$ the value of β becomes indeterminate.

Suppose in this case $q = n - 2$, so that equation (3) of Art. 274 gives

$$\phi\left(\mu_1 + \frac{\beta}{x}\right) + \frac{1}{x}\psi\left(\mu_1 + \frac{\beta}{x}\right) + \frac{1}{x^2}\chi\left(\mu_1 + \frac{\beta}{x}\right) + \ldots = 0.$$

Since $\phi(\mu_1) = 0$ and $\phi'(\mu_1) = 0$, we have, by Art. 92,

$$\phi\left(\mu_1 + \frac{\beta}{x}\right) = \frac{\beta^2}{2x^2}\phi''\left(\mu_1 + \frac{\theta\beta}{x}\right);$$

also

$$\psi\left(\mu_1 + \frac{\beta}{x}\right) = \frac{\beta}{x}\psi'\left(\mu_1 + \frac{\theta_1\beta}{x}\right).$$

Substitute these values in the equation above, multiply by x^2, and then proceed to the limit, and we have for determining the limiting values of β, the quadratic equation

$$\frac{\beta^2}{2}\phi''(\mu_1) + \beta\psi'(\mu_1) + \chi(\mu_1) = 0.$$

If the values of β be possible, we thus obtain two parallel asymptotes.

If this quadratic assume an indeterminate form, we may proceed in the same manner to form a cubic equation in β.

In the case where $\phi'(\mu_1)$ is zero and $\psi(\mu_1)$ is not zero, there is no rectilinear asymptote for the root μ_1 of the equation $\phi(\mu) = 0$, as we have already stated at the beginning of this Article. In this case we may in general obtain a *parabolic* asymptote, as we will now shew.

By Art. 92, since $\phi(\mu_1) = 0$, and $\phi'(\mu_1) = 0$,

$$\phi\left(\mu_1 + \frac{\beta}{x}\right) = \frac{1}{2}\frac{\beta^2}{x^2}\phi''\left(\mu_1 + \frac{\theta\beta}{x}\right).$$

Hence equation (3) of Art. 274 becomes

$$\frac{1}{2}\frac{\beta^2}{x^2}\phi''\left(\mu_1 + \frac{\theta\beta}{x}\right) + \frac{1}{x}\psi\left(\mu_1 + \frac{\beta}{x}\right) + \ldots = 0;$$

as x increases indefinitely this equation approaches to the form $\dfrac{1}{2}\dfrac{\beta^2}{x^2} = -\dfrac{\psi(\mu_1)}{x\phi''(\mu_1)}$, so that $\dfrac{\beta}{x} = \left\{-\dfrac{2\psi(\mu_1)}{x\phi''(\mu_1)}\right\}^{\frac{1}{2}}$.

Hence we have a *parabolic asymptote* determined by the equation

$$y - \mu_1 x = x \left\{ \frac{-2\psi(\mu_1)}{x\phi''(\mu_1)} \right\}^{\frac{1}{2}},$$

that is,

$$(y - \mu_1 x)^2 = \frac{-2x\psi(\mu_1)}{\phi''(\mu_1)}.$$

Next suppose p *less* than $n - 1$.

Then since $\phi'(\mu_1) = 0$ equation (4) of Art. 274 will not determine β; and instead of this equation we have ultimately in the manner just shewn

$$\frac{1}{2}\frac{\beta^2}{x^2} = -\frac{\psi(\mu_1)}{x^{n-p}\phi''(\mu_1)}.$$

If $n - p = 2$, we obtain

$$\beta^2 = -\frac{2\psi(\mu_1)}{\phi''(\mu_1)},$$

so that if $\psi(\mu_1)$ and $\phi''(\mu_1)$ are of different signs we have two possible values of β, and therefore two parallel asymptotes which are equidistant from the origin.

If $n - p$ is not equal to 2, we have a curvilinear asymptote determined by the equation

$$(y - \mu_1 x)^2 = -\frac{2\psi(\mu_1)}{x^{n-p-2}\phi''(\mu_1)}.$$

276. We have assumed in Article 274, that the limit of $\frac{y}{x}$ is *finite;* if it be not, the limit of $\frac{x}{y}$ will be zero, and we must examine if there exists an asymptote parallel to the axis of y. This can generally be easily ascertained in any particular example. Or we may put the given equation in the form

$$y^n \phi_1\left(\frac{x}{y}\right) + y^p \psi_1\left(\frac{x}{y}\right) + \ldots = 0,$$

and proceed as above.

277. If a curve be given by an algebraical equation we may determine the asymptotes which are parallel to the

axis of y thus. Arrange the equation according to powers of y; suppose it to be

$$y^n f(x) + y^{n-a} f_1(x) + y^{n-\beta} f_2(x) + \ldots = 0,$$

where a, β, ... are all positive, then the asymptotes parallel to the axis of y will be given by the real roots of the equation

$$f(x) = 0.$$

For the equation to the curve may be written

$$f(x) + \frac{f_1(x)}{y^a} + \frac{f_2(x)}{y^\beta} + \ldots = 0,$$

and it is obvious that this is satisfied by supposing $y = \infty$ and $f(x) = 0$; and that when y is ∞ no other value of x except those derived from $f(x) = 0$ will satisfy it. Hence the asymptotes parallel to the axis of y are found *by equating to zero the coefficient of the highest power of y* in the equation to the curve.

Similarly the asymptotes parallel to the axis of x may be found by equating to zero the coefficient of the highest power of x in the equation to the curve.

When a curve is given by a rational integral algebraical equation, it will be convenient to determine by the preceding method the asymptotes parallel to the axes, and then proceed for the other asymptotes according to the following rule; we suppose the equation of the nth degree. Substitute for y in the given equation $\mu x + \beta$ and arrange the terms of the equation according to powers of x. Equate to zero the coefficient of x^n; this will give an equation for determining μ; suppose μ_1 one of the real values of μ. Then examine the coefficient of x^{n-1}, and give to μ if it occurs in this coefficient the value μ_1. If we can determine β so as to make this coefficient vanish, then $y = \mu_1 x + \beta$ will be the equation to an asymptote; if the coefficient cannot be made to vanish there is no corresponding asymptote. If the coefficient vanishes whatever be the value of β, then put the coefficient of x^{n-2} equal to zero substituting μ_1 for μ in it; we shall thus have generally a quadratic equation to determine the values of β, and if these values are real, we obtain two parallel asymptotes. If the coefficient of x^{n-2} vanishes, whatever be the value of β, we must equate to zero the coefficient of x^{n-3} and so on.

This rule can be easily shewn to agree with Arts. 274 and 275. Equation (1) of Art. 274, may be supposed the equation to the curve in which n is an integer, $p = n - 1$, $q = n - 2$, ... Then if we put $\mu x + \beta$ for y, and arrange the terms according to powers of x, we shall obtain the expression

$$x^n \phi(\mu) + x^{n-1}\{\psi(\mu) + \beta \phi'(\mu)\} + x^{n-2}\{\chi(\mu) + \beta \psi'(\mu) + \frac{\beta^2}{2} \phi''(\mu)\} + \dots$$

Thus by equating to zero the coefficient of x^n we arrive at the equation for determining μ given in Art. 274. Then by equating to zero the coefficient of x^{n-1} we shall obtain the same value of β as that found in Art. 274; or if the coefficient of x^{n-1} vanishes, whatever β may be, then by equating to zero the coefficient of x^{n-2} we arrive at the quadratic equation given in Art. 275.

Example (1). $y^3 + x^3 - 3axy = 0$.

Put $\mu x + \beta$ for y, then

$$(\mu x + \beta)^3 + x^3 - 3ax (\mu x + \beta) = 0;$$

therefore $(\mu^3 + 1) x^3 + 3x^2 (\mu^2 \beta - a\mu) + Mx + N = 0$.

Hence, $\mu^3 + 1 = 0$,

$$\mu^2 \beta - a\mu = 0,$$

are the equations from which μ and β are to be found; they give $\mu = -1$, $\beta = -a$; therefore

$$y = -x - a$$

is the equation to an asymptote.

Example (2). $x^3 (x + y) = a^2 (x - y)$.

Put $\mu x + \beta$ for y, then

$$x^3(x + \mu x + \beta) = a^2 (x - \mu x - \beta);$$

therefore $x^3 (1 + \mu) + \beta x^2 - xa^2 (1 - \mu) + a^2 \beta = 0$.

Hence, $1 + \mu = 0$ and $\beta = 0$;

therefore $y = -x$ is the equation to an asymptote.

Example (3).　$xy (y - x) (y - x + 3a) + 4a^2x - a^4 = 0$.

Here the term containing the highest power of y is xy^2; thus $x = 0$ gives one asymptote, namely the axis of y. Similarly, the term containing the highest power of x is yx^3; therefore $y = 0$ gives one asymptote, namely the axis of x. Then put $\mu x + \beta$ for y, and we obtain the expression

$$x (\mu x + \beta) \{(\mu - 1) x + \beta\} \{(\mu - 1) x + 3a + \beta\} + 4a^2x - a^4.$$

Arranging this according to powers of x, we have

$$x^4\mu (\mu - 1)^2 + x^3 (\mu - 1) \{3\mu a + (3\mu - 1) \beta\}$$
$$+ x^2 \{\beta^2 (3\mu - 2) + 3a\beta (2\mu - 1)\} + \dots$$

Put $\mu (\mu - 1)^2 = 0$; we have then $\mu = 0$, or $\mu = 1$; the former value of μ will lead to the asymptote coinciding with the axis of x which we have already found. The value $\mu = 1$ makes the coefficient of x^3 in the above expression vanish; we therefore equate to zero the coefficient of x^2, putting $\mu = 1$ in it. We thus obtain $\beta^2 + 3a\beta = 0$; hence, $\beta = 0$, or $\beta = -3a$. Therefore we have for the equations to asymptotes $y = x$, and $y = x - 3a$.

It will be observed that the conclusions of this Chapter all hold whether the axes be rectangular or oblique.

EXAMPLES.

Find the asymptotes of the following curves:

1.　$y^2 (x - 2a) = x^3 - a^3$.　　　　*Result.*　$x = 2a$; $y = \pm (x + a)$.

2.　$y^3 = x^2 (2a - x)$.　　　　　*Result.*　$y = -x + \dfrac{2a}{3}$.

3.　$y (a^2 + x^2) = a^2 (a - x)$.　　　　*Result.*　$y = 0$.

4.　$y^2 (ay + bx) = a^2y^2 + b^2x^2$.　　*Result.*　$y = -\dfrac{b}{a}x + 2a$.

5.　$y^3 = (x - a)^2 (x - c)$.　　　*Result.*　$y = x - \frac{1}{3} (2a + c)$.

6.　$xy^2 + yx^2 = a^3$.　　　*Result.*　$x = 0$; $y = 0$; $y = -x$.

7.　$x^2y^2 = a^2 (x^2 - y^2)$.　　　　*Result.*　$y = \pm a$.

8. $4x^3 = (a + 3x)(x^2 + y^2)$.

> Result. $y = \pm \left(\dfrac{x}{\sqrt{3}} - \dfrac{2a}{3\sqrt{3}} \right)$ and $x = -\dfrac{a}{3}$.

9. $(x + a)y^2 = (y + b)x^2$.

> Result. $x + a = 0$, $y + b = 0$, $y = x + b - a$.

10. $(y - 2x)(y^2 - x^2) - a(y - x)^2 + 4a^2(x + y) = a^3$.

> Result. $y = x$, $y + x = \dfrac{2a}{3}$, $y - 2x = \dfrac{a}{3}$.

11. $y^2(x - y)^2 + ax^2(x - y) - 3a^2y^2 - a^4 = 0$.

> Result. $y = x + \dfrac{a}{2}(1 \pm \sqrt{13})$.

12. $x(x^2 - a^2) - 2y(y^2 - a^2) = 3xy^2 + a^3$.

> Result. $2y = x$, $y + x - a = 0$, $y + x + a = 0$.

13. $x^2(x - y)^2 - a^2(x^2 + y^2) = 0$.

> Result. $x = \pm a$, $y = x \pm a\sqrt{2}$.

14. $(y - x)^2(x^2 - a^2) = a^4$.

15. $y^3 - 3y^2x + 4x^3 + ay^2 + axy - 6ax^2 + 2b^2x - b^2y + c^3 = 0$.

16. If a curve of the third degree be referred to two asymptotes as axes, shew that its equation will be of the form

$$xy(ax + by + c) + a'x + b'y + c' = 0,$$

and that the equation to the third asymptote will be

$$ax + by + c = 0.$$

CHAPTER XX.

TANGENTS AND ASYMPTOTES OF CURVES REFERRED TO POLAR CO-ORDINATES.

278.   IF we have the equation to a curve expressed in terms of x and y, we may transform it to one between polar co-ordinates by assuming $x = r \cos \theta$ and $y = r \sin \theta$.   Thus r becomes a function of θ, and the equation to a curve in polar co-ordinates takes the form $r = f(\theta)$, or $F(r, \theta) = 0$.   In this case the curve is called a *polar curve* or *spiral; r* is called the *radius vector* and θ the *vectorial angle*.

The angle (ψ) which the tangent to a curve makes with the axis of x is given by the equation

$$\tan \psi = \frac{dy}{dx}, \quad \text{(Art. 257).}$$

Hence, by Art. 201,

$$\tan \psi = \frac{\sin \theta \dfrac{dr}{d\theta} + r \cos \theta}{\cos \theta \dfrac{dr}{d\theta} - r \sin \theta}.$$

279.   *Expression for the angle included between the radius vector at any point of a curve, and the tangent to the curve at that point.*

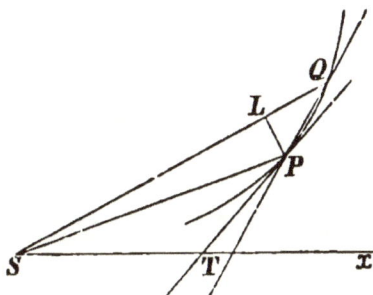

Let P be a point in a curve, the polar co-ordinates of which are r and θ, S being the pole.

Let Q be another point, the co-ordinates of which are

$$r + \Delta r, \text{ and } \theta + \Delta \theta.$$

Draw PL perpendicular to SQ, then

$$PL = r \sin \Delta \theta,$$

$$LQ = r + \Delta r - r \cos \Delta \theta;$$

therefore $\qquad \tan LQP = \dfrac{r \sin \Delta \theta}{r + \Delta r - r \cos \Delta \theta}.$

Let Q move along the curve to P; the limiting position of QP is by definition the tangent to the curve at P; let this be PT. The limit of the angle LQP will be the angle SPT; call this angle ϕ, then

$$\tan \phi = \text{the limit of } \frac{r \sin \Delta \theta}{r + \Delta r - r \cos \Delta \theta}$$

when $\Delta \theta$ and Δr are indefinitely diminished.

Now $\qquad \dfrac{r \sin \Delta \theta}{r + \Delta r - r \cos \Delta \theta} = \dfrac{\dfrac{r \sin \Delta \theta}{\Delta \theta}}{\dfrac{2r \sin^2 \dfrac{\Delta \theta}{2}}{\Delta \theta} + \dfrac{\Delta r}{\Delta \theta}}.$

The limit of $\dfrac{\sin \Delta \theta}{\Delta \theta}$ is 1.

The limit of $\dfrac{\Delta r}{\Delta \theta}$ is denoted by $\dfrac{dr}{d\theta}$.

The limit of $\dfrac{2 . \sin^2 \dfrac{\Delta \theta}{2}}{\Delta \theta}$, that is, of $\dfrac{\sin \dfrac{\Delta \theta}{2}}{\dfrac{\Delta \theta}{2}} \sin \dfrac{\Delta \theta}{2}$, is zero.

Therefore $\qquad \tan \phi = r \dfrac{d\theta}{dr}.$

280. The result of the last Article may also be obtained thus:

$$\tan PTx = \frac{\sin\theta\,\dfrac{dr}{d\theta} + r\cos\theta}{\cos\theta\,\dfrac{dr}{d\theta} - r\sin\theta}, \quad \text{(Art. 278)},$$

$$PSx = \theta\,; \text{ therefore}$$

$$\tan SPT = \frac{\dfrac{\sin\theta\,\dfrac{dr}{d\theta} + r\cos\theta}{\cos\theta\,\dfrac{dr}{d\theta} - r\sin\theta} - \tan\theta}{1 + \dfrac{\tan\theta\left(\sin\theta\,\dfrac{dr}{d\theta} + r\cos\theta\right)}{\cos\theta\,\dfrac{dr}{d\theta} - r\sin\theta}} = r\,\frac{d\theta}{dr} \text{ by reduction.}$$

281. *To find the polar equation to the tangent to a curve.*

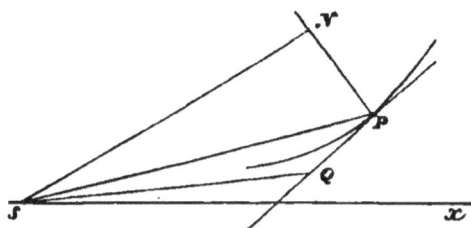

Let $SP = r$, $PSx = \theta$, be the polar co-ordinates of the point of contact.

Let $SQ = r'$, $QSx = \theta'$, be the polar co-ordinates of a point Q in the tangent line. From the triangle SPQ, we have, putting $SPQ = \phi$,

$$\frac{r}{r'} = \frac{\sin SQP}{\sin SPQ} = \frac{\sin(\theta - \theta' + \phi)}{\sin\phi}$$

$$= \sin(\theta - \theta')\cot\phi + \cos(\theta - \theta').$$

But $\tan\phi = r\,\dfrac{d\theta}{dr}\,;$

therefore $\dfrac{r}{r'} = \dfrac{1}{r}\dfrac{dr}{d\theta}\sin(\theta - \theta') + \cos(\theta - \theta')$ ....(1).

This result may be written,

$$r' \frac{d}{d\theta} r \sin(\theta - \theta') = r^2 \ldots\ldots\ldots\ldots\ldots (2).$$

If we put $\frac{1}{r} = u$, and $\frac{1}{r'} = u'$, then

$$-\frac{1}{r^2} \frac{dr}{d\theta} = \frac{du}{d\theta}.$$

Hence, dividing both sides of (1) by r, we obtain

$$u' = u \cos(\theta - \theta') - \frac{du}{d\theta} \sin(\theta - \theta'),$$

or $\qquad u' = u \cos(\theta' - \theta) + \dfrac{du}{d\theta} \sin(\theta' - \theta).$

282. *To find the polar equation to the normal at any point of a curve.*

Let $\qquad SP = r, \quad PSx = \theta,$
$$SN = r', \quad NSx = \theta',$$

N being any point in the normal; then

$$\frac{SP}{SN} = \frac{\sin SNP}{\sin SPN} = \frac{\sin\left(\theta' - \theta + \dfrac{\pi}{2} - \phi\right)}{\sin\left(\dfrac{\pi}{2} - \phi\right)};$$

therefore $\qquad \dfrac{r}{r'} = \sin(\theta' - \theta)\tan\phi + \cos(\theta' - \theta)$

$$= \sin(\theta' - \theta)\frac{r d\theta}{dr} + \cos(\theta' - \theta).$$

This may be written

$$r' \frac{d}{d\theta} r \cos(\theta - \theta') = r \frac{dr}{d\theta},$$

and may be transformed into

$$u' = u \cos(\theta' - \theta) - u^2 \frac{d\theta}{du} \sin(\theta' - \theta).$$

283. The polar equations in Arts. 281 and 282, may also be derived from the rectangular equations to the tangent and normal of Arts. 257 and 258, by transforming these to polar co-ordinates, using the value of $\dfrac{dy}{dx}$ given in Art. 278.

284. From S draw SY perpendicular to the tangent PT; then

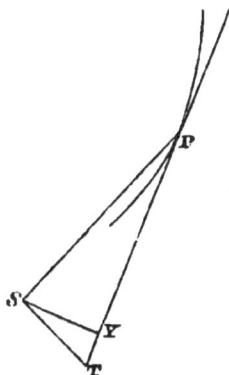

$$SY = r \sin SPT = \frac{r \tan SPT}{\sqrt{(1 + \tan^2 SPT)}}.$$

Hence, if $SY = p$, we have

$$\frac{1}{p^2} = \frac{1}{r^2} + \frac{1}{r^2} \cot^2 SPT = \frac{1}{r^2} + \frac{1}{r^4} \left(\frac{dr}{d\theta}\right)^2$$

$$= u^2 + \left(\frac{du}{d\theta}\right)^2 \text{ if } u = \frac{1}{r}.$$

285. From S draw ST at right angles to the radius vector SP, then ST is called the *polar subtangent*; its value is

$$r \tan SPT, \text{ that is } r^2 \frac{d\theta}{dr}.$$

286. Since an asymptote is a tangent which remains at a finite distance from the origin when the point of contact moves off to an infinite distance, if a polar curve has an asymptote, SP or r must be infinite while ST remains finite. Hence to determine the asymptotes to a polar curve, we must first find those values of θ, if any, which make r infinite.

Suppose α such a value of θ; if for this value of θ the polar subtangent $r^2 \dfrac{d\theta}{dr}$ is *infinite*, there is no corresponding asymptote. If $r^2 \dfrac{d\theta}{dr}$ be *finite* there is an asymptote which may be constructed thus: conceive a straight line drawn from S at an angle α to the initial line; draw from S a second straight line at right angles to the first, to the *right* of it, if $r^2 \dfrac{d\theta}{dr}$ be positive, and to the left of it, if $r^2 \dfrac{d\theta}{dr}$ be negative, and equal in length to $r^2 \dfrac{d\theta}{dr}$; through the end of this second straight line draw a straight line parallel to the first, and it will be the required asymptote.

The terms *right* and *left* in the above rule are to be understood with respect to the straight line first drawn, the eye being supposed to look *along* that line *from S*. The *reason* of the rule must be collected from the figure of Art. 284 and the general principle of the interpretation of signs; that figure makes r increase with θ, and therefore $r^2 \dfrac{d\theta}{dr}$ is positive. If we draw a figure in which r diminishes when θ increases, so that $\dfrac{dr}{d\theta}$ and the polar subtangent are *negative*, we shall find that ST falls to the *left* of SP.

287. Example. $\qquad r = \dfrac{a\theta}{\sin \theta}.$

Here r is infinite when θ is any multiple of π.

Also $\qquad \dfrac{dr}{d\theta} = \dfrac{a\,(\sin\theta - \theta\cos\theta)}{\sin^2\theta};$

therefore $\qquad r^2 \dfrac{d\theta}{dr} = \dfrac{a\theta^2}{\sin\theta - \theta\cos\theta}.$

Hence, when $\sin\theta = 0$, the value of the polar subtangent is $-\dfrac{a\theta}{\cos\theta}.$

When $\theta = \pi$, the polar subtangent $= a\pi$.

When $\theta = 2\pi$, the polar subtangent $= -2a\pi$,
and generally when $\theta = n\pi$, the polar subtangent $= (-1)^{n-1}na\pi$.

To draw the first asymptote, for which $\theta = \pi$, the eye must be supposed to look from S along the direction *opposite* to Sx, and then measure from S at right angles to Sx and towards the right, a straight line in length $a\pi$; a straight line drawn parallel to the initial line and at a distance $a\pi$ from it is the required asymptote.

To draw the second asymptote, for which $\theta = 2\pi$, the eye must be supposed to look along Sx, and then measure to the *left* (since the subtangent is negative) a length $2a\pi$. Hence the asymptote is parallel to the initial line at a distance $2a\pi$ from it, and *above* the initial line.

Proceeding in this way we find an infinite number of asymptotes parallel and equidistant, and all *above* Sx.

If we ascribe to θ negative values, we shall in like manner obtain a series of asymptotes all parallel to Sx, and equidistant, lying below Sx.

EXAMPLES.

1. In the curve $r = a \sin \theta$, shew that $\phi = \theta$.

2. Determine the points in the curve $r = a(1 + \cos \theta)$ at which the tangent is parallel to the initial line.

3. Shew that in the curve $r\theta = a$ the polar subtangent is of constant length.

4. In the curve $r(ae^{\theta} + be^{-\theta}) = ab$, the length of the polar subtangent is $-\dfrac{ab}{ae^{\theta} - be^{-\theta}}$.

5. In any conic section, the focus being the pole, the locus of the extremities of the polar subtangents is a straight line at right angles to the axis major.

6. Find the angle between the radius vector and tangent at any point of an ellipse, (1) the focus being the pole, (2) the centre being the pole. Determine in each case when the angle is a maximum.

7. If $r = a (1 - \cos \theta)$, then $\phi = \dfrac{\theta}{2}$, $p = 2a \sin^3 \dfrac{\theta}{2}$, and the polar subtangent $= 2a \sin^2 \dfrac{\theta}{2} \tan \dfrac{\theta}{2}$.

8. If $r^2 \cos 2\theta = a^2$, shew that $\sin \phi = \dfrac{a^2}{r^2}$.

9. If $r^2 = a^2 \cos 2\theta$, shew that $\phi = \dfrac{\pi}{2} + 2\theta$.

10. If $r = a \sec^3 \dfrac{\theta}{3}$, shew that the locus of Y is a parabola. See the figure in Art. 284.

11. If $r = a (1 + \cos \theta)$, shew that the locus of Y is determined by $r = 2a \left(\cos \dfrac{\theta}{3} \right)^3$.

12. If $r^2 = a^2 \cos 2\theta$, shew that the locus of Y is determined by $r^2 = a^2 \left(\cos \dfrac{2\theta}{3} \right)^3$.

13. Shew that the curve $r \cos \theta = a \cos 2\theta$ has an asymptote having for its equation $r \cos \theta = - a$.

14. Shew that the curve $(r - a) \sin \theta = b$ has an asymptote having for its equation $r \sin \theta = b$.

15. Determine the asymptotes of the curve $r \cos 2\theta = a$.

16. Determine the asymptotes of the curve
$$r \sin 4\theta = a \sin 3\theta.$$

CHAPTER XXI.

CONCAVITY AND CONVEXITY.

288. THE terms 'concave' and 'convex' are commonly not defined in works on the Differential Calculus, but are used in their ordinary sense. The following definition however has been given : "A curve is said to be concave at one of its points with respect to a given straight line, when in passing from that point its two branches are initially included *within* the acute angle formed by the given straight line and the tangent to the curve at that point. When, on the contrary, the two branches are initially *outside* this angle, the curve is said to be convex at this point with respect to the straight line."

289. *To find a test of the convexity or concavity of a curve.*

Let P be a point in a curve whose co-ordinates are x, y.

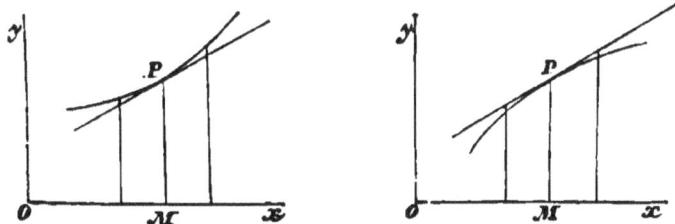

Draw the tangent at P; then if at the point P the curve be *convex* to the axis of x, the ordinates of the curve corresponding to the abscissæ $x \pm h$ must be greater than the corresponding ordinates of the tangent at P, when h has any value contained between some finite limit and zero : if the curve be *concave*, the ordinates of the curve must be less than the ordinates of the tangent. This may be deduced from the definition of Art. 288 ; or if we omit that definition it must still be taken as a consequence of the meaning of the terms *concave* and *convex*.

Let y_1 denote the ordinate of the curve corresponding to the abscissa $x + h$, and y_2 the corresponding ordinate to the tangent at P. If $y = \phi(x)$ be the equation to the curve, we have

$$y_1 = \phi(x) + h\phi'(x) + \frac{h^2}{2} \phi''(x + \theta h).$$

And since the equation to the tangent at P is

$$Y - y = \phi'(x)(X - x),$$

we have

$$y_2 = \phi(x) + h\phi'(x) ;$$

therefore

$$y_1 - y_2 = \frac{h^2}{2} \phi''(x + \theta h).$$

This, if we take h small enough, will have the same sign as $\phi''(x)$; and therefore the curve is convex to the axis of x if $\phi''(x)$ be positive, and concave if $\phi''(x)$ be negative.

We have supposed in the figures that the curve is *above* the axis of x. If it be *below* the axis of x, then $-y_1$ and $-y_2$ are the *numerical* values of the ordinates, and the curve is convex if $-y_1 + y_2$ be positive, that is, if $\phi''(x)$ be negative, and concave if $\phi''(x)$ be positive.

Both cases may be included in one enunciation, thus, "A curve is convex or concave to the axis of x according as $y \dfrac{d^2y}{dx^2}$ is positive or negative."

290. DEFINITION. A point of inflexion is a point at which a curve cuts its tangent at that point.

To find the conditions for the existence of a point of inflexion. Let $y = \phi(x)$ be the equation to a curve; let x, y, be the co-ordinates of a point in a curve, and $x + h$, y_1, the co-ordinates of an adjacent point. Let the tangent of the curve at the point (x, y) be drawn, and let y_2 be the ordinate of this tangent corresponding to the abscissa $x + h$. Then

$$y_1 = \phi(x) + h\phi'(x) + \frac{h^2}{2}\phi''(x + \theta h),$$

$$y_2 = \phi(x) + h\phi'(x);$$

therefore $$y_1 - y_2 = \frac{h^2}{2}\phi''(x + \theta h).$$

Hence, if $\phi''(x)$ be not zero, the sign of $y_1 - y_2$ will, if h be small enough, be the same as that of $\phi''(x)$, whether h be positive or negative, and the curve cannot cut its tangent. Therefore if there be a point of inflexion, we must have $\phi''(x) = 0$. Suppose this condition satisfied, then

$$y_1 - y_2 = \frac{h^3}{\lfloor 3}\phi'''(x + \theta h):$$

and this expression changes its sign when h does, provided $\phi'''(x)$ be not zero. If $\phi'''(x)$ be zero, it may be shewn that $\phi''''(x)$ must also vanish; and generally if for a certain value of x several of the successive differential coefficients of y vanish, beginning with the second, there is a point of inflexion if the first differential coefficient that does *not* vanish is of an *odd* order.

Since generally at a point of inflexion $\frac{d^2y}{dx^2}$ *vanishes* while $\frac{d^3y}{dx^3}$ *is finite*, $\frac{d^2y}{dx^2}$ changes its sign. For $\frac{d^3y}{dx^3}$ is the differential coefficient of $\frac{d^2y}{dx^2}$; therefore, by Art. 89, if $\frac{d^3y}{dx^3}$ be positive $\frac{d^2y}{dx^2}$ increases with x, and if $\frac{d^3y}{dx^3}$ be negative $\frac{d^2y}{dx^2}$ decreases as x increases. Hence $\frac{d^2y}{dx^2}$ must pass from negative to positive if $\frac{d^3y}{dx^3}$ be positive, and from positive to negative if $\frac{d^3y}{dx^3}$ be negative.

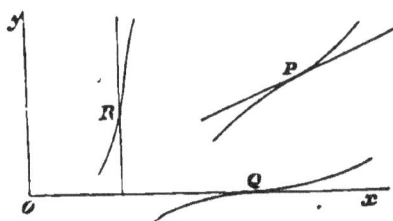

291. In the above figure P, Q, R, are points of inflexion for the curves passing through them. At P there is a change from concavity to convexity with respect to the axis of x. At Q there is a point of inflexion, but the curve on both sides of Q is convex to the axis of x. This agrees with Art. 289; since, if y and $\dfrac{d^2 y}{dx^2}$ both change sign, no change occurs in the sign of their product. At R we have a point of inflexion at which $\dfrac{dy}{dx}$ is infinite and therefore also $\dfrac{d^2 y}{dx^2}$ is infinite by Art. 113, a case which the investigation in Art. 290 does not include. We should therefore in any example ascertain if $\dfrac{d^2 y}{dx^2}$ can become infinite, and if so we must examine that case specially. We may trace the curve in the neighbourhood of that point, or we may examine the sign of $\dfrac{d^2 y}{dx^2}$ for values of x differing slightly from that which gives rise to the infinite value, and thus determine if the curve is concave or convex near the point in question.

If we consider y as the independent variable, we may shew in the manner of the preceding Articles, that a curve is convex or concave to the axis of y, according as $x \dfrac{d^2 x}{dy^2}$ is positive or negative, and that at a point of inflexion $\dfrac{d^2 x}{dy^2}$ must vanish and change its sign. This is often useful in cases in which $\dfrac{d^2 y}{dx^2}$ becomes infinite.

292. The connexion between $\dfrac{d^2 y}{dx^2}$ and the concavity or convexity of a curve, may also be shewn thus.

Let PL, QM, RN, be three equidistant ordinates. Draw the chord PR meeting QM at H. Let $y = \phi(x)$ be the equation to the curve; x, y, the co-ordinates of P; $LM = MN = h$. If the curve be *concave* to the axis of x, QM is *greater* than HM; and therefore $2QM$ greater than $2HM$, that is, greater than $PL + RN$. Hence

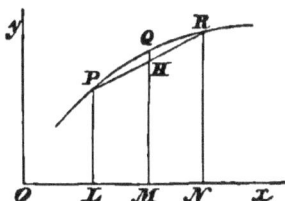

$$\phi(x + 2h) - 2\phi(x + h) + \phi(x) \text{ is negative,}$$

and therefore also $\dfrac{\phi(x + 2h) - 2\phi(x + h) + \phi(x)}{h^2}$ is negative.

Let h diminish indefinitely, and it follows by Art. 127, that $\phi''(x)$ is negative. Similarly, if the curve be convex to the axis of x, then $\phi''(x)$ is positive.

293. We will briefly indicate another method by which the results of this Chapter are sometimes obtained. It is either deduced from some definition of concavity and convexity, or given as the definition of those words, that y being supposed positive, a curve is *convex* to the axis of x, if $\dfrac{dy}{dx}$ be increasing, that is, if $\dfrac{d^2y}{dx^2}$ be positive, and *concave* if $\dfrac{dy}{dx}$ be decreasing, that is, if $\dfrac{d^2y}{dx^2}$ be negative.

Also a point of inflexion may be defined as a point where the curve changes from being concave to being convex, or *vice versa*. Hence $\dfrac{d^2y}{dx^2}$ must change sign at a point of inflexion.

A point of inflexion may also be defined as a point at which the inclination of the tangent to the axis has a maximum or minimum value. Since when this angle has a maximum or minimum value, so also has its tangent, we must have $\dfrac{dy}{dx}$ a maximum or minimum at a point of inflexion.

Hence $\dfrac{d^2y}{dx^2}$ must change sign.

294. A curve referred to polar co-ordinates is said to be concave or convex to the pole at any point, according as the curve in the neighbourhood of that point does, or does not, lie on the same side of the tangent as the pole.

If p be the perpendicular from the pole on the tangent at a point whose co-ordinates are r, θ, it will be seen from a figure, that if the curve be *concave* to the pole, p increases if r increases, and decreases if r decreases; hence $\dfrac{dp}{dr}$ must be *positive*. Similarly if the curve be *convex* to the pole $\dfrac{dp}{dr}$ must be *negative*. Thus at a point of inflexion $\dfrac{dp}{dr}$ must change sign.

295. Since $\qquad \dfrac{1}{p^2} = u^2 + \left(\dfrac{du}{d\theta}\right)^2$, Art. 284,

therefore $\qquad -\dfrac{1}{p^3}\dfrac{dp}{d\theta} = \left(u + \dfrac{d^2u}{d\theta}\right)\dfrac{du}{d\theta}$;

therefore $\qquad \dfrac{dp}{du} = -p^3\left(u + \dfrac{d^2u}{d\theta^2}\right).$

But $\qquad \dfrac{dp}{dr} = \dfrac{dp}{du}\dfrac{du}{dr} = -\dfrac{1}{r^2}\dfrac{dp}{du}$

$$= \dfrac{p^3}{r^2}\left(u + \dfrac{d^2u}{d\theta^2}\right).$$

Hence, at a point of inflexion we must have generally $u + \dfrac{d^2u}{d\theta^2}$ changing its sign.

EXAMPLES.

1. If $y = \dfrac{x^3}{a^2 + x^2}$, there is a point of inflexion at the origin, and also when $x = \pm a\sqrt{3}$.

2. If $y = \dfrac{x^2(x+a)}{a(x-a)}$, there is a point of inflexion when $x = -a(\sqrt[3]{2} - 1)$.

3. If $y(a^4 - b^4) = x(x - a)^4 - xb^4$, there is a point of inflexion when $x = \dfrac{2a}{5}$. Is there a point of inflexion when $x = a$?

4. If $\dfrac{y}{a} = \sqrt{\left(\dfrac{a - x}{x}\right)}$, there is a point of inflexion when $x = \dfrac{3a}{4}$.

5. If $\dfrac{y}{a} = \dfrac{x^2}{a^2} + \left(\dfrac{x - a}{a}\right)^{\frac{3}{5}}$, there is a point of inflexion when $x = a$.

6. If $x^{\frac{1}{3}} = \log y$, there is a point of inflexion when $x = 8$.

7. If $ax^2 - x^2y - a^2y = 0$, there is a point of inflexion when $x = \pm \dfrac{a}{\sqrt{3}}$.

8. If $\dfrac{y}{a} = \sqrt{\left(\dfrac{x}{2a - x}\right)}$, there is a point of inflexion when $x = \dfrac{a}{2}$.

9. If $xy = a^2 \log \dfrac{x}{a}$, there is a point of inflexion when $x = ae^{\frac{3}{2}}$.

10. Find the point of inflexion on the curve,
$$\{y - 2\sqrt[3]{(a^2x)}\}^2 = 4ax. \qquad \textit{Result.} \quad x = \left(\dfrac{8}{9}\right)^6 a.$$

11. If $y(x^2 + a^2) = a^2(a - x)$, there are three points of inflexion which lie on a straight line.

12. If $r = \dfrac{a\theta^2}{\theta^2 - 1}$, there is a point of inflexion when $r = \dfrac{3a}{2}$.

13. If $r = b \cdot \theta^n$, there is a point of inflexion when
$$r = b\{-n(n + 1)\}^{\frac{n}{2}}.$$

14. If $x = a(1 - \cos\phi)$, and $y = a(n\phi + \sin\phi)$, there is a point of inflexion when $\cos\phi = -\dfrac{1}{n}$.

CHAPTER XXII.

SINGULAR POINTS.

296. UNDER the common title of " Singular Points " are included all those points on a curve which offer any singularity depending on the curve itself and independent of the position of the co-ordinate axes. We proceed to define the different singular points and to investigate the conditions of their existence.

Points of Inflexion.

297. These points have been considered in Arts. 288...295 ; the condition for their existence is that $\frac{d^2y}{dx^2}$ should change sign.

Multiple Points.

298. DEFINITION. A multiple point is a point through which two or more branches of a curve pass.

Let $\phi(x, y) = 0$ be an equation in a *rational* form; by Art. 177

$$\frac{d\phi}{dx} + \frac{d\phi}{dy}\frac{dy}{dx} = 0 \dots\dots\dots\dots\dots\dots (1).$$

Now since two or more branches of a curve pass through a multiple point, it will be possible to draw more than one tangent to the curve at that point; hence $\frac{dy}{dx}$ must admit of more than one value. But since the equation $\phi(x, y) = 0$ is supposed *rational*, $\frac{d\phi}{dx}$ and $\frac{d\phi}{dy}$ will each have but *one* value for the given values of x and y. Hence from equation (1) it follows that $\dfrac{dy}{dx}$ cannot have more than one value unless both

$$\frac{d\phi}{dx} = 0, \quad \text{and} \quad \frac{d\phi}{dy} = 0.$$

These then are the conditions for the existence of a multiple point. If values of x and y can be found which satisfy these equations and the equation to the curve, then for such values of x and y we have, by Art. 181,

$$\frac{d^2\phi}{dx^2} + 2\frac{d^2\phi}{dx\,dy}\frac{dy}{dx} + \frac{d^2\phi}{dy^2}\left(\frac{dy}{dx}\right)^2 = 0 \dots\dots\dots (2).$$

From this quadratic equation we can find two values of $\dfrac{dy}{dx}$, and thus determine two tangents which can be drawn through the multiple point. In this case the multiple point is called a double point.

If the above equation assumes an indeterminate form by the vanishing of $\dfrac{d^2\phi}{dx^2}$, $\dfrac{d^2\phi}{dx\,dy}$, and $\dfrac{d^2\phi}{dy^2}$, for the values of x and y under consideration, we have, by Art. 184,

$$\frac{d^3\phi}{dx^3} + 3\frac{d^3\phi}{dx^2\,dy}\frac{dy}{dx} + 3\frac{d^3\phi}{dx\,dy^2}\left(\frac{dy}{dx}\right)^2 + \frac{d^3\phi}{dy^3}\left(\frac{dy}{dx}\right)^3 = 0 \dots\dots (3).$$

This cubic equation gives three values of $\dfrac{dy}{dx}$; if they are all real, *three* tangents to the curve pass through the point under consideration; the point is then called a triple point.

If the equation (3) assumes an indeterminate form by the vanishing of the coefficients of the different powers of $\dfrac{dy}{dx}$, we must proceed to the *fourth* derived equation from $\phi(x, y) = 0$, and we thus obtain a biquadratic equation for determining $\dfrac{dy}{dx}$.

299. If the two values of $\dfrac{dy}{dx}$ furnished by equation (2) of Art. 298 are equal, the two branches which pass through the point in question have a common tangent at that point. In this case, however, the method by which we have arrived at equation (2) is not satisfactory, because in obtaining it we

have assumed $\dfrac{dy}{dx}$ to have more than *one* value. But as in this case two different branches of the curve pass through the same point, there will generally be *two different values* of $\dfrac{d^2y}{dx^2}$; by Art. 181,

$$\frac{d^2\phi}{dx^2} + 2\frac{d^2\phi}{dx\,dy}\frac{dy}{dx} + \frac{d^2\phi}{dy^2}\left(\frac{dy}{dx}\right)^2 + \frac{d\phi}{dy}\frac{d^2y}{dx^2} = 0,$$

and since $\phi(x, y)$ is rational, each of the differential coefficients of ϕ has only one value; hence if $\dfrac{d\phi}{dy}$ be different from zero $\dfrac{d^2y}{dx^2}$ can have *only one value*. But, by supposition $\dfrac{d^2y}{dx^2}$ has more than one value; therefore $\dfrac{d\phi}{dy} = 0$ is the condition that must hold at the point where two branches touch. Since $\dfrac{d\phi}{dy} = 0$, it follows from (1) of Art. 298 that $\dfrac{d\phi}{dx}$ also $= 0$.

If $\dfrac{d^2y}{dx^2}$ should have two *equal* values, then the reasoning of this Article may be applied to $\dfrac{d^3y}{dx^3}$ and the *third* derived equation of $\phi(x, y) = 0$; and the same result as before may be deduced.

Points where two or more values of $\dfrac{dy}{dx}$ are equal are called " Points of Osculation."

300. **Example.** Let $y^2 - x^2(1 - x^2) = 0$.

Here $\dfrac{d\phi}{dy} = 2y,$ $\qquad \dfrac{d\phi}{dx} = -2x(1 - x^2) + 2x^3.$

Hence $x = 0$, $y = 0$, are the co-ordinates of a point which *may* be a double point. Equation (2) of Art. 298 becomes

$$1 - \left(\frac{dy}{dx}\right)^2 = 0,$$

therefore $\dfrac{dy}{dx} = \pm 1$, and there is a double point.

We may in this case put the given equation in the form
$$y = \pm x\sqrt{(1 - x^2)},$$

T. D. C. Y

and from this we see that for values of x comprised between 0 and 1, both positive and negative, y is possible, and that there are *two* values of y for every value of x. When $x = 0$ the two values of y become equal; but since

$$\frac{dy}{dx} = \pm \sqrt{(1 - x^2)} \mp \frac{x^2}{\sqrt{(1 - x^2)}},$$

we see that when $x = 0$ there are *two* values of $\frac{dy}{dx}$. Hence, instead of clearing an equation of radicals so as to bring it into a *rational* form, and then applying the method of Art. 298, we may often detect a multiple point more easily by observing what values of x make *one of the radicals in the value of y vanish.*

Cusps.

301. DEFINITION. A cusp is a point of a curve at which two branches meet a common tangent and stop at that point. If the two branches lie on *opposite* sides of the common tangent, the cusp is said to be of the *first* species; if on the *same* side, the cusp is said to be of the *second* species.

Since a cusp is really a multiple point, if a cusp exist in the curve $\phi(x, y) = 0$ at any point, we must have

$$\frac{d\phi}{dx} = 0, \quad \text{and} \quad \frac{d\phi}{dy} = 0,$$

at that point. To distinguish a cusp from an ordinary multiple point, we must trace the curve in the vicinity of the point in question.

Example. Let $(cy - bx)^2 - \dfrac{(x - a)^3}{a} = 0$ ....(1).

Here when $x = a$ and $y = \dfrac{ab}{c}$ we have the equation to the curve satisfied and also

$$\frac{d\phi}{dx} = 0, \quad \text{and} \quad \frac{d\phi}{dy} = 0.$$

Putting the given equation in the form

$$y = \frac{bx}{c} \pm \frac{1}{c} \sqrt{\left\{ \frac{(x - a)^3}{a} \right\}} \quad ....(2),$$

we see that y is impossible so long as x is less than a, and that when x is greater than a there are two values of y for every value of x. When $x = a$ the radical in y vanishes, and the two values of y become equal; at the same time $\dfrac{dy}{dx}$ has only *one* value, namely $\dfrac{b}{c}$. Hence there is a cusp.

In the figure, A represents the cusp; the straight line OA has for its equation $y = \dfrac{bx}{c}$; and since of the two values of y given by equation (2), one is *greater* and the other *less* than $\dfrac{bx}{c}$, it is obvious that the two branches lie on *opposite* sides of OA, and the cusp at A is of the *first* species. Generally the cusp is of the *first* species if the two values of $\dfrac{d^2y}{dx^2}$ indefinitely near to the point are of *contrary* signs, and of the second species if they are of the *same* sign.

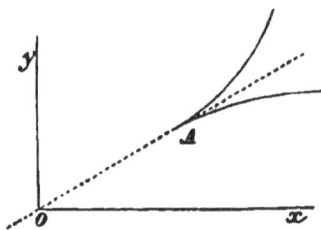

Cusps of the first species have been called "keratoid cusps," and of the second "rhamphoid cusps."

Conjugate Points.

302. DEFINITION. A conjugate point is an *isolated* point the co-ordinates of which satisfy the equation to the curve. For example, let

$$y^2 = \frac{x^2}{a^2}\,(x^2 - a^2).$$

Here the values $x = 0$, $y = 0$, satisfy the equation to the curve, but no branch of the curve passes through the point thus determined, y being impossible for all other values of x comprised between $-a$ and a. Hence the origin of co-ordinates is a conjugate point in this curve.

In the above example, since

$$y = \pm \frac{x}{a}\,\sqrt{(x^2 - a^2)},$$

we find that the value of $\dfrac{dy}{dx}$ is impossible when $x = 0$; but $\dfrac{dy}{dx}$

Y 2

may be possible at a conjugate point; for example, suppose

$$y = \pm \frac{x^2}{a^2} \sqrt{(x^2 - a^2)}.$$

Here, when $x = 0$, we have $\frac{dy}{dx} = 0$; but the origin is a conjugate point, since $x = 0$, $y = 0$, satisfy the equation, and y
is impossible for all other values of x between $-a$ and a. In
like manner $\frac{d^2y}{dx^2}$ or any number of the differential coefficients
of y *may* be possible at a conjugate point, but they cannot be
all possible, for if they were we should have nothing to distinguish the point in question from an ordinary point of the
curve.

To find the condition for the existence of a conjugate point.
Since at a conjugate point the values of the differential
coefficients of y cannot be all possible, let the nth differential
coefficient of y be the first that is impossible. Suppose the
equation to the curve to be put in a rational form, and
denoted by $\phi(x, y) = 0$. Take the nth derived equation; we
have

$$\frac{d\phi}{dy}\frac{d^ny}{dx^n} + \ldots\ldots + \frac{d^n\phi}{dx^n} = 0,$$

where the terms not written down contain differential coefficients of ϕ with respect to x and y, and also differential
coefficients of y with respect to x of orders inferior to the nth.
If then $\frac{d\phi}{dy}$ be not zero the value of $\frac{d^ny}{dx^n}$ furnished by the
above equation will be possible; hence $\frac{d\phi}{dy} = 0$ is a necessary
condition for the existence of a conjugate point; but

$$\frac{d\phi}{dx} + \frac{d\phi}{dy}\frac{dy}{dx} = 0,$$

therefore also $\frac{d\phi}{dx} = 0.$

303. It appears from the preceding Articles that if
$\phi(x, y) = 0$ be the equation to a curve, we must have at

an ordinary multiple point, at a cusp, and at a conjugate point,

$$\frac{d\phi}{dx} = 0, \quad \text{and} \quad \frac{d\phi}{dy} = 0.$$

Hence, whenever we have found values of x and y which satisfy these three equations, we must, by examining the particular curve, and tracing it in the vicinity of the point in question, determine what species of singular point exists.

We now pass to some other singular points which occur but rarely, and, as the student will find by experience, never present themselves in curves the equations to which are of an *algebraical* form. See Art. 6.

Points d'arrêt.

304. A *point d'arrêt* is a point at which a single branch of a curve suddenly stops.

Example. Let $y = x \log x$.

Here when $x = 0$ we have $y = 0$; but if x be negative, y becomes impossible. Hence the origin is a *point d'arrêt*.

Again, suppose $y = e^{-\frac{1}{x}}$.

Here if x be made indefinitely small and *positive*, we have y approaching the limit zero; but if x be *negative* and indefinitely small, y is indefinitely great.

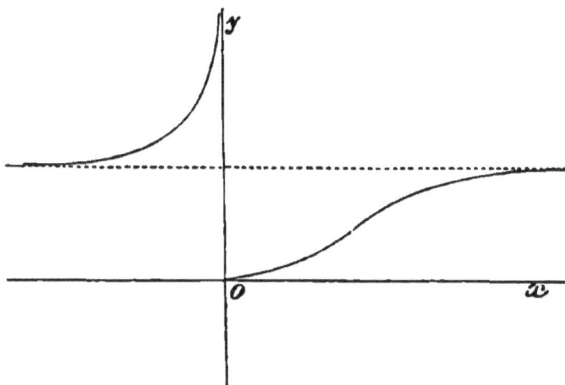

Hence the curve has the form represented in the figure, the origin being a *point d'arrêt*; the dotted line is an asymptote having for its equation $y = 1$.

305. A *point saillant* is a point at which two branches of a curve meet and stop without having a common tangent.

Example. Let
$$y = \frac{x}{1 + e^{\frac{1}{x}}},$$

therefore
$$\frac{dy}{dx} = \frac{1}{1 + e^{\frac{1}{x}}} + \frac{e^{\frac{1}{x}}}{x\,(1 + e^{\frac{1}{x}})^2}.$$

Here, if x be *positive* and approach zero as its limit, we have ultimately $y = 0$ and $\frac{dy}{dx} = 0$; but if x be *negative*, we have ultimately $y = 0$ and $\frac{dy}{dx} = 1$. Hence at the origin two branches meet, one having the axis of x as its tangent, and the other inclined to the axis of x at an angle of 45°.

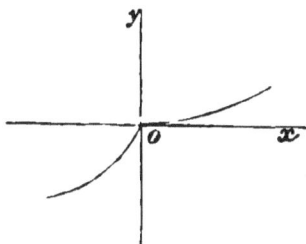

Branches Pointillées.

306. If a curve has an *infinite* number of conjugate points, that series of points is called a *branche pointillée*.

For example, suppose $y^2 = x \sin^2 x$; for all positive values of x there are two possible values of y, but when x is negative y is impossible, unless x be a multiple of π. Hence we have an *infinite number* of conjugate points lying on the axis of x and forming a *branche pointillée*.

EXAMPLES.

1. If $a^2 y^2 = a^2 x^2 - x^4$ there is a multiple point at the origin.

2. In the following curves there is a point of inflexion at the origin:
$$y = \sin x; \quad y = x \cos x; \quad y = \tan x; \quad y = x^2 \tan x.$$

3. The following curves have cusps at the origin:
$$y^2 = x^3; \quad (y - x)^2 = x^3; \quad (y - x^2)^2 = x^5.$$

4. If $y = \phi(x) + (x-a)^{\frac{2p+1}{2q}} F(x)$, when $x = a$, there is a cusp of the first kind if $\dfrac{2p+1}{2q}$ be greater than 1 and less than 2, and a cusp of the second kind if $\dfrac{2p+1}{2q}$ be greater than 2.

5. The curve $y^3 = (x-a)^2 (x-c)$ has a cusp of the first kind at the point $x = a$.

6. The curve $(xy+1)^2 + (x-1)^3 (x-2) = 0$ has a cusp of the first kind at the point $x = 1$.

7. The curve $y - b = (x-a)^{\frac{3}{2}} + (x-a)^{\frac{5}{2}}$ has a cusp of the second kind at the point $x = a$.

8. The curve $x^4 - 2ax^2 y - axy^2 + a^2 y^2 = 0$ has a cusp of the second kind at the origin.

9. The curve $x^4 - ax^2 y - axy^2 + a^2 y^2 = 0$ has a conjugate point at the origin.

10. The curve $x^4 - 2ay^3 - 3a^2 y^2 - 2a^2 x^2 + a^4 = 0$ has a double point when $x = \pm a$, and $\dfrac{dy}{dx}$ then $= \pm\sqrt{\frac{4}{3}}$; also a double point when $y = -a$, and $\dfrac{dy}{dx}$ then $= \pm\sqrt{\frac{2}{3}}$.

11. If $ay^2 = (x-a)^2 (x-b)$, when $x = a$ there is a conjugate point if a be less than b, a double point if a be greater than b, and a cusp if $a = b$.

12. Shew that the curve $ay^2 - x^3 + bx^2 = 0$ has a conjugate point at the origin, and a point of inflexion when $x = \dfrac{4b}{3}$.

13. Find the points of inflexion in the following curves:
$$y^2(1 + x^2) = (1 - x + x^2)^2; \quad r^2\theta = a^2; \quad r\theta \sin\theta = a.$$

14. Find the singular points in the following curves:
$$(y + x + 1)^2 = (1 - x)^5; \quad y^4 - axy^2 + x^4 = 0;$$
$$y^2 = x^3 - x^4; \quad y^4 + xy^3 + x^2(ay - bx) = 0.$$

CHAPTER XXIII.

DIFFERENTIAL COEFFICIENTS OF AN ARC, AN AREA, A VOLUME, AND A SURFACE.

307.   THE length of the arc of a curve APQ, reckoned from any fixed point A to the point P, is evidently a function of the abscissa x of the point P. This function is often very difficult to determine, but its differential coefficient with respect to x can always be assigned.

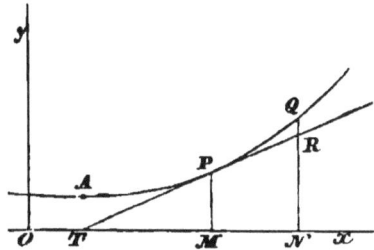

Let P, Q, be two points on a curve ;

x, y, the co-ordinates of P ;

$x + \Delta x$, $y + \Delta y$, the co-ordinates of Q.

Draw the ordinates PM, QN, and the tangent at P meeting QN at R and Ox at T.

Let $\qquad AP = s, \qquad AQ = s + \Delta s.$

We assume as an axiom, that the length Δs is greater than the chord PQ, and less than $PR + RQ$.

The chord $PQ = \sqrt{\{(\Delta x)^2 + (\Delta y)^2\}}$,

$$PR = MN \sec PTM = MN \sqrt{(1 + \tan^2 PTM)}$$

$$= \Delta x \sqrt{\left\{1 + \left(\frac{dy}{dx}\right)^2\right\}},$$

$$QR = y + \Delta y - RN$$
$$= y + \Delta y - (PM + \Delta x \tan PTM)$$
$$= \qquad \Delta y - \Delta x \frac{dy}{dx};$$

therefore Δs lies between $\sqrt{\{(\Delta x)^2 + (\Delta y)^2\}}$ and

$$\Delta x \sqrt{\left\{1 + \left(\frac{dy}{dx}\right)^2\right\}} + \Delta y - \Delta x \frac{dy}{dx};$$

therefore $\dfrac{\Delta s}{\Delta x}$ lies between $\sqrt{\left\{1 + \left(\dfrac{\Delta y}{\Delta x}\right)^2\right\}}$ and

$$\sqrt{\left\{1 + \left(\frac{dy}{dx}\right)^2\right\}} + \frac{\Delta y}{\Delta x} - \frac{dy}{dx}.$$

Now the limit of $\sqrt{\left\{1 + \left(\dfrac{\Delta y}{\Delta x}\right)^2\right\}}$, when Δx is indefinitely diminished, is

$$\sqrt{\left\{1 + \left(\frac{dy}{dx}\right)^2\right\}}.$$

The limit of $\sqrt{\left\{1 + \left(\dfrac{dy}{dx}\right)^2\right\}} + \dfrac{\Delta y}{\Delta x} - \dfrac{dy}{dx}$ is

$$\sqrt{\left\{1 + \left(\frac{dy}{dx}\right)^2\right\}} + \frac{dy}{dx} - \frac{dy}{dx}, \text{ or } \sqrt{\left\{1 + \left(\frac{dy}{dx}\right)^2\right\}}.$$

The limit of $\dfrac{\Delta s}{\Delta x}$ is, by definition, $\dfrac{ds}{dx}$; hence

$$\frac{ds}{dx} = \sqrt{\left\{1 + \left(\frac{dy}{dx}\right)^2\right\}} \quad \dots\dots\dots\dots \dots\dots\dots (1).$$

Square and multiply by $\left(\dfrac{dx}{ds}\right)^2$, then

$$1 = \left(\frac{dx}{ds}\right)^2 + \left(\frac{dy}{ds}\right)^2 \quad \dots\dots\dots\dots\dots\dots (2).$$

If x and y are each functions of a third variable t, since

$$\frac{dx}{ds} = \frac{\dfrac{dx}{dt}}{\dfrac{ds}{dt}}, \text{ and } \frac{dy}{ds} = \frac{\dfrac{dy}{dt}}{\dfrac{ds}{dt}},$$

we have from (2), $\left(\dfrac{ds}{dt}\right)^2 = \left(\dfrac{dx}{dt}\right)^2 + \left(\dfrac{dy}{dt}\right)^2 \quad \dots\dots\dots\dots (3).$

308. Of the axioms on which the preceding demonstration is founded, the first will probably be readily granted ; the second is more difficult, and will not be necessarily true if the arc be not *concave towards the chord PQ throughout its extent.* It must be understood therefore, in stating it, that the arc *PQ* must be taken so small that it is always concave towards its chord.

There is another mode of arriving at the results given in Art 307, which is preferred by some writers: they assert that no precise idea can be formed of the *length of an arc*, except by regarding it as the *limit of the perimeter of a polygon inscribed in that arc, when the length of each side of the polygon is indefinitely diminished.* If we adopt this definition of the length of an arc, we must shew that the limit mentioned does exist, and is the same in whatever manner we suppose the polygon inscribed, provided that each side is ultimately indefinitely diminished.

Draw two chords dividing the whole arc we are considering into two portions; then in each of these subdivisions place two chords dividing the whole arc into four portions ; in each of the last subdivisions place two chords, and so on. The perimeters of the polygons thus formed constitute a series continually increasing ; and as it is easy to see they cannot increase without limit, we prove the first point, namely, that there is a *limit to the perimeter of the inscribed polygon when the length of each side is indefinitely diminished.*

Suppose now two polygons with indefinitely small sides inscribed in the curve, one of them being one of the series just considered, and the other described after any other law. Draw tangents to the curve at the angular points of *both* polygons, thus forming *one* polygon circumscribing the arc. Then it is easy to see that any chord of either polygon bears to the corresponding portion of the circumscribing figure, a ratio which can be made as near to unity as we please by sufficiently diminishing the length of each chord. Hence the perimeter of each inscribed figure bears to that of the circumscribed figure a ratio which is ultimately one of equality, and consequently the ratio of the perimeter of one inscribed figure to that of the other inscribed figure is ultimately one of equality. This proves the second point involved in the definition of the length of an arc, namely, that the limit obtained is the same according to whatever law the polygons be inscribed.

From this definition of the length of an arc it follows that the ultimate ratio of the length of an indefinitely small arc to its chord is one of equality, that is,

$$\frac{\Delta s}{\sqrt{\{(\Delta x)^2 + (\Delta y)^2\}}} \quad \text{or} \quad \frac{\dfrac{\Delta s}{\Delta x}}{\sqrt{\left\{1 + \left(\dfrac{\Delta y}{\Delta x}\right)^2\right\}}} = 1, \text{ ultimately,}$$

therefore $\qquad \dfrac{ds}{dx} = \sqrt{\left\{1 + \left(\dfrac{dy}{dx}\right)^2\right\}}.$

309.   Since secant $PTx = \sqrt{\left\{1 + \left(\dfrac{dy}{dx}\right)^2\right\}}$,

we have $\qquad \cos PTx = \dfrac{1}{\sqrt{\left\{1 + \left(\dfrac{dy}{dx}\right)^2\right\}}} = \dfrac{dx}{ds},$

and $\qquad \sin PTx = \cos PTx \tan PTx$

$$= \frac{dx}{ds}\frac{dy}{dx} = \frac{dy}{ds}.$$

310.   If x and y be expressed in terms of θ from the equations

$$x = r \cos \theta, \quad y = r \sin \theta,$$

we have $\qquad \dfrac{ds}{d\theta} = \dfrac{ds}{dx}\dfrac{dx}{d\theta}$

$$= \sqrt{\left\{1 + \left(\frac{dy}{dx}\right)^2\right\}}\frac{dx}{d\theta}$$

$$= \sqrt{\left\{\left(\frac{dx}{d\theta}\right)^2 + \left(\frac{dy}{d\theta}\right)^2\right\}}.$$

But $\qquad \dfrac{dx}{d\theta} = \cos \theta \dfrac{dr}{d\theta} - r \sin \theta,$

$$\frac{dy}{d\theta} = \sin \theta \frac{dr}{d\theta} + r \cos \theta;$$

therefore
$$\frac{ds}{d\theta} = \sqrt{\left\{\left(\frac{dr}{d\theta}\right)^2 + r^2\right\}}.$$

Also
$$\frac{ds}{dr} = \frac{ds}{d\theta}\frac{d\theta}{dr} = \sqrt{\left\{1 + r^2\left(\frac{d\theta}{dr}\right)^2\right\}}.$$

We have shewn in Art. 279, that

$$\tan\phi = r\frac{d\theta}{dr},$$

where ϕ is the angle between the radius vector at the point whose polar co-ordinates are r, θ, and the tangent at that point. Hence

$$\sin\phi = \frac{r\dfrac{d\theta}{dr}}{\sqrt{\left\{1 + r^2\left(\dfrac{d\theta}{dr}\right)^2\right\}}} = \frac{r\dfrac{d\theta}{dr}}{\dfrac{ds}{dr}} = r\frac{d\theta}{ds}.$$

Similarly
$$\cos\phi = \frac{dr}{ds}.$$

These results may also be deduced immediately from the figure in Art. 279; for $\sin\phi$ is the limiting value of $\dfrac{PL}{PQ}$, that is, of $\dfrac{PL}{\Delta s}\cdot\dfrac{\Delta s}{PQ}$ or of $\dfrac{r\sin\Delta\theta}{\Delta s}\cdot\dfrac{\Delta s}{PQ}$. The limit of $\dfrac{r\sin\Delta\theta}{\Delta s}$ is $\dfrac{rd\theta}{ds}$; and the limit of $\dfrac{\Delta s}{PQ}$ is unity; hence $\sin\phi = \dfrac{rd\theta}{ds}$. Similarly the value of $\cos\phi$ may be found.

311.   The value of $\dfrac{ds}{d\theta}$, in Art. 310, may also be obtained thus:

Let P, Q, be points on a curve, and suppose

$$SP = r, \qquad PSx = \theta,$$
$$SQ = r + \Delta r, \qquad QSx = \theta + \Delta\theta.$$

Draw PL perpendicular to SQ, then

$$PL = r\sin\Delta\theta,$$

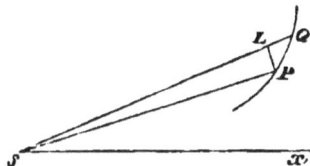

$$LQ = r + \Delta r - r \cos \Delta\theta$$
$$= \Delta r + 2r \sin^2 \frac{\Delta\theta}{2}.$$

Also the chord $PQ = \sqrt{(PL^2 + LQ^2)}$.

From this, if we proceed according to the method of the preceding Articles, we shall arrive at

$$\frac{ds}{d\theta} = \sqrt{\left\{ r^2 + \left(\frac{dr}{d\theta}\right)^2 \right\}}.$$

312. If A denote the area contained between a curve and the axis of x, we have shewn in Art. 43 that

$$\frac{dA}{dx} = y.$$

313. To find the differential coefficient of the area of a curve referred to polar co-ordinates.

Let A denote the area contained between the radius SP, the radius SC drawn to some fixed point C on the curve, and the curve CP. Let ΔA denote the area PSQ. With centre S and radius SP describe an arc meeting SQ at L, and with centre S and radius SQ describe an arc meeting SP produced at M. Then ΔA lies between PSL and QSM, that is, between

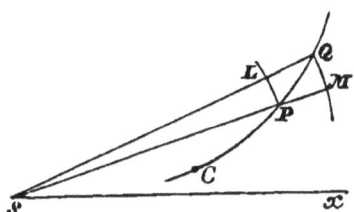

$$\frac{r^2 \Delta\theta}{2} \quad \text{and} \quad \frac{(r + \Delta r)^2}{2} \Delta\theta;$$

therefore $\frac{\Delta A}{\Delta\theta}$ lies between $\frac{r^2}{2}$ and $\frac{(r + \Delta r)^2}{2}$.

Hence, proceeding to the limit, we have

$$\frac{dA}{d\theta} = \frac{r^2}{2}.$$

314. *Differential coefficient of the volume of a solid of revolution.*

Suppose the curve APQ in the figure of Art. 307 to revolve round the axis of x, and thus to generate a solid.

Let V denote the volume of a portion of this solid contained between two planes perpendicular to the axis Ox, one drawn through a fixed point A and the other through P. Let ΔV denote the volume of the solid contained between planes through P and Q perpendicular to the axis. The volume of a cylinder having MN for its axis and for its base the circular area formed by the revolution of PM round the axis Ox, is $\pi y^2 \Delta x$. The volume of a cylinder having MN for its axis and for its base the circular area formed by the revolution of QN round Ox, is $\pi (y + \Delta y)^2 \Delta x$. Hence ΔV lies between $\pi y^2 \Delta x$ and $\pi (y + \Delta y)^2 \Delta x$. Therefore $\dfrac{\Delta V}{\Delta x}$ lies between πy^2 and $\pi (y + \Delta y)^2$. Hence, proceeding to the limit, we have

$$\frac{dV}{dx} = \pi y^2.$$

315. *Differential coefficient of the surface of a solid of revolution.*

Let P, Q, be two points in a curve which by revolving round the axis Ox generates a solid of revolution. Let A be a fixed point on the curve, and suppose $AP = s$, $PQ = \Delta s$. Let S denote the area of the surface formed by the revolution of AP, and ΔS the area of the surface formed by the revolution of PQ. Draw PR and QT each equal to Δs and each parallel to Ox. If PR revolve round Ox it generates a cylinder, the surface of which is $2\pi y \Delta s$. If QT revolve round Ox it generates a cylinder, the surface of which is $2\pi (y + \Delta y) \Delta s$. We assume as an axiom that the surface generated by the arc PQ lies between the former and the latter. Hence ΔS lies between $2\pi y \Delta s$ and $2\pi (y + \Delta y) \Delta s$, and proceeding to the limit, we have

$$\frac{dS}{ds} = 2\pi y;$$

therefore $$\frac{dS}{dx} = 2\pi y \frac{ds}{dx}.$$

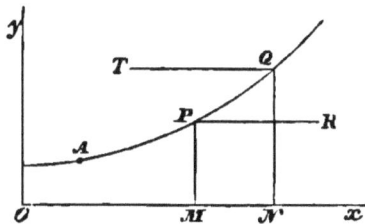

EXAMPLES.

1. In the ellipse $\dfrac{ds}{dx} = \sqrt{\left(\dfrac{a^2 - e^2 x^2}{a^2 - x^2}\right)}$; and if $x = a \sin \phi$,

$$\frac{ds}{d\phi} = a \sqrt{(1 - e^2 \sin^2 \phi)}.$$

2. In the parabola $y^2 = 4ax$, $\dfrac{ds}{dx} = \sqrt{\left(\dfrac{a + x}{x}\right)}$.

3. In the circle $\dfrac{ds}{dx} = \dfrac{a}{y}$.

4. Find the differential coefficient of the arc of the curve
$e^y (e^x - 1) = e^x + 1$.

$$\textit{Result.} \quad \frac{ds}{dx} = \frac{e^{2x} + 1}{e^{2x} - 1}.$$

5. In the curve $x^{\frac{2}{3}} + y^{\frac{2}{3}} = a^{\frac{2}{3}}$, $\dfrac{ds}{dx} = \dfrac{a^{\frac{1}{3}}}{x^{\frac{1}{3}}}$.

6. In the curve $r = a (1 + \cos \theta)$, $\dfrac{ds}{d\theta} = 2a \cos \dfrac{\theta}{2}$.

7. In the curve $r = a^\theta$, $\dfrac{ds}{d\theta} = r \sqrt{\{1 + (\log a)^2\}}$.

8. In the curve $r^2 = a^2 \cos 2\theta$, $\dfrac{ds}{d\theta} = \dfrac{a^2}{r}$.

9. In the curve $r = a\theta$, $\dfrac{ds}{dr} = \dfrac{\sqrt{(a^2 + r^2)}}{a}$.

10. If $e^{-y} = \cos x$, $\dfrac{dx}{ds} = \cos x$.

CHAPTER XXIV.

CONTACT. CURVATURE. EVOLUTES AND INVOLUTES.

316. LET $y = \phi(x)$ be the equation to one curve, and $y = \psi(x)$ the equation to another; then if $\phi(a) = \psi(a)$ the curves *intersect* at the point whose abscissa is a. If moreover $\phi'(a) = \psi'(a)$ the curves have a common tangent at the common point; in this case they are said to have a contact of the *first* order. If moreover $\phi''(a) = \psi''(a)$ the curves are said to have a contact of the *second* order. If $\phi(a) = \psi(a)$, $\phi'(a) = \psi'(a)$, $\phi''(a) = \psi''(a)$, $\phi'''(a) = \psi'''(a)$, and so on up to $\phi^n(a) = \psi^n(a)$, the curves are said to have a contact of the nth order at the common point. When we speak of two curves having contact of the nth order we imply that they have not contact of a higher order; that is, with the preceding notation we imply that $\phi^{n+1}(a)$ is *not* equal to $\psi^{n+1}(a)$.

317. If two curves have at any point a contact of the nth order, then in the vicinity of the common point no curve can pass between them unless it has with both of them a contact of an order not lower than the nth. For let $y = \phi(x)$ and $y = \psi(x)$ be the equations to two curves which have contact of the nth order at the point $x = a$; and let y_1 denote the ordinate in the former curve corresponding to the abscissa $a + h$, and y_2 the ordinate in the latter curve corresponding to the same abscissa; then, by Art. 92,

$$y_1 = \phi(a) + h\phi'(a) + \frac{h^2}{\lfloor 2} \phi''(a) \ldots + \frac{h^n}{\lfloor n} \phi^n(a) + \frac{h^{n+1}}{\lfloor n+1} \phi^{n+1}(a + \theta h),$$

$$y_2 = \psi(a) + h\psi'(a) + \frac{h^2}{\lfloor 2} \psi''(a) \ldots + \frac{h^n}{\lfloor n} \psi^n(a) + \frac{h^{n+1}}{\lceil n+1} \psi^{n+1}(a + \theta h).$$

Hence, since the curves have contact of the nth order,

$$y_1 - y_2 = \frac{h^{n+1}}{\underline{|n+1}} \left\{ \phi^{n+1}(a + \theta h) - \psi^{n+1}(a + \theta h) \right\}.$$

Now suppose $y = \chi(x)$ to be the equation to a third curve which has contact of the mth order with the first curve at the point $x = a$; then if $y_3 = \chi(a + h)$, we have

$$y_1 - y_3 = \frac{h^{n+1}}{\underline{|m+1}} \left\{ \phi^{m+1}(a + \theta h) - \chi^{m+1}(a + \theta h) \right\}.$$

If m be less than n we can give such a value to h as will render $y_1 - y_2$ less than $y_1 - y_3$ for that value of h and all numerically inferior values both positive and negative. Hence in the vicinity of the common point the second curve is nearer to the first than the third is.

In the above expressions θ denotes merely a proper fraction, and it is not necessarily the *same* proper fraction in the different cases.

318. The expression for $y_1 - y_2$ in Art. 317, when h is sufficiently diminished, has the same sign as

$$h^{n+1} \{ \phi^{n+1}(a) - \psi^{n+1}(a) \},$$

and therefore changes sign with h if n be *even;* therefore if two curves have contact of an *even* order they cross each other at the common point. If two curves have contact of an *odd* order they do not cross each other at the common point.

319. In order that a curve may have contact of the nth order with a given curve, it appears from Art. 316 that $n + 1$ equations must be satisfied. Hence, if the equation to a species of curves contain $n + 1$ constants, we may, by giving suitable values to those constants, find the particular curve of the species that has contact of the nth order with a given curve at a given point. For example, the equation to a straight line is of the form $y = mx + c$; since there are two constants, m and c, we may, by properly determining them, find the straight line which has contact of the *first* order with a given curve at a given point. If the

T. D. C.   z

given curve be $y = \phi(x)$, and the given point that whose co-ordinates are $x = a$, $y = \phi(a)$, then we must have

$$ma + c = \phi(a),$$

and $m = \phi'(a).$

Hence m and c are determined.

If $y = \phi(x)$ be the equation to a curve, then

$$y = \phi(a) + (x - a)\,\phi'(a) + \frac{(x-a)^2}{\lfloor 2} \phi''(a) \ldots + \frac{(x-a)^n}{\lfloor n} \phi^n(a)$$

is the equation to a curve which has a contact of the nth order with the given curve at the point $x = a$. This may be easily verified.

320. *Circle of curvature.* The general equation to a circle involves three constants ; hence at any point of a curve a circle may be found which has contact of the *second* order with the curve at that point. We proceed to determine the radius and the centre of such a circle.

DEFINITION. The circle of curvature at any point of a curve is a circle which has at that point a contact of the second order with the curve.

Let $(X - a)^2 + (Y - b)^2 = \rho^2 \ldots\ldots\ldots\ldots(1)$

be the equation to a circle, so that a, b, are the co-ordinates of its centre and ρ its radius. From (1) by differentiating we have

$$\left. \begin{array}{l} X - a + (Y - b)\dfrac{dY}{dX} = 0 \\[2mm] 1 + \left(\dfrac{dY}{dX}\right)^2 + (Y - b)\dfrac{d^2Y}{dX^2} = 0 \end{array} \right\} \ldots\ldots\ldots\ldots (2).$$

If this circle is the circle of curvature at the point (x, y) of a given curve, we must have

$$\left. \begin{array}{l} X = x \\ Y = y \\ \dfrac{dY}{dX} = \dfrac{dy}{dx} \\[2mm] \dfrac{d^2Y}{dX^2} = \dfrac{d^2y}{dx^2} \end{array} \right\} \ldots\ldots\ldots\ldots\ldots (3).$$

Hence, from (2),

$$x - a + (y - b) \frac{dy}{dx} = 0$$

$$1 + \left(\frac{dy}{dx}\right)^2 + (y - b) \frac{d^2y}{dx^2} = 0 \qquad \Bigg\} \quad \ldots\ldots\ldots(4).$$

Therefore

$$y - b = - \frac{1 + \left(\frac{dy}{dx}\right)^2}{\frac{d^2y}{dx^2}}$$

$$x - a = \frac{\frac{dy}{dx}\left\{1 + \left(\frac{dy}{dx}\right)^2\right\}}{\frac{d^2y}{dx^2}} \qquad \Bigg\} \quad \ldots\ldots\ldots\ldots(5).$$

By (1) and (5) we have

$$\rho = \frac{\left\{1 + \left(\frac{dy}{dx}\right)^2\right\}^{\frac{3}{2}}}{\frac{d^2y}{dx^2}}.$$

Hence the values of a, b, ρ, are found, and thus the position and the radius of the circle of curvature at any point of a curve are determined.

In the value of ρ it will be proper in any particular example to give to the radical in the numerator the same sign as $\frac{d^2y}{dx^2}$ has, so as to make ρ positive. Hence if y be positive and the curve *concave* to the axis of x we should put

$$\rho = - \frac{\left\{1 + \left(\frac{dy}{dx}\right)^2\right\}^{\frac{3}{2}}}{\frac{d^2y}{dx^2}}.$$

From the first of equations (4) we see that the point (a, b) is on the normal to the given curve at the point (x, y).

The centre of the circle of curvature at any point is called for shortness the "centre of curvature." Also the radius of the circle of curvature is called the "radius of curvature."

If a straight line be drawn from any point of a curve in any direction the portion of this straight line which is intercepted by the circle of curvature at the assumed point is called the *chord of curvature* at the assumed point in the assumed direction. By the nature of a circle the length of the chord of curvature will be obtained by multiplying the diameter of the circle of curvature by the cosine of the angle between the chord of curvature and the common normal to the curve and the circle at the assumed point.

321. If p be the perpendicular from the origin on the tangent at the point (x, y) of a curve, we have

$$p = \frac{x \dfrac{dy}{dx} - y}{\sqrt{\left\{1 + \left(\dfrac{dy}{dx}\right)^2\right\}}},$$

therefore $\dfrac{dp}{dx} =$

$$\frac{x \dfrac{d^2y}{dx^2}\left\{1 + \left(\dfrac{dy}{dx}\right)^2\right\} - \dfrac{dy}{dx}\dfrac{d^2y}{dx^2}\left(x \dfrac{dy}{dx} - y\right)}{\left\{1 + \left(\dfrac{dy}{dx}\right)^2\right\}^{\frac{3}{2}}}$$

$$= \frac{\left(x + y\dfrac{dy}{dx}\right)\dfrac{d^2y}{dx^2}}{\left\{1 + \left(\dfrac{dy}{dx}\right)^2\right\}^{\frac{3}{2}}}.$$

Also, if $\qquad\qquad r^2 = x^2 + y^2,$

$$r \frac{dr}{dx} = x + y\frac{dy}{dx}.$$

From these values of $\dfrac{dp}{dx}$ and $\dfrac{dr}{dx}$, and the value of ρ given in Art. 320, we see that,

$$\frac{dp}{dx} = \frac{1}{\rho} r \frac{dr}{dx},$$

and $\qquad\qquad \rho = r \dfrac{dr}{dp}.$

322. If x and y be each a function of a third variable t, we have

$$\frac{dy}{dx} = \frac{\dfrac{dy}{dt}}{\dfrac{dx}{dt}}, \quad \frac{d^2y}{dx^2} = \frac{\dfrac{d^2y}{dt^2}\dfrac{dx}{dt} - \dfrac{d^2x}{dt^2}\dfrac{dy}{dt}}{\left(\dfrac{dx}{dt}\right)^3}.$$

Using these values, we deduce

$$\rho = \frac{\left\{\left(\dfrac{dx}{dt}\right)^2 + \left(\dfrac{dy}{dt}\right)^2\right\}^{\frac{3}{2}}}{\dfrac{d^2y}{dt^2}\dfrac{dx}{dt} - \dfrac{d^2x}{dt^2}\dfrac{dy}{dt}}.$$

For example, if $t = s$ the arc of the curve measured from some fixed point, then

$$\rho = \frac{1}{\dfrac{d^2y}{ds^2}\dfrac{dx}{ds} - \dfrac{d^2x}{ds^2}\dfrac{dy}{ds}} \quad\text{....................}(1),$$

since by Art. 307

$$\left(\frac{dx}{ds}\right)^2 + \left(\frac{dy}{ds}\right)^2 = 1 \quad\text{....................}(2).$$

Hence

$$\frac{1}{\rho} = \frac{d^2y}{ds^2}\frac{dx}{ds} - \frac{d^2x}{ds^2}\frac{dy}{ds} \quad\text{....................}(3).$$

By differentiating (2) we obtain

$$0 = \frac{dx}{ds}\frac{d^2x}{ds^2} + \frac{dy}{ds}\frac{d^2y}{ds^2} \quad\text{....................}(4).$$

Square (3) and (4), and add; thus

$$\frac{1}{\rho^2} = \left(\frac{d^2x}{ds^2}\right)^2 + \left(\frac{d^2y}{ds^2}\right)^2.$$

From (3), by means of (4), we may also deduce

$$\frac{1}{\rho} = \frac{\dfrac{d^2y}{ds^2}}{\dfrac{dx}{ds}} = -\frac{\dfrac{d^2x}{ds^2}}{\dfrac{dy}{ds}}.$$

323. If we put $x = r\cos\theta$, and $y = r\sin\theta$, we have from

Art. 201 the values of $\dfrac{dy}{dx}$ and $\dfrac{d^2y}{dx^2}$. Substitute these values in the expression for ρ in Art. 320, and we find

$$\rho = \frac{\left\{ r^2 + \left(\dfrac{dr}{d\theta}\right)^2 \right\}^{\frac{3}{2}}}{r^2 + 2\left(\dfrac{dr}{d\theta}\right)^2 - r\dfrac{d^2r}{d\theta^2}}.$$

If $r = \dfrac{1}{u}$, then $\dfrac{dr}{d\theta} = -\dfrac{1}{u^2}\dfrac{du}{d\theta}$, and

$$\frac{d^2r}{d\theta^2} = \frac{2}{u^3}\left(\frac{du}{d\theta}\right)^2 - \frac{1}{u^2}\frac{d^2u}{d\theta^2}.$$

Substitute these in the above value of ρ; then

$$\rho = \frac{\left\{ u^2 + \left(\dfrac{du}{d\theta}\right)^2 \right\}^{\frac{3}{2}}}{u^3\left(u + \dfrac{d^2u}{d\theta^2}\right)}.$$

This result may also be found thus:

By Art. 321 $\qquad \rho = r\dfrac{dr}{dp} = -\dfrac{1}{u^3}\dfrac{du}{dp}.$

By Art. 284 $\qquad \dfrac{1}{p^2} = u^2 + \left(\dfrac{du}{d\theta}\right)^2,$

therefore $\qquad -\dfrac{1}{p^3}\dfrac{dp}{d\theta} = \left(u + \dfrac{d^2u}{d\theta^2}\right)\dfrac{du}{d\theta},$

and $\qquad \dfrac{dp}{du} = -p^3\left(u + \dfrac{d^2u}{d\theta^2}\right).$

Hence $\qquad \rho = \dfrac{1}{u^3p^3\left(u + \dfrac{d^2u}{d\theta^2}\right)}$

$$= \frac{\left\{ u^2 + \left(\dfrac{du}{d\theta}\right)^2 \right\}^{\frac{3}{2}}}{u^3\left(u + \dfrac{d^2u}{d\theta^2}\right)}.$$

The chord of curvature passing through the origin will be obtained by multiplying 2ρ by the cosine of the angle between the radius vector and the normal to the curve at the point considered. (Art. 320.) Hence the chord of curvature through the origin

$$= 2\rho \frac{p}{r} = 2p \frac{dr}{dp};$$

$$= \frac{2u^2 + 2 \left(\frac{du}{d\theta}\right)^2}{u^2 \left(u + \frac{d^2u}{d\theta^2}\right)}.$$

324. If ψ be the angle which the tangent at the point (x, y) of a curve makes with the axis of x, we have

$$\psi = \tan^{-1} \frac{dy}{dx},$$

therefore $\quad \dfrac{d\psi}{ds} = \dfrac{\dfrac{d^2y}{dx^2}}{1 + \left(\dfrac{dy}{dx}\right)^2} \dfrac{dx}{ds} = \dfrac{\dfrac{d^2y}{dx^2}}{\left\{1 + \left(\dfrac{dy}{dx}\right)^2\right\}^{\frac{3}{2}}};$

therefore $\quad \rho = \dfrac{ds}{d\psi}.$

325. If two polar curves have a common point the polar co-ordinates of that point must satisfy the equations to both curves. If they have contact of the first order at that point the value of $\frac{dy}{dx}$ is the same for both curves at that point, and hence, by Art. 201, the value of $\frac{dr}{d\theta}$ is the same for both curves. If the curves have contact of the second order the value of $\frac{d^2y}{dx^2}$ also is the same for both curves at the common point, and hence, by Art. 201, the value of $\frac{d^2r}{d\theta^2}$ is the same for both curves at that point. Proceeding in this way, we see that if two curves have contact of the nth order at any point, if they are referred to polar co-ordinates, the values of

$\dfrac{dr}{d\theta}$, $\dfrac{d^2r}{d\theta^2}$, ... $\dfrac{d^nr}{d\theta^n}$ will be the same for both curves at the common point.

326. Since $\quad \dfrac{1}{p^2} = \dfrac{1}{r^2} + \dfrac{1}{r^4}\left(\dfrac{dr}{d\theta}\right)^2,$

it follows from the last Article, that if two curves have contact of the first order the value of p will be the same for both curves at the common point. Also, since

$$\dfrac{dp}{dr} \text{ or } \dfrac{\dfrac{dp}{d\theta}}{\dfrac{dr}{d\theta}} \text{ involves only } r,\ \dfrac{dr}{d\theta},\ \text{and } \dfrac{d^2r}{d\theta^2},$$

it follows that if two curves have contact of the second order the value of $\dfrac{dp}{dr}$ must also be the same for both curves at the common point.

327. We may apply the preceding Article to establish the equation proved in Art. 321 as follows.

If R be the radius vector of a point in a circle,

$\quad P$ the perpendicular on the tangent,

$\quad c$ the radius of the circle,

$\quad b$ the distance of the centre from the origin,

we have, from the properties of a circle,

$$2cP = R^2 + c^2 - b^2.$$

Differentiating, $\qquad c = R\dfrac{dR}{dP}.$

If this circle be the circle of curvature at a point in a curve having r for its radius vector and p for the perpendicular on the tangent, we have by the last Article,

$$R = r,$$
$$P = p,$$
$$\dfrac{dR}{dP} = \dfrac{dr}{dp},$$

therefore
$$c = r\frac{dr}{dp};$$

that is, the radius of curvature $= r\dfrac{dr}{dp}$.

328. *At a point where the radius of curvature is a maximum or a minimum the circle of curvature has contact of the third order with the curve.*

Since
$$\rho = \frac{\left\{1 + \left(\frac{dy}{dx}\right)^2\right\}^{\frac{3}{2}}}{\frac{d^2y}{dx^2}},$$

we have, when $\dfrac{d\rho}{dx} = 0$,

$$3\left(\frac{d^2y}{dx^2}\right)^2\frac{dy}{dx} - \frac{d^3y}{dx^3}\left\{1 + \left(\frac{dy}{dx}\right)^2\right\} = 0.$$

If in Art. 320 we differentiate the second of equations (2), we have

$$3\frac{dY}{dX}\frac{d^2Y}{dX^2} + (Y - b)\frac{d^3Y}{dX^3} = 0.$$

Hence
$$\frac{d^3Y}{dX^3} = -\frac{3\dfrac{dY}{dX}\dfrac{d^2Y}{dX^2}}{Y - b}$$

$$= \frac{3\left(\dfrac{d^2y}{dx^2}\right)^2\dfrac{dy}{dx}}{1 + \left(\dfrac{dy}{dx}\right)^2},$$

by equations (3) and (5) of that Article. In order that the circle of curvature may have contact of the *third* order with the curve at the proposed point, we must have

$$\frac{d^3Y}{dX^3} = \frac{d^3y}{dx^3},$$

therefore
$$\frac{d^3y}{dx^3}\left\{1 + \left(\frac{dy}{dx}\right)^2\right\} = 3\left(\frac{d^2y}{dx^2}\right)^2\frac{dy}{dx}.$$

This is the relation we have already shewn to hold at points where the radius of curvature is a maximum or minimum.

329. In the figure of Art. 284 let $SP = r$ and $SY = p$; if p_1 denote the perpendicular from S on the tangent at Y to the locus of Y, then will

$$p_1 = \frac{p^2}{r}.$$

Let x, y, be the co-ordinates of P,

x', y', the co-ordinates of Y;

then
$$p_1 = \frac{x' \dfrac{dy'}{dx'} - y'}{\sqrt{\left\{ 1 + \left(\dfrac{dy'}{dx'}\right)^2 \right\}}} \quad \dots\dots\dots\dots\dots\dots (1).$$

The equation to the tangent at P is

$$\eta - y = \frac{dy}{dx}(\xi - x),$$

η and ξ being the variable co-ordinates.

Since the point Y is on the tangent,

$$y' - y = \frac{dy}{dx}(x' - x) \quad \dots\dots\dots\dots\dots (2).$$

The equation to SY is $\qquad \eta = \dfrac{y'}{x'} \xi \quad \dots\dots\dots\dots\dots (3).$

But SY is perpendicular to PY, therefore

$$\frac{y'}{x'} = -\frac{dx}{dy} \quad \dots\dots\dots\dots\dots (4).$$

Combining (2) and (4),

$$(y' - y)\, y' = -x'\,(x' - x)\,;$$

therefore $\qquad yy' + xx' = y'^2 + x'^2 \quad \dots\dots\dots\dots\dots (5).$

Differentiate (5), thus

$$y'\frac{dy}{dx} + y\frac{dy'}{dx} + x' + x\frac{dx'}{dx} = 2y'\frac{dy'}{dx} + 2x'\frac{dx'}{dx}.$$

This by (4) reduces to

$$(2x' - x)\frac{dx'}{dx} + (2y' - y)\frac{dy'}{dx} = 0,$$

therefore

$$\frac{dy'}{dx} = -\frac{2x - x}{2y' - y}.$$

Substitute in (1), and we obtain

$$p_1 = \frac{x'^2 + y'^2}{\sqrt{(x^2 + y^2)}} = \frac{p^2}{r}.$$

330. DEFINITION. The *evolute* of a plane curve is the locus of the centre of curvature; a curve when considered with respect to its evolute is called an *involute*.

If x', y', be the co-ordinates of the centre of curvature at the point (x, y) of a curve, we have by Art. 320,

$$x - x' + (y - y')\frac{dy}{dx} = 0 \dots\dots\dots\dots (1),$$

$$1 + \left(\frac{dy}{dx}\right)^2 + (y - y')\frac{d^2y}{dx^2} = 0 \dots\dots\dots\dots (2).$$

By means of the equation to the curve y, $\frac{dy}{dx}$, and $\frac{d^2y}{dx^2}$ can be expressed in terms of x; hence from the above equations we can, by eliminating x, obtain a relation between x' and y' which is the equation to the evolute. From the above equations, x' and y' may be considered functions of x; differentiating the first, we have

$$1 + \left(\frac{dy}{dx}\right)^2 + (y - y')\frac{d^2y}{dx^2} - \frac{dx'}{dx} - \frac{dy'}{dx}\frac{dy}{dx} = 0.$$

By means of (2) this gives

$$\frac{dx'}{dx} + \frac{dy'}{dx}\frac{dy}{dx} = 0 \dots\dots\dots\dots (3),$$

therefore

$$1 + \frac{dy'}{dx'}\frac{dy}{dx} = 0 \dots\dots\dots\dots (4).$$

Hence (1) may be written

$$y - y' = \frac{dy'}{dx'}(x - x'),$$

which shews that the point (x, y) is situated on the *tangent* to the *evolute* at the point (x', y'). Also (1) shews that the point (x', y') is on the *normal* to the *curve* at the point (x, y). Hence the normal at any point of an involute is a tangent at the corresponding point of the evolute.

331. If ρ be the length of the radius of curvature at the point (x, y) of a curve, and x', y' the co-ordinates of the centre of curvature, we have

$$\rho^2 = (x - x')^2 + (y - y')^2.$$

As x' and y' are functions of x, so also is ρ; hence differentiating we have

$$(x - x')\left(1 - \frac{dx'}{dx}\right) + (y - y')\left(\frac{dy}{dx} - \frac{dy'}{dx}\right) = \rho\frac{d\rho}{dx}.$$

By means of equation (1) of the preceding Article this gives

$$(x - x')\frac{dx'}{dx} + (y - y')\frac{dy'}{dx} = -\rho\frac{d\rho}{dx} \quad\dotsc\dotsc\dotsc\dotsc (1).$$

From equations (1) and (3) of the preceding Article we obtain

$$\frac{\dfrac{dx'}{dx}}{x' - x} = \frac{\dfrac{dy'}{dx}}{y' - y} = \pm\left\{\frac{\left(\dfrac{dx'}{dx}\right)^2 + \left(\dfrac{dy'}{dx}\right)^2}{(x' - x)^2 + (y' - y)^2}\right\}^{\frac{1}{2}} = \pm\frac{1}{\rho}\frac{ds'}{dx},$$

s' being the length of the arc of the evolute. See Art. 307. Hence, by (1),

$$\frac{1}{\rho}\{(x' - x)^2 + (y' - y)^2\}\frac{ds'}{dx} = \pm\rho\frac{d\rho}{dx},$$

therefore

$$\frac{ds'}{dx} = \pm\frac{d\rho}{dx} \quad\dotsc\dotsc\dotsc\dotsc\dotsc\dotsc\dotsc (2).$$

Since $\dfrac{d\,(s' \mp \rho)}{dx} = 0$, we have, by Art. 102,

$$s' \mp \rho = \text{some constant, say } l.$$

Let ABC be the given curve, and $A'B'C'$ the evolute,

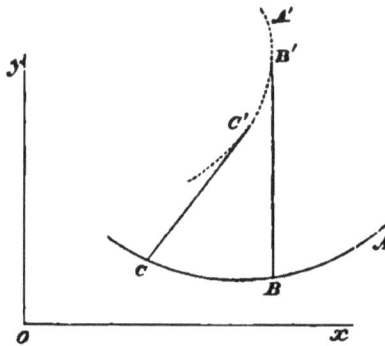

BB' being the radius of curvature of the given curve at B, and CC' at C. Then if A' be the fixed point on the evolute from which the arc is measured, we have

$$A'B' + B'B = l,$$

$$A'B'C' + C'C = l,$$

therefore $$B'C' = BB' - CC'.$$

Hence, if a flexible string of length l be fastened at A' and placed in contact with the evolute $A'B'C'$, then, as the string is unwound from the *evolute*, the free end of it will describe the *involute* CBA. From this property the names evolute and involute are obtained.

In the figure as s' increases ρ diminishes and we have $s' + \rho =$ a constant; if s' be measured in the direction from C' towards A', then s' and ρ increase together and we have $s' - \rho =$ a constant.

It will be observed that a curve has only *one* evolute ; but a curve has an infinite number of involutes, for in the equation $s' \mp \rho =$ some constant, the constant may have any value we please.

332. The following polar formulæ for determining the evolute of a curve are sometimes useful.

Let O be the centre of curvature corresponding to the point P of a curve referred to polar co-ordinates. Let SY be the perpendicular on the tangent at P.

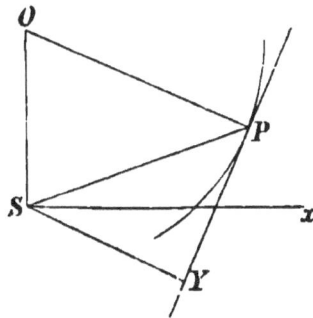

Let $\qquad SP = r, \quad PO = \rho, \quad SY = p,$

$\qquad\qquad SO = r', \quad p' = $ perpendicular from S on PO.

From the triangle SOP we have

$$r'^2 = \rho^2 + r^2 - 2r\rho \cos SPO$$
$$= \rho^2 + r^2 - 2r\rho \sin SPY$$
$$= \rho^2 + r^2 - 2\rho p \dots\dots\dots\dots\dots\dots\dots\dots (1).$$

Also $\qquad p'^2 = r^2 - p^2 \dots\dots\dots\dots\dots\dots\dots\dots\dots (2).$

$$\rho = r \frac{dr}{dp} \dots\dots\dots\dots\dots\dots\dots\dots\dots (3).$$

From the given equation to the curve we can find p in terms of r, and then between (1) and (2) we can eliminate r, and thus we have an equation between p' and r' to determine the locus of O. Since PO is a tangent to the locus of O, p' is the perpendicular from the origin on the tangent to the evolute at O.

In the figure the curve is drawn *concave* to the pole.

If the curve be *convex* to the pole $\dfrac{dr}{dp}$ is negative (Art. 294),

and we should take $\rho = - r \dfrac{dr}{dp}$; in this case we shall find instead of (1) the equation

$$r'^2 = \rho^2 + r^2 + 2\rho p.$$

Thus in both cases we have

$$r'^2 = \rho^2 + r^2 - 2pr\,\frac{dr}{dp}.$$

333. *Involute of a circle.*

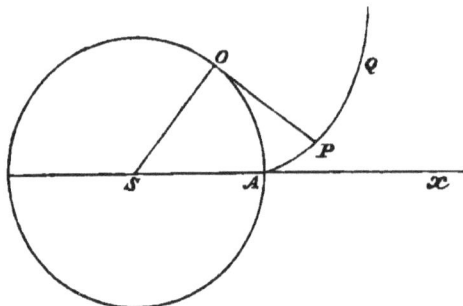

Let S be the centre of a circle, APQ a portion of the involute, $OP = OA$ the portion of the string unwound. Let $SO = a$, $OSA = \phi$, and let x, y be the co-ordinates of P, the origin being at S, and SA the direction of the axis of x.

Then
$$OP = a\phi,$$
$$x = a\cos\phi + a\phi\sin\phi,$$
$$y = a\sin\phi - a\phi\cos\phi.$$

Let $AP = s$, then
$$\frac{ds}{d\phi} = \sqrt{\left\{\left(\frac{dx}{d\phi}\right)^2 + \left(\frac{dy}{d\phi}\right)^2\right\}},\quad \text{Art. 307,}$$
$$= a\phi.$$

Hence, as we shall see in the Integral Calculus,
$$s = \frac{a\phi^2}{2}.$$

EXAMPLES.

1. In the curve
$$y = \frac{c}{2}\,(e^{\frac{x}{c}} + e^{-\frac{x}{c}}),$$

the ordinate at any point is a mean proportional between the radius of curvature there and at the lowest point.

2.   In the curve
$$y = x^4 - 4x^3 - 18x^2,$$
the radius of curvature at the origin $= \frac{1}{36}.$

3.   In the curve
$$y = x^3 + 5x^2 + 6x,$$
the radius of curvature at the origin $= 22.506\ldots$
Find at what point the radius of curvature is infinite.

4.   If $\phi(x, y) = 0$ be the equation to a curve, then
$$\rho = \frac{\left\{\left(\frac{d\phi}{dx}\right)^2 + \left(\frac{d\phi}{dy}\right)^2\right\}^{\frac{3}{2}}}{\left(\frac{d\phi}{dy}\right)^2 \frac{d^2\phi}{dx^2} - 2\frac{d\phi}{dx}\frac{d\phi}{dy}\frac{d^2\phi}{dx\,dy} + \left(\frac{d\phi}{dx}\right)^2 \frac{d^2\phi}{dy^2}}.$$

5.   Find the parabola whose axis is parallel to that of y which has the closest possible contact with the curve
$y = \dfrac{x^3}{a^2}$ at the point where $x = a$.

$$\text{Result.} \quad \left(x - \frac{a}{2}\right)^2 = \frac{a}{3}\left(y - \frac{a}{4}\right).$$

6.   If $r = a(1 - \cos\theta),$ $\quad \rho = \dfrac{4a}{3}\sin\dfrac{\theta}{2}.$

7.   If $r = a(2\cos\theta - 1),$ $\quad \rho = \dfrac{a(5 - 4\cos\theta)^{\frac{3}{2}}}{9 - 6\cos\theta}.$

8.   If the curves $f(x, y) = 0$ and $\phi(x, y) = 0$ touch, shew that at the point of contact
$$\frac{df}{dx}\frac{d\phi}{dy} - \frac{df}{dy}\frac{d\phi}{dx} = 0.$$

9.   Apply the last result to find if the straight line
$$\frac{x}{a} + \frac{y}{b} - 1 = 0$$
touches the curve
$$x^{\frac{2}{3}} + y^{\frac{2}{3}} - (a^2 + b^2)^{\frac{1}{3}} = 0.$$

10. When the angle between the radius vector and the perpendicular on the tangent has a maximum or minimum value, shew that $p\rho = r^2$.

11. If at every point of a curve $2a \dfrac{dx}{ds} = \dfrac{b^2}{y} + y$, then

 $\rho = \dfrac{2ay^2}{y^2 - b^2}$. Shew also that $\dfrac{1}{n} + \dfrac{1}{\rho} = \dfrac{1}{a}$, where n is the portion of the normal intercepted by the axis of x.

12. Find the value of ρ when $r = a \cos \theta$.

13. If $x = \sqrt{(c^2 + s^2)}$, find ρ.

14. The equations which determine the co-ordinates a, b, of the centre of curvature of a curve may be put in the following form, where $r^2 = x^2 + y^2$:

$$2a \frac{d^2x}{dy^2} = \frac{d^2r^2}{dy^2}, \qquad 2b \frac{d^2y}{dx^2} = \frac{d^2r^2}{dx^2}.$$

15. In the parabola $y^2 = 4mx$,

$$a = 2m + 3x, \qquad b = -\frac{2x^{\frac{3}{2}}}{\sqrt{m}}, \qquad \rho = \frac{2(m+x)^{\frac{3}{2}}}{\sqrt{m}}.$$

 Shew that the circle of curvature at any point of a parabola, except the vertex, cuts the axis at two points on opposite sides of the vertex.

16. If $\qquad\qquad Ax^2 + By^2 + C = 0$,

 then $a = \dfrac{A(A-B)}{BC} x^3, \qquad b = \dfrac{B(B-A)}{AC} y^3.$

17. If $\dfrac{dy}{dx} = \dfrac{a}{s}$, then $\rho = \dfrac{a^2 + s^2}{a}$.

18. The radius of curvature of the curve $y^2 = \dfrac{ax(x - 3a)}{x - 4a}$, at one of the points where $y = 0$ is $\dfrac{3a}{8}$, and at the other $\dfrac{3a}{2}$.

19. If $s = a \sin^n \psi$, find ρ. See Art. 324.

20. Find the equation to the circle of curvature of the curve $y^4 = 4a^2x^2 - x^4$, at the origin.

T. D. C. $\qquad\qquad\qquad\qquad\qquad\qquad\qquad\qquad$ A A

21.   If $y + ae^{-\frac{x}{a}} = 0$, then $\rho = \dfrac{(a^2 + y^2)^{\frac{3}{2}}}{ay}$.

22.   Shew that the circle $\left(x - \dfrac{3a}{4}\right)^2 + \left(y - \dfrac{3a}{4}\right)^2 = \dfrac{a^2}{2}$ and the curve $\sqrt{x} + \sqrt{y} = \sqrt{a}$ have contact of the *third* order at the point $x = y = \dfrac{a}{4}$.

23.   If $r = a \sec^2 \dfrac{\theta}{2}$, find ρ.   *Result.* $\rho = 2a \sec^3 \dfrac{\theta}{2}$.

24.   Find the two parabolas which, having their axes parallel to the co-ordinate axes respectively, have a contact of the second order with the circle $x^2 + y^2 = 5a^2$, at the point $x = a$, $y = 2a$.

Results. $\left(y - \dfrac{8a}{5}\right)^2 = \dfrac{2a}{5}\left(\dfrac{7a}{5} - x\right)$, $\left(x - \dfrac{a}{5}\right)^2 = \dfrac{16a}{5}\left(\dfrac{11a}{5} - y\right)$.

25.   In the curve $\dfrac{y}{c} = \dfrac{1}{2}\left(e^{\frac{x}{c}} + e^{-\frac{x}{c}}\right)$, shew that the co-ordinates of the centre of curvature are

$$Y = 2y, \quad X = x - y\sqrt{\left(\dfrac{y^2}{c^2} - 1\right)},$$

and find the equation to the evolute.

26.   Find the equation to the evolute of the ellipse, and the whole length of the evolute.

Results. $(ax)^{\frac{2}{3}} + (by)^{\frac{2}{3}} = (a^2 - b^2)^{\frac{2}{3}}$; $4\left(\dfrac{a^2}{b} - \dfrac{b^2}{a}\right)$.

27.   If $r = f(p)$ be the polar equation to a curve, shew that the equation to the locus of the foot of the perpendicular drawn from the pole on the tangent is $p' = \dfrac{r'^2}{f(r')}$. Find the locus when $p^2 = \dfrac{b^2 r}{2a - r}$, and shew that it is a circle.

28. Find the evolute of the curve $p^2 = r^2 - a^2$.

29. If A be the area between a curve, its radius of curvature, and its evolute, then

$$\frac{dA}{dx} = \frac{\left\{1 + \left(\frac{dy}{dx}\right)^2\right\}^2}{2\dfrac{d^2y}{dx^2}}.$$

30. If ρ be the radius of curvature of a curve, then the radius of curvature of the evolute at the corresponding point is $\rho \dfrac{d\rho}{ds}$.

31. If x', y' be the co-ordinates of the centre of curvature of the curve $y^3 = a^2x$, shew that

$$x' = \frac{a^4 + 15y^4}{6ya^2}, \qquad y' = \frac{a^4y - 9y^5}{2a^4}.$$

32. Shew that in a parabola the radius of curvature at any point is equal to twice the portion of the normal which is intercepted between the point and the directrix.

33. Investigate the following expressions for the radius of curvature at any point of an ellipse:

(1) $\dfrac{(rr')^{\frac{3}{2}}}{ab}$,

(2) $\dfrac{b^2}{a(1 - e^2 \sin^2 \phi)^{\frac{3}{2}}}$,

where r and r' are the focal distances of the point and ϕ is the angle which the normal at the point makes with the major axis.

34. The locus of the centres of all ellipses having the directions of their axes given, and having a contact of the second order with a given curve at a given point, is a rectangular hyperbola passing through that point.

35. Find the asymptotes of the evolute of the curve $y = a \tan x$.

36. Shew that corresponding to the portion of the curve $a^3y^2 = x^5$ near the origin, the evolute is approximately a curve whose equation is $xy^2 = c^3$.

37. Shew that corresponding to the portion of the curve $a^{\frac{4}{3}}y = a^{\frac{1}{3}}x^2 + x^{\frac{5}{3}}$ near the origin, the evolute is approximately a curve whose equation is

$$(y - a)^3 + \beta^2 x = 0.$$

38. Shew that the chord of curvature parallel to the axis of x of the curve $\sec\dfrac{y}{a} = e^{\frac{x}{a}}$ is constant; and that the evolute of this curve for the portion near the origin is approximately a curve whose equation is

$$\sec\left(\frac{3y}{a}\right)^{\frac{1}{3}} = e^{\frac{x-a}{a}}.$$

39. If along a curve and its circle of curvature at any point equal arcs (δs) be measured from the point of contact and on the same side of it, shew that the distance between their extremities will be ultimately $\dfrac{1}{6}\dfrac{d\rho}{ds}\dfrac{(\delta s)^3}{\rho^2}$.

40. Shew that in general a conic section may be found which has a contact of the *fourth* order with a given curve at a proposed point, and shew how to find it when the length of the curve is given in terms of the angle which the normal makes with a fixed line.

   If the curve be an equiangular spiral, and α be the angle between the radius vector and the tangent at any point, shew that the conic section is an ellipse, the major axis of which makes with the normal to the curve an angle ω given by the equation

$$\tan 2\omega + 3\tan\alpha = 0.$$

CHAPTER XXV

334. SUPPOSE

$$F(x, y, a) = 0 \quad\quad\quad\quad (1)$$

to be the equation to a curve, a being some constant quantity. By changing a into $a + h$, we obtain another curve of the same species as (1), the equation to which is

$$F(x, y, a + h) = 0 \quad\quad\quad\quad (2).$$

The point of intersection of (1) and (2) will be found by combining the equations. Now (2) may be written

$$F(x, y, a) + hF'(x, y, a + \theta h) = 0 \quad\quad\quad (3),$$

the accent denoting that $F(x, y, a)$ is to be differentiated with respect to a, and in the result a changed into $a + \theta h$. Hence, combining (3) and (1), we have the point of intersection determined by

$$F(x, y, a) = 0, \text{ and } F'(x, y, a + \theta h) = 0 \quad\quad (4).$$

If we diminish h indefinitely, the equations (4) become

$$F(x, y, a) = 0, \text{ and } F'(x, y, a) = 0 \quad\quad\quad (5).$$

The point determined by equations (5) is the *limit of the intersection of* (1) *and* (2).

If between equations (5) we eliminate a, we obtain the equation to a curve which is called the *locus of the ultimate intersections of the curves formed by varying* a *continuously in the equation* $F(x, y, a) = 0$.

The quantity a is called the parameter of the curve.

335. *The locus of the ultimate intersections of a series of curves touches each of the series of intersecting curves.*

Let $F(x, y, a) = 0$ be the equation which gives the series of curves by varying continuously the quantity a. Then the locus of the ultimate intersections is found by eliminating a between

$$F(x, y, a) = 0 \dots\dots\dots\dots\dots(1),$$

and

$$F'(x, y, a) = 0 \dots\dots\dots\dots\dots(2).$$

Suppose from (2) we obtain a in terms of x and y, say $a = \phi(x, y)$; then if we substitute in (1) we have

$$F\{x, y, \phi(x, y)\} = 0 \dots\dots\dots\dots(3),$$

which is therefore the equation to the locus of the ultimate intersections. Now if for any assigned value of a the equations (1) and (2) give possible values to x and y, then the curve represented by (1) when a has this assigned value, will meet the curve represented by (3).

The value of $\dfrac{dy}{dx}$ for the curve (1) is found by the equation

$$\frac{dF(x, y, a)}{dx} + \frac{dF(x, y, a)}{dy} \frac{dy}{dx} = 0 \dots\dots\dots\dots(4).$$

The value of $\dfrac{dy}{dx}$ for the curve (3) is found by the equation

$$\frac{dF(x, y, \phi)}{dx} + \frac{dF(x, y, \phi)}{dy} \frac{dy}{dx}$$
$$+ \frac{dF(x, y, \phi)}{d\phi} \left\{ \frac{d\phi}{dx} + \frac{d\phi}{dy} \frac{dy}{dx} \right\} = 0 \dots\dots\dots\dots(5).$$

But $\dfrac{dF}{d\phi}$ only differs from $\dfrac{dF}{da}$ in having $\phi(x, y)$ in the place of a; hence by (2) we have *at the point where* (1) *and* (3) *meet,* $\dfrac{dF}{d\phi} = 0$. Thus (5) becomes at that point

$$\frac{dF(x, y, \phi)}{dx} + \frac{dF(x, y, \phi)}{dy} \frac{dy}{dx} = 0 \dots\dots\dots\dots(6).$$

Since at the point of intersection of (1) and (3) we have $a = \phi(x, y)$, equation (6) gives for $\dfrac{dy}{dx}$ at that point the same value as equation (4). Hence (1) and (3) *touch* at their common point.

From this property the locus of the ultimate intersections of a series of curves is called the *envelop* of the series of curves.

336. Example. Required the locus of the ultimate intersections of a series of parabolas found by varying a in the equation

$$y = ax - \frac{1+a^2}{2p} x^2.$$

Here $\quad F(x, y, a) = y - ax + \frac{1+a^2}{2p} x^2 = 0 \ldots\ldots\ldots(1),$

$$F'(x, y, a) = \frac{ax^2}{p} - x = 0 \ldots\ldots\ldots(2).$$

From (2) $\qquad\qquad a = \frac{p}{x}.$

Substitute in (1) and we have

$$y - p + \frac{p^2 + x^2}{2p} = 0,$$

or $\qquad\qquad x^2 + 2py - p^2 = 0,$

which is the equation to a parabola.

337. Required the locus of the ultimate intersections of a series of normals drawn at different points of a given curve.

Let x, y be co-ordinates of a point in the given curve, then

$$x' - x + (y' - y)\frac{dy}{dx} = 0 \ldots\ldots\ldots\ldots(1)$$

is the equation to the normal at that point; $x', y',$ being the variable co-ordinates. From the equation to the given curve y and $\frac{dy}{dx}$ can be expressed as functions of x; thus x is the *parameter* in (1), by varying which the series of normals is obtained. Hence the required locus is to be found by eliminating x between (1) and the equation obtained from (1) by differentiating it with respect to x, which is

$$- 1 + (y' - y) \frac{d^2y}{dx^2} - \left(\frac{dy}{dx}\right)^2 = 0 \dots\dots\dots\dots(2).$$

It appears from (1) and (2), compared with Art. 320, that x', y' will be the co-ordinates of the centre of curvature at the point (x, y) of the given curve. *Hence the locus of the ultimate intersections of the normals to a curve is the evolute of that curve.*

338. It may happen that the envelop does not touch *all* the curves of the series, as will appear from an example.

Suppose the centre of a circle of variable radius to move along the axis of x, so that the abscissa OP of its centre and its radius PM are the abscissa and ordinate of an ellipse AMB which has for its equation

$$\frac{x^2}{m^2} + \frac{y^2}{n^2} = 1 :$$

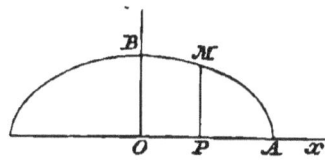

required the envelop of the system of circles.

If $OP = a$, the equation to the circle will be

$$(x - a)^2 + y^2 - \frac{n^2}{m^2}(m^2 - a^2) = 0 \dots\dots\dots \dots(1).$$

Hence differentiating with respect to a, we have

$$x - a - \frac{n^2 a}{m^2} = 0 ;$$

therefore $$a = \frac{m^2 x}{m^2 + n^2} \dots\dots\dots\dots\dots\dots(2).$$

Substitute in (1) and we obtain

$$\frac{x^2}{m^2 + n^2} + \frac{y^2}{n^2} = 1 \dots\dots\dots\dots\dots\dots (3),$$

which is the equation to the envelop.

For all values of a comprised between $\dfrac{m^2}{\sqrt{(m^2 + n^2)}}$ and m, the circles do not ultimately intersect, and are *not* touched by the envelop: for the value of y found from (2) and (3) is

$$y = \pm\, n \sqrt{\left\{ 1 - \frac{(m^2 + n^2)\, a^2}{m^4} \right\}},$$

which is impossible when a is greater than $\dfrac{m^2}{\sqrt{(m^2 + n^2)}}$.

Therefore in the enunciation of Art. 335 we do not assert that the envelop touches each of the series of curves, but that it touches each of the series of *intersecting* curves. The demonstration in that Article assumes that the equations (1) and (2) lead to *possible* values of x and y; or in other words, that one curve of the series ultimately intersects the adjacent curve.

339. The method of Art. 334 may be extended to the case in which there are n parameters connected by $n - 1$ equations. For example, suppose

$$F(x, y, a, b, c) = 0 \dots\dots\dots\dots\dots\dots\dots (1)$$

to be the equation to a curve, the parameters a, b, c, being connected by the equations

$$\left.\begin{array}{l} \phi_1 (a, b, c) = 0 \\ \phi_2 (a, b, c) = 0 \end{array}\right\} \dots\dots\dots\dots\dots\dots\dots (2),$$

and that we require the locus of the ultimate intersections of the curves obtained by giving to the parameters in (1) all possible values consistent with (2). If from equations (2) we find the values of b and c in terms of a and substitute them in (1), we may then proceed as in Art. 334. If however the solution of equations (2) be difficult we may proceed thus. Regarding b and c in (1) as implicit functions of a, we have, if we differentiate with respect to a, and put the result equal to zero as in Art. 334,

$$\frac{dF}{da} + \frac{dF}{db}\frac{db}{da} + \frac{dF}{dc}\frac{dc}{da} = 0 \dots \dots\dots\dots\dots\dots(3).$$

To find $\dfrac{db}{da}$ and $\dfrac{dc}{da}$, we have by differentiating (2),

$$\left.\begin{array}{l} \dfrac{d\phi_1}{da} + \dfrac{d\phi_1}{db}\dfrac{db}{da} + \dfrac{d\phi_1}{dc}\dfrac{dc}{da} = 0 \\[2mm] \dfrac{d\phi_2}{da} + \dfrac{d\phi_2}{db}\dfrac{db}{da} + \dfrac{d\phi_2}{dc}\dfrac{dc}{da} = 0 \end{array}\right\} \quad\ldots\ldots\ldots\ldots(4).$$

If the values of $\dfrac{db}{da}$ and $\dfrac{dc}{da}$ from (4) be substituted in (3), and then a, b, c, be eliminated between (1), (2), and (3), the resulting equation between x and y will determine the required locus.

This process may be rendered more symmetrical by supposing a, b, c all functions of a third variable, say t; then using Da, Db, Dc for $\dfrac{da}{dt}$, $\dfrac{db}{dt}$, $\dfrac{dc}{dt}$ respectively, we have instead of (3) and (4) the equations

$$\left.\begin{array}{l} \dfrac{dF}{da} Da + \dfrac{dF}{db} Db + \dfrac{dF}{dc} Dc = 0 \\[2mm] \dfrac{d\phi_1}{da} Da + \dfrac{d\phi_1}{db} Db + \dfrac{d\phi_1}{dc} Dc = 0 \\[2mm] \dfrac{d\phi_2}{da} Da + \dfrac{d\phi_2}{db} Db + \dfrac{d\phi_2}{dc} Dc = 0 \end{array}\right\} \quad\ldots\ldots\ldots\ldots(5).$$

And the solution of the problem will be facilitated by the use of indeterminate multipliers. Thus multiply the second of equations (5) by λ, the third by μ, and add to the first; this gives

$$\left(\dfrac{dF}{da} + \lambda\dfrac{d\phi_1}{da} + \mu\dfrac{d\phi_2}{da}\right) Da + \left(\dfrac{dF}{db} + \lambda\dfrac{d\phi_1}{db} + \mu\dfrac{d\phi_2}{db}\right) Db$$
$$+ \left(\dfrac{dF}{dc} + \lambda\dfrac{d\phi_1}{dc} + \mu\dfrac{d\phi_2}{dc}\right) Dc = 0 \ldots\ldots\ldots\ldots (6).$$

Now since λ and μ are at present undetermined, we may take them such that

$$\left.\begin{array}{l} \dfrac{dF}{da} + \lambda\dfrac{d\phi_1}{da} + \mu\dfrac{d\phi_2}{da} = 0 \\[2mm] \dfrac{dF}{db} + \lambda\dfrac{d\phi_1}{db} + \mu\dfrac{d\phi_2}{db} = 0 \end{array}\right\} \quad\ldots\ldots\ldots\ldots\ldots(7),$$

from which it follows by (6) that

$$\frac{dF}{dc} + \lambda \frac{d\phi_1}{dc} + \mu \frac{d\phi_2}{dc} = 0 \dots\dots\dots\dots\dots (8).$$

Hence we have to eliminate a, b, c, λ and μ from equations (1), (2), (7) and (8) ; the result is the equation to the envelop required.

Example. A straight line moves so that the length intercepted between the co-ordinate axes is constant: required the envelop of the moving straight line.

Let the equation to the straight line be

$$\frac{x}{a} + \frac{y}{b} = 1 \dots\dots\dots\dots\dots\dots (9),$$

so that $\quad a^2 + b^2 = $ a constant $= k^2$, say $\dots\dots\dots\dots(10).$

From (9) $\quad \dfrac{x}{a^2} Da + \dfrac{y}{b^2} Db = 0,$

from (10) $\quad a Da + b Db = 0 ;$

thus $\quad \left(\dfrac{x}{a^2} + \lambda a\right) Da + \left(\dfrac{y}{b^2} + \lambda b\right) Db = 0,$

therefore $\quad \dfrac{x}{a^2} + \lambda a = 0,$ and $\dfrac{y}{b^2} + \lambda b = 0\dots\dots\dots\dots(11) ;$

multiply the first of these equations by a and the second by b and add ; thus

$$\frac{x}{a} + \frac{y}{b} + \lambda (a^2 + b^2) = 0,$$

that is, $\quad 1 + \lambda k^2 = 0,$ therefore $\lambda = -\dfrac{1}{k^2}.$

Then from (11)

$$a^3 = k^2 x, \text{ and } b^3 = k^2 y.$$

Therefore by (9)

$$\frac{x}{\sqrt[3]{(k^2 x)}} + \frac{y}{\sqrt[3]{(k^2 y)}} = 1,$$

or $\quad\quad\quad\quad x^{\frac{2}{3}} + y^{\frac{2}{3}} = k^{\frac{2}{3}}.$

This equation determines the envelop.

EXAMPLES.

1. Find the envelop of the series of straight lines $\frac{x}{a} + \frac{y}{b} = 1$,
   where $\sqrt{a} + \sqrt{b} = \sqrt{k}$ a constant.

   Result. $x^{\frac{1}{3}} + y^{\frac{1}{3}} = k^{\frac{1}{3}}$.

2. Ellipses are described with coincident centre and axes, and having the sum of the semiaxes $= c$. Shew that the equation to the locus of ultimate intersections is

   $$x^{\frac{2}{3}} + y^{\frac{2}{3}} = c^{\frac{2}{3}}.$$

3. Find the envelop of all ellipses having a constant area, the axes being coincident.

   Result. $4x^2y^2 = c^4$ where πc^2 is the given area.

4. A straight line cuts off from the co-ordinate axes distances AB, AC, such that $nAB + AC = c$, shew that the envelop of the straight lines is

   $$(y + nx - c)^2 = 4nxy.$$

5. Find the evolute of a parabola $y^2 = 4ax$, by the method of Art. 337, taking the equation to the normal in the form

   $$y = m(x - 2a) - am^3.$$

   Result. $27ay^2 = 4(x - 2a)^3$.

6. Find the evolute of the curve $x^{\frac{2}{3}} + y^{\frac{2}{3}} = a^{\frac{2}{3}}$. See Example 9, to Chapter XVIII.

   Result. $(x + y)^{\frac{2}{3}} + (x - y)^{\frac{2}{3}} = 2a^{\frac{2}{3}}$.

7. Shew that the envelop of the series of parabolas

   $$\sqrt{\left(\frac{x}{a}\right)} + \sqrt{\left(\frac{y}{b}\right)} = 1,$$

   under the condition $ab = c^2$, is an hyperbola having its asymptotes coinciding with the axes.

8. Find the locus of the ultimate intersections of the straight lines drawn at right angles to normals to the parabola $y^2 = 4ax$, at the points where they cut the axis.

   Result. $y^2 = 4a(2a - x)$.

9. Straight lines drawn at right angles to the tangents of a parabola at the points where they meet a given straight line perpendicular to the axis, are in general tangents to a confocal parabola.

10. Find the envelop of the curves $\left(\dfrac{x-a}{h}\right)^2 + \left(\dfrac{y-b}{k}\right)^2 = 1$, the variable parameters a, b, being connected by the equation $\left(\dfrac{a}{h}\right)^2 + \left(\dfrac{b}{k}\right)^2 = 1$.

$Result.$ $\dfrac{x^2}{h^2} + \dfrac{y^2}{k^2} = 4.$

11. Circles are described on successive double ordinates of a parabola as diameters: shew that their envelop is an equal parabola. Find what part of this system of circles does not admit of an envelop.

12. A circle moves with its centre on a parabola whose equation is $y^2 - 4ax = 0$, and always passes through the vertex of the parabola: shew that the circle always touches the curve $y^2(x+2a) + x^3 = 0$.

13. A series of parabolas of latus rectum l is described with their vertices in a given parabola of latus rectum l'. Shew that the locus of the ultimate intersections is a parabola with latus rectum $l + l'$, the concavities being in the same direction and the axes parallel.

14. Find the envelop of all ellipses having the same centre and in which the straight line joining the ends of the axes is of constant length.

$Result.$ $x \pm y = \pm c.$

15. From any point of the ellipse $\dfrac{x^2}{a^2} + \dfrac{y^2}{b^2} = 1$, perpendiculars are drawn to the axes, and the feet of these perpendiculars are joined: shew that the straight line thus formed always touches the curve $\left(\dfrac{x}{a}\right)^{\frac{2}{3}} + \left(\dfrac{y}{b}\right)^{\frac{2}{3}} = 1$.

16. From every point of the ellipse $\frac{x^2}{h^2} + \frac{y^2}{k^2} - 1 = 0$ pairs of tangents are drawn to the ellipse $\frac{x^2}{a^2} + \frac{y^2}{b^2} - 1 = 0$: shew that the locus of the ultimate intersections of the chords of contact is $\frac{h^2 x^2}{a^4} + \frac{k^2 y^2}{b^4} = 1$.

17. Circles are drawn passing through the origin having their centres on the curve $a^2 y^2 - b^2 (2ax - x^2) = 0$: shew that the locus of the ultimate intersections of these circles is $(x^2 + y^2 - 2ax)^2 - 4a^2 x^2 - 4b^2 y^2 = 0$.

18. The circle whose equation is $x^2 + y^2 + 2ax + 2by + 2c = 0$, is cut by another circle which passes through the origin and whose centre is on the curve $\frac{x^2}{\alpha^2} + \frac{y^2}{\beta^2} = 1$: shew that the chord joining the points of intersection touches the curve $\alpha^2 x^2 + \beta^2 y^2 = (ax + by + c)^2$.

19. Find the locus of the ultimate intersections of the straight lines
$$y \cos \theta - x \sin \theta = c - c \sin \theta \log \tan \left(\frac{\pi}{4} + \frac{\theta}{2} \right),$$
where θ is the variable parameter.

Result. $2y = c \left(e^{\frac{x}{c}} + e^{-\frac{x}{c}} \right)$.

20. The equation to a spiral is $r^n \cos n\theta = a^n$; straight lines are drawn through the extremities of the radii vectores at right angles to them: shew that the envelop of these straight lines is the curve
$$r^m \cos m\theta = a^m, \text{ where } m = \frac{n}{n+1}.$$

21. A series of ellipses has the same centre and directrix: shew that the envelop is a pair of parabolas, but that the envelop will not meet those ellipses whose excentricity is less than $\frac{1}{\sqrt{2}}$.

22. Find the locus of the ultimate intersections of an ellipse which touches a given straight line at a given point at the extremity of the axis minor, the excentricity varying as the axis major. Find the limits of the excentricity in order that two consecutive ellipses may intersect.

23. A straight line is drawn from the focus to any point of a conic section, and a circle is described on it as a diameter : shew that the locus of the ultimate intersections of all such circles is a circle, except, in a certain case, where it is a straight line.

24. Shew that the locus of the ultimate intersections of all the chords of an ellipse which join the points of contact of pairs of tangents at right angles to one another is a confocal ellipse.

25. Find the locus of the ultimate intersections of the straight lines $x \cos 3\theta + y \sin 3\theta = a (\cos 2\theta)^{\frac{3}{2}}$, where θ is the variable parameter.

$$Result. \quad (x^2 + y^2)^2 = a^2 (x^2 - y^2).$$

26. Find the envelop of the circles described on the radii of an ellipse, drawn from the centre, as diameters.

$$Result. \quad (x^2 + y^2)^2 = a^2 x^2 + b^2 y^2.$$

27. On any radius vector of the curve $r = c \sec^n \dfrac{\theta}{n}$ as diameter is described a circle : shew that the envelop of all such circles is the curve $r = c \sec^{n-1} \dfrac{\theta}{n-1}$.

28. Find the locus of the ultimate intersections of a family of parabolas of which the pole of a given equiangular spiral is the focus, and its tangents directrices.

$$Result. \quad \text{A similar equiangular spiral.}$$

29. Perpendiculars are drawn from the pole of an equi-
    angular spiral on the tangents to the curve : find the
    envelop of the circles described on these perpendiculars
    as diameters.

    Result.   A similar equiangular spiral.

30. From every point of a parabola as centre a circle is
    described with a radius exceeding the focal distance
    of the point by a constant quantity : find the envelop
    of the circles.

    Result. $(x + c + a)\{y^2 + (x - a)^2 - c^2\} = 0$; where c is
    the constant quantity.

31. Find the envelop of the straight lines obtained by vary-
    ing θ in the equation $ax \sec \theta - by \csc \theta = a^2 - b^2$.

    Result. $(ax)^{\frac{2}{3}} + (by)^{\frac{2}{3}} = (a^2 - b^2)^{\frac{2}{3}}.$

32. From a fixed point A in the circumference of a circle
    any chord AP is drawn and bisected at H, and on
    PH as diameter a circle is described : find the locus
    of the ultimate intersections of the system of circles
    described according to this law.

    Result. $a^2 (x^2 + y^2) = (2x^2 + 2y^2 - 3ax)^2$;

    where $x^2 + y^2 = 2ax$ is the equation to the given circle.

CHAPTER XXVI.

TRACING OF CURVES.

340. IN this Chapter we shall give some examples of tracing curves from their equations.

Example (1). Let $y^2 = \dfrac{x^2(x^2 - 4a^2)}{x^2 - a^2}$ .... (1).

First find the value of $\dfrac{dy}{dx}$: taking the logarithms of both sides of the equation and differentiating, we have

$$\frac{1}{y}\frac{dy}{dx} = \frac{1}{x} + \frac{x}{x^2 - 4a^2} - \frac{x}{x^2 - a^2};$$

therefore $\dfrac{dy}{dx} = \pm \dfrac{x\sqrt{(x^2 - 4a^2)}}{\sqrt{(x^2 - a^2)}}\left\{\dfrac{1}{x} + \dfrac{x}{x^2 - 4a^2} - \dfrac{x}{x^2 - a^2}\right\}$ .... (2).

Next find the asymptotes: since

$$y^2 = \frac{x^2\left(1 - \dfrac{4a^2}{x^2}\right)}{1 - \dfrac{a^2}{x^2}},$$

therefore $y = \pm x\left(1 - \dfrac{4a^2}{x^2}\right)^{\frac{1}{2}}\left(1 - \dfrac{a^2}{x^2}\right)^{-\frac{1}{2}}$

$$= \pm x\left\{1 - \frac{2a^2}{x^2} - \frac{2a^4}{x^4} - \ldots\right\}\left\{1 + \frac{a^2}{2x^2} + \frac{3a^4}{8x^4} + \ldots\right\}$$

$$= \pm x\left\{1 - \frac{3a^2}{2x^2}\ldots\right\}$$

$$= \pm\left\{x - \frac{3a^2}{2x}\ldots\right\} \ldots\ldots\ldots\ldots\ldots\ldots\ldots\ldots (3).$$

Hence $\qquad\qquad y = x$

and $\qquad\qquad y = -x$

are asymptotes.

T. D. C.

Also when $x = \pm a$ we see that y is infinite.

Hence $x = \quad a$

and $x = -a$

are asymptotes.

We may now assign different values to x, and note the corresponding values of y and $\dfrac{dy}{dx}$ obtained from (1) and (2). Since the curve is symmetrical with respect to the axis of x, we may confine our attention to the positive values of y.

When $x = 0$, $y = 0$, $\dfrac{dy}{dx} = \pm 2$.

From $x = 0$ to $x = a$, y is possible.

When $x = a$, $y = \infty$, $\dfrac{dy}{dx} = \infty$.

From $x = a$ to $x = 2a$, y is impossible.

When $x = 2a$, $y = 0$, $\dfrac{dy}{dx} = \infty$.

When x is greater than $2a$, y is possible.

It is not necessary to give negative values to x in this example, because the curve is symmetrical with respect to the axis of y.

If we draw the asymptotes and make use of the above list of particular values of y and $\dfrac{dy}{dx}$, we shall have sufficient materials for ascertaining the general form of the curve. If necessary, in any example, we may find $\dfrac{d^2y}{dx^2}$, in order to determine the points of inflexion ; also by examining when $\dfrac{dy}{dx}$ vanishes, we can determine the maxima and minima values of y.

If we take the upper sign in equation (3), we have for the asymptote

$$y = x \dotfill (4) ;$$

and for the curve $y = x - \dfrac{3a^2}{2x}$ &c. $\dotfill (5)$.

When x is very large the terms included in the &c. of equation (5) will be very small compared with $\dfrac{3a^2}{2x}$. Hence comparing (4) and (5) we see that corresponding to the same abscissa the ordinate of the curve is *less* than that of the asymptote, and therefore the curve lies *below* the asymptote as represented in the figure.

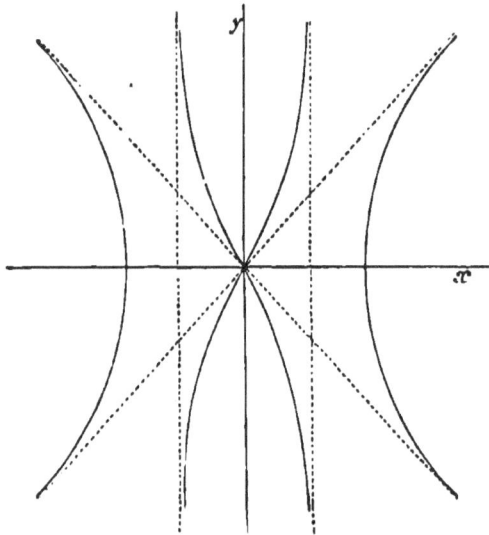

Example (2). Suppose

$$y^2 = \frac{x\,(x-a)\,(x-2a)}{x+3a} \quad\ldots\ldots\ldots\ldots\ldots (1)\,;$$

therefore $\quad \dfrac{2}{y}\dfrac{dy}{dx} = \dfrac{1}{x} + \dfrac{1}{x-a} + \dfrac{1}{x-2a} - \dfrac{1}{x+3a},$

$$\frac{dy}{dx} = \frac{1}{2}\left\{\frac{x\,(x-a)\,(x-2a)}{x+3a}\right\}^{\frac{1}{2}}\left\{\frac{1}{x} + \frac{1}{x-a} + \frac{1}{x-2a} - \frac{1}{x+3a}\right\}$$

$$\ldots\ldots\ldots\ldots\ldots (2).$$

Also from (1) we have

$$y = \pm\, x\left(1 - \frac{a}{x}\right)^{\frac{1}{2}}\left(1 - \frac{2a}{x}\right)^{\frac{1}{2}}\left(1 + \frac{3a}{x}\right)^{-\frac{1}{2}}$$

$$x\left(1 - \frac{a}{2x} - \frac{a^2}{8x^2}\ldots\right)\left(1 - \frac{a}{x} - \frac{a^2}{2x^2}\ldots\right)\left(1 - \frac{3a}{2x} + \frac{27a^2}{8x^2}\ldots\right).$$

If the three series be multiplied together we have

$$y = \pm x \left(1 - \frac{3a}{x} + \frac{11a^2}{2x^2} \cdots \right)$$

$$= \pm \left(x - 3a + \frac{11a^2}{2x} \cdots \right) \dots \dots \dots (3).$$

Hence $y = x - 3a$

and $y = -x + 3a$

are asymptotes.

Also from (1) $x = -3a$

is an asymptote.

From (1) and (2) we have the following results, confining ourselves to the positive values of y.

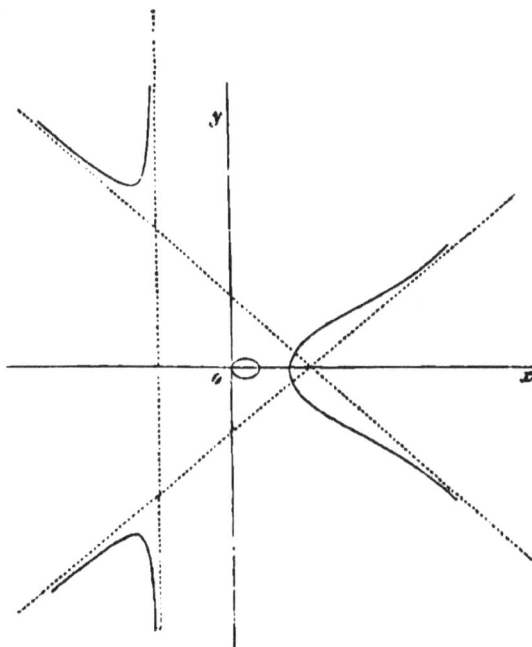

When $x = 0$, $y = 0$, $\dfrac{dy}{dx} = \infty$.

From $x = 0$ to $x = a$, y is possible.

When $x = a$, $y = 0$, $\dfrac{dy}{dx} = \infty$.

From $x = a$ to $x = 2a$, y is impossible.

When $\qquad x = 2a, \qquad y = 0, \qquad \dfrac{dy}{dx} = \infty$.

When x is greater than $2a$, y is possible.

When x is negative and between 0 and $-3a$, y is impossible.

When $\qquad x = -3a, \qquad y = \infty, \qquad \dfrac{dy}{dx} = \infty$.

When x lies between $-3a$ and $-\infty$, y is possible.

From (3) we see that the equation to the curve when x is very great is approximately

$$y = x - 3a + \frac{11a^2}{2x} ;$$

and whether x be positive or negative $x - 3a + \dfrac{11a^2}{2x}$ is numerically greater than $x - 3a$. Hence the curve lies *above* the asymptote.

342. In the above examples the value of y is given explicitly in terms of x. In a similar manner we may proceed if x is given explicitly in terms of y. But if the equation connecting x and y does not admit of easy solution we must abandon this method. In such cases we may find the asymptotes by Art. 277: we may determine the nature of the curve near the origin by a method exemplified in the next two Articles; from these results we may obtain some idea of the form of the curve. By transforming the equation to polar co-ordinates we shall sometimes be enabled to trace it more accurately.

343. To determine the form of the curve

$$x^4 - ayx^2 + by^3 = 0 \dots\dots\dots\dots\dots (1)$$

near the origin.

First, suppose that near the origin the term by^3 can be neglected in comparison with the other two terms in (1); in that case we should have

$$x^4 - ayx^2 = 0,$$

therefore $\qquad\qquad\qquad x^2 = ay.$

This makes y vary as x^2, and therefore y^3 vary as x^6. Hence the neglected term by^3 varies as x^6, while the terms retained, x^4 and ayx^2, vary as x^4. But by taking x small enough x^6 can be made as small as we please compared with x^4, and therefore near the origin one branch of the curve may be found approximately by neglecting by^3. The branch we thus obtain, being determined by the equation $x^2 = ay$, is a portion of a parabola having its axis coincident with that of y.

Next, assume that near the origin the term ayx^2 may be neglected in comparison with the others. We thus find

$$x^4 + by^3 = 0 ;$$

therefore y varies as $x^{\frac{4}{3}}$.

Hence the neglected term ayx^2 would vary as $x^{2+\frac{4}{3}}$; that is as $x^{\frac{10}{3}}$, while the terms retained would vary as x^4. But since $x^{\frac{10}{3}}$ can be made as *great* as we please compared with x^4 by taking x small enough, we do *not* obtain an approximate branch near the origin by neglecting ayx^2.

Again, assume that x^4 may be neglected near the origin; then

$$by^3 - ax^2y = 0,$$

therefore $$by^2 - \ ax^2 = 0.$$

Hence y varies as x; the terms retained vary as x^3 and the rejected term varies as x^4; and thus an approximation to the curve near the origin is given by

$$by^2 - ax^2 = 0, \quad \text{or } y = \pm\, x \sqrt{\dfrac{a}{b}}\,.$$

The figure shews the nature of the curve near the origin; AB is the parabolic branch, and CD, $C'D'$, are the two branches found by neglecting x^4.

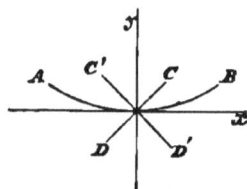

The conclusions in this case may be verified by solving the given equation with respect to x^2. We thus find

$$x^2 = \frac{y}{2} \{a \pm \sqrt{(a^2 - 4by)}\}.$$

Expand $\sqrt{(a^2 - 4by)}$ in powers of y by the Binomial Theorem, and take the upper sign, then

$$x^2 = ay \text{ approximately};$$

with the lower sign

$$x^2 = \frac{b}{a} y^2 \text{ approximately.}$$

In this manner, or by transforming the equation into a polar form, we may complete the tracing of the curve. It will be found that the branches extending from the origin to C and B respectively, unite, thus forming a loop. The branch from the origin to D' extends to infinity, and has no rectilinear asymptote. The curve is obviously symmetrical with respect to the axis of y.

344. Determine the nature of the curve

$$y^4 + ay^2x - x^4 = 0$$

near the origin.

First, if we neglect x^4 we have

$$y^4 + ay^2x = 0,$$

therefore $\qquad y^2 = -ax.$

Hence x varies as y^2; the rejected term varies as y^8, while the terms retained vary as y^4, and therefore we have in the parabola $y^2 = -ax$ an approximation to the given curve near the origin.

Next, reject the term ay^2x; thus

$$y^4 - x^4 = 0,$$

therefore $\qquad y = \pm x.$

Hence y varies as x; the rejected term varies as x^3, and the terms retained vary as x^4; hence this does *not* give us an approximate branch.

Again, reject y^4; thus

$$ay^2x - x^4 = 0,$$

therefore

$$y^2 = \frac{x^3}{a}.$$

Hence y^2 varies as x^3; the rejected term varies as x^4, and the terms retained vary as x^4, and consequently we obtain an approximate branch.

The branch to the left of the axis of y is that given by $y^2 = -ax$, and the cusp to the right of the axis of y is that given by $y^2 = \frac{x^3}{a}$.

In this example, y^2 may be found in terms of x and the whole curve traced.

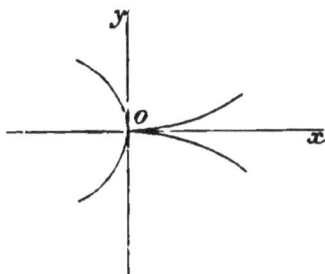

345. We may observe that in the examples of the preceding Articles, the supposition which was found inadmissible near the origin, will be admissible for points at a very great distance from the origin. Thus if

$$y^4 + ay^2x - x^4 = 0,$$

when x and y are indefinitely great, ay^2x may be neglected in comparison with y^4 and x^4; and $y^4 = x^4$, or $y = \pm x$, will be an approximation at points remote from the origin. If we find the asymptotes by Art. 277, we shall have

$$y = \pm \left(x - \frac{a}{4} \right);$$

to which

$$y = \pm x$$

may be considered an approximation when x and y are indefinitely great.

. 346. Required the nature of the curve

$$y^4 + xy^3 + ax^2y - bx^3 = 0$$

near the origin.

Assume

$$ax^2y - bx^3 = 0$$

as an approximation near the origin. Hence

$$ay = bx,$$

therefore $\qquad y$ varies as x,

the terms retained vary as x^3, and those rejected vary as x^4, and we have therefore an approximation to the curve at the origin. If we examine all the six cases which present themselves by retaining two of the terms of the given equation and rejecting the other two, we shall find that the only other allowable supposition is, that xy^3 and bx^3 can be rejected, and we obtain for an approximation

$$y^4 + ax^2 y = 0,$$

or $\qquad y^2 = - ax^2.$

It will be easy to draw the branches we have found; the equation $y^2 = - ax^2$ gives us a cusp, the two branches being on the two sides of the negative part of the axis of y.

347. If in any examples we wish only to find the *directions of the tangents* at the origin, we may arrive at them immediately, as shewn in Art. 195.

Suppose $\qquad y^4 + xy^3 + ax^2 y - bx^3 = 0,$

therefore $\qquad (y + x) \left(\dfrac{y}{x}\right)^3 + a\dfrac{y}{x} - b = 0.$

Hence, when x and y vanish, we have

$$\text{the limit of } \frac{y}{x} = \frac{b}{a}.$$

Besides this, the limit of $\dfrac{y}{x}$ *may* have an infinite value, and this can be determined by examining if $\dfrac{x}{y}$ has zero for a limit. The given equation may be put in the form

$$y + x + \left(\frac{x}{y}\right)^3 \left\{a - b\frac{x}{y}\right\} = 0.$$

Hence one of the limiting values of $\dfrac{x}{y}$ is zero.

In like manner, if $y^4 + ay^2x - x^4 = 0$,

we have
$$y\left(\frac{y}{x}\right)^3 + a\left(\frac{y}{x}\right)^2 - x = 0.$$

Hence $\dfrac{y}{x}$ has zero for one of its limiting values. Also from the given equation we may deduce

$$y + a\frac{x}{y} - x\left(\frac{x}{y}\right)^3 = 0.$$

Hence $\dfrac{x}{y}$ has zero for one of its limiting values. Thus $\dfrac{y}{x}$ may be zero or infinity when x and y are indefinitely diminished, and therefore the axes of x and y are tangents to the branches through the origin.

In connexion with the subject of tracing curves from equations of the form $\phi(x, y) = 0$ the student may with advantage consult Chapter XXIII. of the treatise on the *Theory of Equations*.

348. We shall now give some examples of polar curves.

Suppose $r = a \sec \dfrac{\theta}{3}$,   therefore $\dfrac{dr}{d\theta} = \dfrac{a}{3}\dfrac{\sin\dfrac{\theta}{3}}{\cos^2\dfrac{\theta}{3}}$,

$$\tan\phi = r\frac{d\theta}{dr} = 3\cot\frac{\theta}{3}. \quad \text{(Art. 279.)}$$

The polar subtangent $= r^2\dfrac{d\theta}{dr} = 3a\operatorname{cosec}\dfrac{\theta}{3}$.

When $\dfrac{\theta}{3} = \dfrac{\pi}{2}$, r is infinite, and the polar subtangent $= 3a$; hence we have an asymptote. As θ increases from 0 to $\dfrac{3\pi}{2}$, $\dfrac{dr}{d\theta}$ is positive, and r is positive and increases with θ. As θ

increases from $\dfrac{3\pi}{2}$ to 3π, r is negative, and $\dfrac{dr}{d\theta}$ is positive so that r numerically diminishes.

To draw the asymptote we proceed thus: since, when

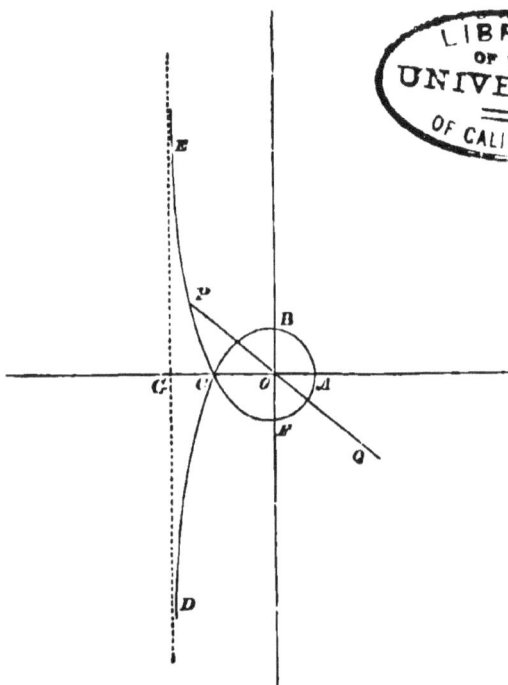

$\theta = \dfrac{3\pi}{2}$, r is infinite, and the polar subtangent is $3a$, the eye must be supposed at O looking along OF, and a distance $OG = 3a$ must be measured to the right of OF and at right angles to it; a straight line drawn through G parallel to OF is the required asymptote.

As θ changes from 0 to $\dfrac{3\pi}{2}$ the branch $ABCD$ is traced out, cutting OA at right angles at A since $\tan\phi = \infty$ when $\theta = 0$. When θ becomes greater than $\dfrac{3\pi}{2}$, r is negative, and according to the usual conventions with respect to sign, must be measured in a direction *opposite* to that which it would have if it were positive. For example, if the angle AOQ

measured in the ordinary way round from OA be $\dfrac{3\pi}{2}+\dfrac{\pi}{4}$ the corresponding value of r is

$$\dfrac{a}{\cos\dfrac{1}{3}\left(\dfrac{3\pi}{2}+\dfrac{\pi}{4}\right)} \text{ or } -\dfrac{a}{\sin\dfrac{\pi}{12}} \text{ or } -a\sqrt{2}\,(\sqrt{3}+1)\,;$$

hence we take $OP = a\sqrt{2}\,(\sqrt{3}+1)$ measuring it along QO produced. In this way, as θ changes from $\dfrac{3\pi}{2}$ to 3π, we obtain the portion $ECFA$ of the curve.

If we suppose θ negative, or positive and greater than 3π, we shall only obtain repetitions of the branches already found.

349. A very common mistake in drawing polar curves is made with respect to the asymptotes. For example, if r is infinite when $\theta = 0$, it is assumed that the initial line is an asymptote. This involves a double error, for in the first place it does not follow that because r is infinite there *is* an asymptote; and secondly, if there be an asymptote it may be *parallel* to the initial line instead of coinciding with it. For example, the polar equation to the parabola from the vertex is

$$r = \frac{4a\cos\theta}{\sin^2\theta}\,.$$

Here when $\theta = 0$, r is ∞, but the curve has no asymptote. In the curve

$$r = \frac{a}{\sin\dfrac{\theta}{3}}\,,$$

when $\theta = 0$, r is infinite; there is an asymptote, but it does not coincide with the initial line; it will be found to be parallel to it and at a distance $3a$ from it.

350. Trace the curve

$$r = \frac{a\sin\theta}{\theta}\,.$$

Here
$$\frac{dr}{d\theta} = \frac{a\,(\theta \cos \theta - \sin \theta)}{\theta^2},$$

$$\tan \phi = \frac{\theta \sin \theta}{\theta \cos \theta - \sin \theta}.$$

As r is never infinite there is no asymptote; r is positive from $\theta = 0$ to $\theta = \pi$, negative from $\theta = \pi$ to $\theta = 2\pi$, and so on.

When $\theta = 0$, $\tan \phi$ assumes the form $\frac{0}{0}$; on examination it will be found infinite.

The curve begins at A, crossing the initial line at right

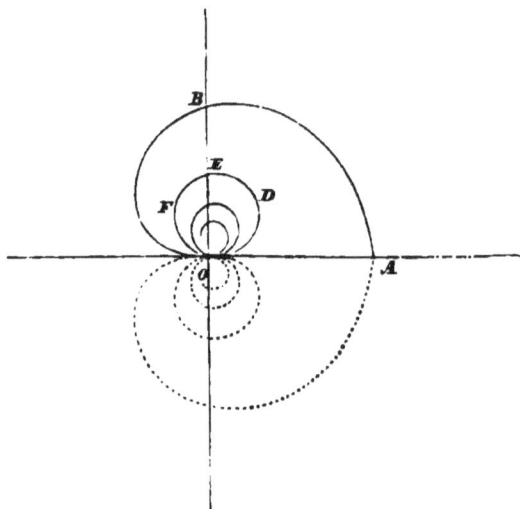

angles, since there $\tan \phi$ is infinite : as θ changes from 0 to π the portion ABO is traced out; as θ changes from π to 2π the portion $ODEFO$ is traced out, and so on. The curve forms an infinite number of loops, each smaller than the preceding and all passing through O.

If we ascribe negative values to θ we obtain the dotted part lying *below* the straight line OA.

351. Trace the curve

$$r = \frac{a\theta^2}{1 + \theta^2}.$$

In this case the curve begins at the pole O and makes

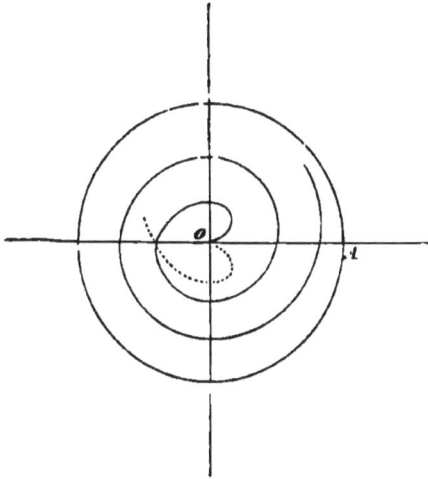

an infinite number of revolutions round it; r can never become so great as a, to which value however it continually approaches. Hence $r = a$ is the equation to *an asymptotic circle*, to which the curve continually approaches as θ increases without limit.

If we give to θ negative values, we have a branch similar to that obtained from positive values of θ. It is represented in the figure by the dotted portion.

352. We shall now give the equations and the figures of a few curves which frequently occur in problems.

The Logarithmic Curve.

The equation to this curve is

$$y = be^{\frac{x}{c}};$$

or, which is an equivalent form,

$$y = ba^x.$$

The curve extends to infinity both in the positive and negative directions of the axis of x. As x

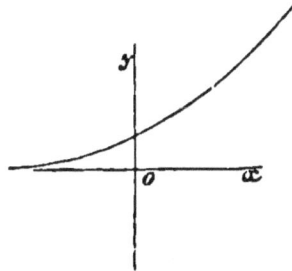

is increased numerically in the negative direction, y tends to the limit zero, so that the axis of x is an asymptote.

353. *The Catenary.*

The equation to this curve is

$$y = \frac{c}{2}\left(e^{\frac{x}{c}} + e^{-\frac{x}{c}}\right).$$

It is the curve in which a flexible string would hang if suspended from two points, as is shewn in works on Statics.

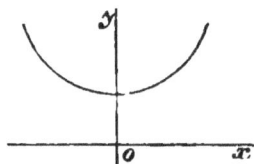

354. *The Logarithmic Spiral.*

The equation to this curve is

$$r = be^{\frac{\theta}{c}},$$

or
$$r = ba^{\theta}.$$

Taking the first form we have

$$\tan \phi = r\,\frac{d\theta}{dr} = c.$$

Since ϕ is thus constant the curve is also called the *equiangular spiral.*

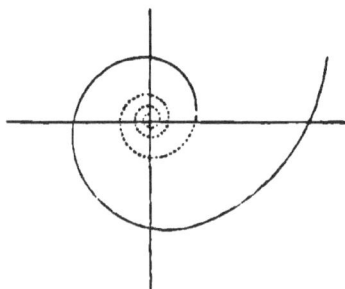

The dotted part arises from negative values of θ.

355.   *The Spiral of Archimedes.*

$$r = a\theta.$$

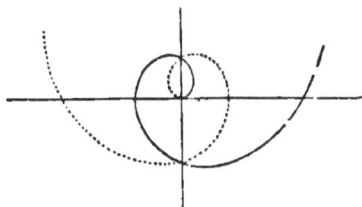

356.   *The Cycloid.*

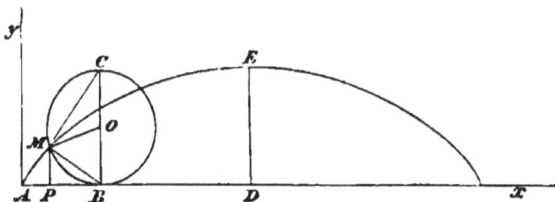

The cycloid is traced out by a fixed point in the circumference of a circle as the circle *rolls* along a straight line.

Let Ax be the straight line along which the circle rolls;

> M the fixed point in the circumference of the circle BMC which traces out the cycloid;

> A the point in the straight line Ax with which M was originally in contact;

> O the centre of the circle:

$$AP = x, \quad MP = y, \quad MOB = \phi, \quad OB = a.$$

The arc $MB = a\phi$, and by the nature of the curve it is equal to AB;

therefore $x = a\phi - PB = a\phi - a \sin \phi,$

$y = a - a \cos \phi.$

If we eliminate ϕ we have

$$x = a \cos^{-1} \frac{a - y}{a} - \sqrt{(2ay - y^2)}.$$

357. From the last equation we have

$$\frac{dx}{dy} = \sqrt{\left(\frac{y}{2a-y}\right)}.$$

Hence the equation to the tangent at M is

$$y' - y = \sqrt{\left(\frac{2a-y}{y}\right)}(x' - x),$$

and the equation to the normal at M is

$$y' - y = -\sqrt{\left(\frac{y}{2a-y}\right)}(x' - x).$$

If in the last equation we put $y' = 0$, we have

$$x' - x = \sqrt{\{y(2a-y)\}} = a\sin\phi = PB.$$

Hence MB is the direction of the normal at M, and therefore MC is the direction of the tangent at M.

If in the equations of Art. 356 we put $\phi = \pi$, we have $y = 2a$ and $x = a\pi$ as the co-ordinates of the vertex E.   Hence

$$PD = a\pi - a\phi + a\sin\phi$$
$$= a(\theta + \sin\theta), \qquad \text{if } \theta = \pi - \phi.$$

Also the distance of M from a straight line through E parallel to Ax is $2a - a(1 - \cos\phi)$ or $a(1 - \cos\theta)$.

358. If we take the vertex as the origin, and the tangent at that point as the axis of y, we have by the last Article

$$\left.\begin{array}{l} y = PN = a(\theta + \sin\theta) \\ x = AN = a(1 - \cos\theta) \end{array}\right\} \quad \cdots\cdots\cdots\cdots\cdots (1).$$

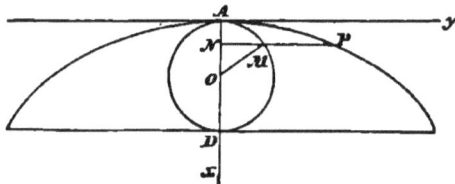

Describe a semicircle on AD as diameter: let PN meet this circle at M, and join M with the centre O; then

$$AN = a(1 - \cos AOM);$$

therefore $AOM = \theta.$

Since the arc $AM = a\theta$, it follows that

$$MP = \text{arc } AM.$$

From (1) we have

$$y = a \cos^{-1}\frac{a - x}{a} + \sqrt{(2ax - x^2)} \ldots\ldots\ldots\ldots(2),$$

therefore

$$\frac{dy}{dx} = \sqrt{\left(\frac{2a - x}{x}\right)}.$$

If s denote the arc AP, we have

$$\frac{ds}{dx} = \sqrt{\left\{1 + \left(\frac{dy}{dx}\right)^2\right\}} = \sqrt{\left(\frac{2a}{x}\right)},$$

therefore

$$s = \sqrt{(8ax)},$$

as will appear from the Integral Calculus.

The normal to the curve at P is parallel to MD, as we may see from Art. 357 or from an independent investigation.

By the property of the circle it follows that

$$MD = 2a \cos \frac{\theta}{2}.$$

If we investigate the value of the radius of curvature at P we shall find it to be twice MD, that is,

$$4a \cos \frac{\theta}{2}, \text{ or } 2\sqrt{(4a^2 - 2ax)}.$$

359. *The evolute of the cycloid is an equal cycloid.*

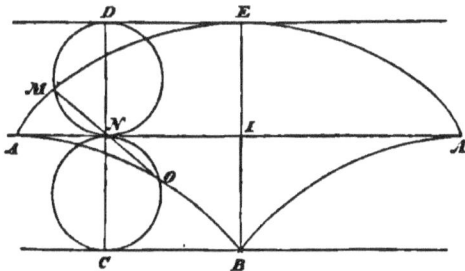

For it appears by Art. 358 that the radius of curvature at a point M of a cycloid is twice MN. Hence if we produce MN to O, making $NO = MN$, the point O is the centre of curvature corresponding to the point M. Draw EIB and make $IB = 2a$; draw BC parallel to ED; the circle described on NC as a diameter will pass through O.

The arc $NO =$ arc MN and therefore $= AN$,

therefore the arc $OC = NI = CB$.

Hence O is a point in a cycloid generated by rolling a circle of radius a along BC. Hence the evolute of the cycloid AEA' is composed of the two semi-cycloids AB and $A'B$.

360. The epicycloid is the curve traced out by a point in the perimeter of a circle which rolls on the outside of a fixed circle.

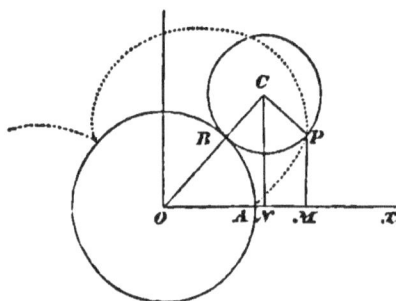

Let O and C be the centres of the fixed and the rolling circles respectively, B the point of contact, P the tracing point, A its initial position. Take OA as the axis of x; draw CN, PM, perpendicular to the axis of x. Let

$$OB = a, \quad BC = b, \quad AOB = \theta, \quad BCP = \phi.$$

Then $x = ON + NM$

$$= (a + b) \cos \theta + b \sin (\phi - \tfrac{1}{2}\pi + \theta)$$

$$= (a + b) \cos \theta - b \cos (\phi + \theta).$$

But the arc $AB =$ the arc BP, by the mode of generation, that is, $a\theta = b\phi$, therefore

$$x = (a + b) \cos \theta - b \cos \frac{a+b}{b} \theta.$$

Similarly $y = (a + b) \sin \theta - b \sin \frac{a+b}{b} \theta.$

The hypocycloid is the curve traced out by a point in the perimeter of a circle which rolls on the *inside* of a fixed circle.

It may be found by a method similar to the above that for the hypocycloid

$$x = (a - b) \cos \theta + b \cos \frac{a-b}{b} \theta,$$

$$y = (a - b) \sin \theta - b \sin \frac{a-b}{b} \theta.$$

361. The radius of the rolling circle may be greater or less than the radius of the fixed circle both in the epicycloid and in the hypocycloid; it is however easy to infer from the figure, that a hypocycloid in which the radius of the rolling circle is greater than the radius of the fixed circle may be counted as an epicycloid. This can also be shewn from the equations. For in the equations to the hypocycloid put $\frac{b-a}{b} \theta = \phi$; then those equations may be written

$$x = (a+b-a) \cos \phi - (b-a) \cos \frac{a+b-a}{b-a} \phi,$$

$$y = (a+b-a) \sin \phi - (b-a) \sin \frac{a+b-a}{b-a} \phi;$$

these are the equations to an epicycloid in which the radius of the fixed circle is a, and the radius of the rolling circle is $b - a$.

Similarly we may shew that a hypocycloid in which the radius of the rolling circle is *greater* than half the radius of the fixed circle may be counted as a hypocycloid in which the radius of the rolling circle is *less* than half the radius of the fixed circle. Thus we can obtain *all* epicycloids and hypocycloids if in addition to epicycloids we take hypocycloids in which the radius of the rolling circle is less than half the radius of the fixed circle.

362. If a and b are in the proportion of two whole numbers we may eliminate θ between the two equations which determine an epicycloid or a hypocycloid, and thus obtain the equation to the curve in an algebraical form. For example, suppose in the hypocycloid that $a = 4b$; then

$$x = 3b \cos \theta + b \cos 3\theta = 4b \cos^3 \theta,$$

$$y = 3b \sin \theta - b \sin 3\theta = 4b \sin^3 \theta ;$$

therefore $\qquad x^{\frac{2}{3}} + y^{\frac{2}{3}} = a^{\frac{2}{3}}.$

If in the hypocycloid we suppose $a = 2b$, we obtain

$$x = 2b \cos \theta \text{ and } y = 0 ;$$

thus y is always zero and x may have any value between $-a$ and $+a$; therefore the curve reduces to a diameter of the fixed circle.

363. If in Art. 360 the describing point, instead of being *on* the perimeter of the rolling circle, is on a fixed radius of that circle, but either *within* or *without* the perimeter, the curve generated is called the epitrochoid when the rolling circle moves on the outside of the fixed circle, and the hypotrochoid when the rolling circle moves on the inside of the fixed circle. In the former case we have

$$x = (a + b) \cos \theta - mb \cos \frac{a + b}{b} \theta,$$

$$y = (a + b) \sin \theta - mb \sin \frac{a + b}{b} \theta,$$

and in the latter case

$$x = (a - b) \cos \theta + mb \cos \frac{a - b}{b} \theta,$$

$$y = (a - b) \sin \theta - mb \sin \frac{a - b}{b} \theta,$$

mb being the distance of the describing point from the centre of the rolling circle.

364. If a circle roll along a straight line the curve traced out by a point *in* the perimeter of the rolling circle is, as we have already stated, called the cycloid. If the describing point be *inside* the perimeter the curve is called the *prolate* cycloid, if *outside* the *curtate* cycloid ; the term *trochoid* is also used to denote both the *prolate* cycloid and the *curtate* cycloid.

The equations

$$x = a \, (1 - m \cos \theta),$$
$$y = a \, (\theta + m \sin \theta),$$

will represent a prolate cycloid, a common cycloid, or a curtate cycloid, according as m is less than unity, equal to unity, or greater than unity. See Art. 358.

EXAMPLES.

Trace the following curves :

1. $y^3 = ax^2 - x^3$.

2. $y^3 = a^3 - x^3$.

3. $y^2 (x - a) = (x + a) x^2$.

4. $x^2 y^2 = a^2 (x^2 - y^2)$.

5. $y^2 (x - 4a) = ax (x - 3a)$.

6. $(x^2 + y^2)^3 = 4a^2 x^2 y^2$.

7. $y^2 (2a - x) = x^3$.   (The cissoid.)

8. $x^2 y^2 = (a^2 - y^2) (b + y)^2$.   (The conchoid.)   Transfer the origin to the point $(0, -b)$ and then change to polar co-ordinates and we have for the equation

$$r = b \, \mathrm{cosec} \, \theta \pm a.$$

9. $(x^2 + y^2)^2 = a^2 (x^2 - y^2)$.   (The lemniscata.)

10. $r = a\theta \sin \theta$.

11. $r = a \, (\theta + \sin \theta)$.

12. $r \sin \theta = a \cos^2 \theta$.

13. $r = \log \sin \theta$.

14. $r^2 \cos \theta = a^2 \sin^2 3\theta$.

15. $r^2 \cos \theta = a^2 \sin^3 \theta$.

16. $r (\theta - \sin \theta) = a (\theta + \sin \theta)$.

17. $r = a (1 - \cos \theta)$.   (The cardioide.)

18. $r\theta = a$.   (The hyperbolic spiral.)

19. Find the equations to the tangent and normal at the point P in the epicycloid. See the Figure to Art. 360. Shew that the normal at P passes through B.

20. Trace the curve determined by the equations

$$x = a \, (1 - \cos \phi), \;\; y = a\phi \, ;$$

this curve is called the *companion to the cycloid*.

21. Obtain in an algebraical form the equation to the epicycloid for which $a = 2b$.

$$\text{Result.} \quad 4\,(x^2 + y^2 - a^2)^3 = 27a^4y^2.$$

22. Shew that when $a = b$ the epicycloid becomes the cardioide.

23. Trace the curve whose equation is $r = a \cos \dfrac{\theta}{3}$; and shew that if A be the point where the curve meets the prime radius produced backwards and $PSQR$ any chord drawn through the pole S meeting the curve at P, Q, and R, the angles PAQ and QAR are each $60°$, and the angle ASQ equal to thrice the angle APS.

24. Shew that the equations

$$r = a - a \tan \theta \quad \text{and} \quad 2a = r - r \tan \theta$$

represent the same curve in different positions, and that the radii vectores to the points of intersection bisect the angles between the tangents at those points.

25. Trace the curve $\dfrac{y}{c} = \sin \dfrac{x}{a} \log \left(m \sin \dfrac{x}{a} \right)$, (1) when m is greater than unity, (2) when m is equal to unity, (3) when m is less than unity and greater than the reciprocal of the base of the Napierian logarithms, (4) when m is less than the reciprocal of the base of the Napierian logarithms.

CHAPTER XXVII.

ON DIFFERENTIALS.

365. IN the preceding pages we have given the propositions commonly found in works on the Differential Calculus,
and have used the method of limits in all the demonstrations.
We now offer a few remarks on another method of treating
the subject.

In the expansion of $f(x+h)$ by Taylor's Theorem, the
coefficient of h was shewn to be that function of x which we
had called the differential coefficient of $f(x)$ with respect to x.
Lagrange proposed to *define* the differential coefficient of $f(x)$
with respect to x as *the coefficient of h in the expansion of*
$f(x+h)$, and thus to avoid all reference to the theory of
limits. Lagrange's views were propounded towards the close
of the last century and were generally adopted by elementary
writers.

One objection to this method is its use of infinite series
without ascertaining that those series are *convergent*, and the
proof that $f(x+h)$ can always be expanded in a series of
ascending powers of h, which is made the foundation of the
Differential Calculus, labours under serious defects. Another
objection is that it is impossible to avoid introducing the
notion of a limit in the applications of the subject to geometry
and mechanics; the definition of the tangent line to a curve
may be given as an example.

366. Nearly all the recent treatises on the Differential
Calculus have followed the method of limits, and the only
point of importance in which a difference exists among them
is with respect to the use of *differentials*. In the present
work $\dfrac{dy}{dx}$ has been defined as *one* symbol, thus: *if* $y = \phi\,(x)$

the limit of $\dfrac{\phi(x+h)-\phi(x)}{h}$ *when h is indefinitely diminished is denoted by* $\dfrac{dy}{dx}$. Some writers add the following words: *the quantities dx and dy are called the differentials of x and y respectively; their absolute values are indeterminate, and they may be either finite or indefinitely small provided their relative magnitudes be such that* $\dfrac{dy}{dx}$ *is equal to the limit above mentioned.*

With this meaning attached to dy and dx such equations may occur as

$$dy = \phi'(x)\, dx,$$

where $\phi'(x)$ is the differential coefficient of $\phi(x)$ or y.

Equations expressed by means of differentials are in general capable of immediate translation into the language of differential coefficients. For example, if x and y be co-ordinates of a point on a curve and be functions of a third variable t, and if s denote the corresponding arc of the curve beginning at some fixed point, we have, by Art. 307,

$$\left(\frac{dx}{dt}\right)^2 + \left(\frac{dy}{dt}\right)^2 = \left(\frac{ds}{dt}\right)^2,$$

and by differentiation

$$\frac{dx}{dt}\frac{d^2x}{dt^2} + \frac{dy}{dt}\frac{d^2y}{dt^2} = \frac{ds}{dt}\frac{d^2s}{dt^2}.$$

A writer who uses *differentials* will express these results thus,

$$dx^2 + dy^2 = ds^2,$$

$$dx\, d^2x + dy\, d^2y = ds\, d^2s.$$

The student may look upon the latter as merely abbreviated methods of writing the previous equations, and may take $dx,\ dy,\ d^2x,\ \dots$ as standing for $\dfrac{dx}{dt},\ \dfrac{dy}{dt},\ \dfrac{d^2x}{dt^2},\ \dots$ respectively.

367. Let u be a function of any number of variables, for example three, so that $u = \phi(x,\, y,\, z)$. If we suppose

x, y, z, all functions of a variable t, and for shortness put

$$\frac{du}{dt} = Du, \quad \frac{dx}{dt} = Dx, \quad \frac{dy}{dt} = Dy, \quad \frac{dz}{dt} = Dz,$$

we have

$$Du = \left(\frac{d\phi}{dx}\right) Dx + \left(\frac{d\phi}{dy}\right) Dy + \left(\frac{d\phi}{dz}\right) Dz \ldots\ldots\ldots (1).$$

In works on the Differential Calculus, which use *differentials*, we find an equation similar to the above occurring at an early period, namely,

$$du = \left(\frac{d\phi}{dx}\right) dx + \left(\frac{d\phi}{dy}\right) dy + \left(\frac{d\phi}{dz}\right) dz \ldots\ldots\ldots (2).$$

The introduction and use of this equation form the principal difference between such works and one which, like the present, uses only *differential coefficients*. To establish (2) the following method is adopted.

Let $\qquad u = \phi (x, y, z)$,

and $\qquad u + \Delta u = \phi (x + \Delta x, y + \Delta y, z + \Delta z)$,

therefore

$$\Delta u = \phi (x + \Delta x, y + \Delta y, z + \Delta z) - \phi (x, y, z)$$

$$= \frac{\phi (x + \Delta x, y + \Delta y, z + \Delta z) - \phi (x, y + \Delta y, z + \Delta z)}{\Delta x} \Delta x$$

$$+ \frac{\phi (x, y + \Delta y, z + \Delta z) - \phi (x, y, z + \Delta z)}{\Delta y} \Delta y$$

$$+ \frac{\phi (x, y, z + \Delta z) - \phi (x, y, z)}{\Delta z} \Delta z \ldots\ldots\ldots\ldots\ldots (3).$$

If Δx, Δy, and Δz diminish without limit, the quantity

$$\frac{\phi (x + \Delta x, y + \Delta y, z + \Delta z) - \phi (x, y + \Delta y, z + \Delta z)}{\Delta x}$$

approaches the limit $\left(\frac{d\phi}{dx}\right)$. If then we put for this quantity $\left(\frac{d\phi}{dx}\right) + a$, we know that a diminishes without limit when

Δx, Δy, and Δz do so. In this manner we may deduce from (3) the equation

$$\Delta u = \left\{ \left(\frac{d\phi}{dx} \right) + \alpha \right\} \Delta x + \left\{ \left(\frac{d\phi}{dy} \right) + \beta \right\} \Delta y + \left\{ \left(\frac{d\phi}{dz} \right) + \gamma \right\} \Delta z \dots (4),$$

where α, β, γ, all diminish without limit when Δx, Δy, Δz do so. If then du, dx, dy, and dz, denote quantities whose absolute magnitudes are undetermined, but whose *relative* magnitudes are those to which Δu, Δx, Δy, and Δz, respectively approach as their limits when they are all indefinitely diminished, we have

$$du = \left(\frac{d\phi}{dx} \right) dx + \left(\frac{d\phi}{dy} \right) dy + \left(\frac{d\phi}{dz} \right) dz.$$

Having thus established (2), we give an example of its application. Since in establishing (2) we had no occasion to consider whether x, y, and z, were independent or not, the result is universally true, whatever relation be given or supposed between the variables. If, for example, $\phi(x, y, z)$ is always $= 0$, we have

$$\left(\frac{d\phi}{dx} \right) dx + \left(\frac{d\phi}{dy} \right) dy + \left(\frac{d\phi}{dz} \right) dz = 0 \dots \dots (5).$$

Now if $\phi(x, y, z) = 0$ is the only equation connecting x, y, and z, we may if we please vary x and z without changing y. Hence in the preceding investigation $\Delta y = 0$ throughout, and therefore in (5) $dy = 0$; thus we have

$$\left(\frac{d\phi}{dx} \right) dx + \left(\frac{d\phi}{dz} \right) dz = 0 \dots \dots (6).$$

Hence $$\frac{dz}{dx} = - \frac{\left(\frac{d\phi}{dx} \right)}{\left(\frac{d\phi}{dz} \right)},$$

where $\frac{dz}{dx}$ is the differential coefficient of z, supposing x to vary and y to be constant. See Art. 188.

368. It would occupy too much space if we were to proceed further with the subject of differentials. Differential

coefficients have been used exclusively in the present work, from the conviction that the subject is thus presented in the clearest form, and that if some of the operations are thus rendered a little longer than they would otherwise be, there is at the same time far less liability to error. The equation (2) is certainly of great use in applications of the Differential Calculus, particularly in the higher parts of the Geometry of Three Dimensions: after the remarks already made, the student will probably find little difficulty in those applications. Perhaps he may be further assisted by referring to the theorem for the expansion of a function of three variables. If $u = \phi(x, y, z)$, we have

$$\phi(x+h, y+k, z+l) - \phi(x, y, z) \text{ or } \Delta u$$

$$= h\frac{du}{dx} + k\frac{du}{dy} + l\frac{du}{dz} + R,$$

where R involves squares and products of h, k, l. Hence the smaller h, k, l, are taken, the smaller is the error contained in the assertion

$$\Delta u \doteq h\frac{du}{dx} + k\frac{du}{dy} + l\frac{du}{dz}.$$

MISCELLANEOUS EXAMPLES.

1. Find $\dfrac{du}{dv}$ if $u = \sin^{-1}\sqrt{x} - \sqrt{(x - x^2)}$,

    and $v = \cos^{-1}(x^{\frac{1}{3}}a^{\frac{1}{6}}) - (x^{\frac{1}{3}}a^{\frac{1}{3}} - x^{\frac{1}{4}}a^{\frac{2}{3}})^{\frac{1}{2}}$.

    Result. $-\dfrac{3x^{\frac{1}{6}}}{2a^{\frac{1}{6}}\sqrt{1-x}\sqrt{1-x^{\frac{1}{3}}a^{\frac{1}{3}}}}.$

2. Find the maxima and minima values of $(\sin x)^{\sin x}$.

3. Find the area of the greatest isosceles triangle that can be inscribed in a given ellipse, the triangle having its vertex coincident with one extremity of the major axis.

4. $APQB$ is a semicircle whose diameter is AB, and PQ is parallel to AB. Draw AQ and BP, and let them meet at R: find the position of P and Q so that the triangle PQR may be a maximum.

    Result. $\dfrac{PQ}{AB}$ must be equal to $\dfrac{\sqrt{17}-1}{4}$.

5. A figure made up of a rectangle and an isosceles triangle is inscribed in a semicircle : determine its dimensions so that its area may be a maximum.

   Result. The height of the rectangle must be half the radius of the circle.

6. Find the cone of least surface, excluding the base, that can surround a given sphere.

   Result. The sine of the semivertical angle $= \sqrt{2} - 1$.

7. Find the cone of least surface, including the base, that can surround a given sphere.

   Result. The sine of the semivertical angle $= \frac{1}{3}$.

8. Find the maximum value of $\cos \theta \cos \phi \cos \psi$, where $\theta + \phi + \psi = \pi$.

9. Transform $\dfrac{d^2u}{dx^2} + \dfrac{d^2u}{dy^2}$ by assuming

   $$x' = l_1 x + m_1 y, \quad y' = l_2 x + m_2 y.$$

Result. $(l_1^2 + m_1^2) \dfrac{d^2u}{dx'^2} + 2 (l_1 l_2 + m_1 m_2) \dfrac{d^2u}{dx' dy'} + (l_2^2 + m_2^2) \dfrac{d^2u}{dy'^2}.$

10. An equation between three variables contains n arbitrary functions of one of them, and $4n^2 - n - 1$ arbitrary constants : shew that generally the equation must be differentiated at least $4n - 2$ times in order that the functions and constants may be eliminated.

11. If V be any function of x, y, z, and V' the value of V when vw is substituted for x, wu for y, and uv for z; then

$$x^2 \frac{d^2V}{dx^2} + y^2 \frac{d^2V}{dy^2} + z^2 \frac{d^2V}{dz^2} + yz \frac{d^2V}{dy\,dz} + zx \frac{d^2V}{dz\,dx} + xy \frac{d^2V}{dx\,dy}$$

$$= \frac{1}{2} \left\{ u^2 \frac{d^2V'}{du^2} + v^2 \frac{d^2V'}{dv^2} + w^2 \frac{d^2V'}{dw^2} \right\}.$$

12. If $y = e^{ux} + e^{-ux}$, and $z + xe^{-2ux} = 0$, shew that the general term in the value of y when expanded in a series is

$$\frac{x^{2n}}{\underline{|n}} \{(2n + 1)^{n-1} - (2n - 1)^{n-1}\}.$$

13. If $y = x + a\psi(y) + \beta\phi(y) + \ldots\ldots$, then

$$F(y) = F(x) + \ldots + \frac{1}{\lfloor n} \frac{d^{n-1}}{dx^{n-1}} \left[F'(x) \{ x\psi(x) + \beta\phi(x) + \ldots \}^n \right] + \ldots$$

14. If $y = z + x\phi(y)$, and $y' = z' + x'\psi(y')$, z and z' being independent variables, shew that the general term in the expansion of $f(y, y')$ in powers and products of x and x' is

$$\frac{d^{m+n-2}}{dz^{m-1} dz'^{n-1}} \left\{ \overline{\phi(z)}\rceil^m \overline{\psi(z')}\rceil^n \frac{d^2 f(z, z')}{dz\, dz'} \right\} \frac{x^m}{\lfloor m} \cdot \frac{x'^n}{\lfloor n} .$$

Find the coefficient of $x^2 x'$ in the expansion of $\cos(ay + a'y')$, when $y = z + x \sin y$, and $y' = z' + x' \sin y'$.

15. In any curve the part of the tangent between the point of contact and the perpendicular from the origin on the tangent is equal to $\dfrac{r\,dr}{ds}$.

16. Shew that the equation to the normal at any point of a curve may be put under the form

$$\frac{x' - x}{\dfrac{d^2 x}{ds^2}} = \frac{y' - y}{\dfrac{d^2 y}{ds^2}} .$$

Shew that this equation is the analytical expression of the fact, that if a tangent be drawn to a curve at any point P, and in the tangent PT be taken equal to the arc PQ and on the same side of P, then the straight line QT is ultimately perpendicular to the tangent.

17. In the ellipse the focal distance cuts the curve at an angle, the tangent of which is a mean proportional between the tangents of the angles at which the corresponding diameter and a parallel through the point to the transverse axis cut the curve.

18. If a curve be referred to axes inclined at an angle a to each other, shew that the radius of curvature is

$$\frac{\left\{ 1 + 2\cos a \dfrac{dy}{dx} + \left(\dfrac{dy}{dx}\right)^2 \right\}^{\frac{3}{2}}}{- \sin a \dfrac{d^2 y}{dx^2}} .$$

19. The equation to a parabola referred to any two tangents being $\left(\dfrac{x}{a}\right)^{\frac{1}{2}}+\left(\dfrac{y}{b}\right)^{\frac{1}{2}}=1$, shew that the radius of curvature is $\dfrac{2}{ab\sin\alpha}\{ax-2\cos\alpha\sqrt{(abxy)}+by\}^{\frac{3}{2}}$, where α is the inclination of the tangents; and thence find the co-ordinates of the vertex assuming that the curvature is a maximum at that point.

20. If a curve pass through the origin and touch the axis of y, the diameter of the circle of curvature is equal to the limit of $\dfrac{y^2}{x}$; if it touch the axis of x the diameter is equal to the limit of $\dfrac{x^2}{y}$.

21. If a curve pass through the origin at an inclination α to the axis of x, shew that the diameter of curvature at the origin is the limit of $\dfrac{x^2+y^2}{x\sin\alpha-y\cos\alpha}$. Hence, shew that the radius of curvature at the origin of the curve $y^2+2ay-2ax=0$ is $2\sqrt{2}a$.

22. If ϕ be the angle between the tangent and the radius vector of a polar curve, shew that the radius of curvature is $\dfrac{r\operatorname{cosec}\phi}{1+\dfrac{d\phi}{d\theta}}$.

23. The equations to an epicycloid being
$$x=a\,(2\cos\theta-\cos 2\theta),$$
$$y=a\,(2\sin\theta-\sin 2\theta),$$
shew that $\rho=\dfrac{8a}{3}\sin\dfrac{\theta}{2}$, and that the evolute is an epicycloid in which the radius of each circle is $\dfrac{a}{3}$.

24. In the curve $y=x^4-4x^3-18x^2$, find the nature of the curve at the points $x=3$, -1, and $\frac{3}{2}(1\pm\sqrt{5})$.

25. Shew that the curve $y=e^{-x^2}$ has points of inflexion when $x=\pm\dfrac{1}{\sqrt{2}}$.

26. In any curve the equation $\dfrac{d\phi}{d\theta} + 1 = 0$ holds at a point of inflexion, θ and ϕ being the angles which the prime radius and tangent make respectively with the radius vector.

27. Is $\dfrac{dr}{d\theta}$ necessarily of the form $\dfrac{0}{0}$ at a multiple point?

28. Find the singular points in the curves

$$4\,(x^2 + y^2) = 1 + 3y^3,$$

and $\qquad y^3 - 2xy + 2x^2 - x^3 = 0.$

29. Find the nature of the curve

$$y + 1 = 2x - x^2 \pm (2 - x)^{\frac{5}{2}}$$

at the point $x = 2$.

30. Determine the point of inflexion in the curve

$$y = x^3 - 9x^2 + 24x - 16.$$

31. From the pole of the curve $r = Aa^\theta$ perpendiculars are drawn upon the tangent; through the points of intersection of the perpendiculars with the tangents, straight lines are drawn parallel to the radii vectores: shew that the equation to the locus of the ultimate intersections of all such straight lines is $r = A \cos\alpha\, a^{\theta + \alpha}$, where $\cot\alpha = \log a$.

32. If radii vectores of an equiangular spiral be diameters of a series of circles, the locus of the ultimate intersections of the circles will be a similar spiral.

CHAPTER XXVIII.

MISCELLANEOUS PROPOSITIONS.

369. In the present Chapter we shall investigate various propositions which afford valuable illustrations of the principles of the subject and lead to important results.

370. The following formula is due to Jacobi:

$$\frac{d^{n-1}(1-x^2)^{n-\frac{1}{2}}}{dx^{n-1}} = (-1)^{n-1}\frac{1 \cdot 3 \cdot 5 \dots (2n-1)\sin n\theta}{n},$$

where $x = \cos\theta$. This we shall now demonstrate.

Put y for $1 - x^2$: we have

$$\frac{d^n y^{n+\frac{1}{2}}}{dx^n} = \frac{d^{n-1}}{dx^{n-1}}\frac{d}{dx}y^{n+\frac{1}{2}} = -(2n+1)\frac{d^{n-1}}{dx^{n-1}}xy^{n-\frac{1}{2}};$$

thus by Art. 80

$$\frac{d^n y^{n+\frac{1}{2}}}{dx^n} = -(2n+1)x\frac{d^{n-1}y^{n-\frac{1}{2}}}{dx^{n-1}} - (n-1)(2n+1)\frac{d^{n-2}y^{n-\frac{1}{2}}}{dx^{n-2}} \dots(1).$$

Also

$$\frac{d^n y^{n+\frac{1}{2}}}{dx^n} = \frac{d^n yy^{n-\frac{1}{2}}}{dx^n};$$

thus by Art. 80

$$\frac{d^n y^{n+\frac{1}{2}}}{dx^n} = y\frac{d^n y^{n-\frac{1}{2}}}{dx^n} - 2nx\frac{d^{n-1}y^{n-\frac{1}{2}}}{dx^{n-1}} - n(n-1)\frac{d^{n-2}y^{n-\frac{1}{2}}}{dx^{n-2}} \dots\dots(2).$$

From (1) and (2) by eliminating $\dfrac{d^{n-2}y^{n-\frac{1}{2}}}{dx^{n-2}}$ we obtain

$$(2n+1-n)\frac{d^n y^{n+\frac{1}{2}}}{dx^n} = (2n+1)y\frac{d^n y^{n-\frac{1}{2}}}{dx^n} - (2n+1)nx\frac{d^{n-1}y^{n-\frac{1}{2}}}{dx^{n-1}} \dots(3).$$

Assume that Jacobi's formula is true for a specific value of n; differentiate both sides with respect to x: thus

$$\frac{d^n y^{n-\frac{1}{2}}}{dx^n} = (-1)^n \frac{1 \cdot 3 \cdot 5 \dots (2n-1) \cos n\theta}{\sin \theta}.$$

Using this result, and also Jacobi's formula, on the right-hand side of (3), we obtain

$$(n+1) \frac{d^n y^{n+\frac{1}{2}}}{dx^n} = (-1)^n 1 \cdot 3 \cdot 5 \dots (2n+1) \cos n\theta \sin \theta$$

$$+ (-1)^n 1 \cdot 3 \cdot 5 \dots (2n+1) \sin n\theta \cos \theta$$

$$= (-1)^n 1 \cdot 3 \cdot 5 \dots (2n+1) \sin (n+1) \theta ;$$

therefore $\dfrac{d^n y^{n+\frac{1}{2}}}{dx^n} = (-1)^n \dfrac{1 \cdot 3 \cdot 5 \dots (2n+1) \sin (n+1) \theta}{n+1}.$

This shews that if Jacobi's formula is true for a specific value of n it is true for that value increased by unity; and it is obviously true when $n = 1$, and when $n = 2$: therefore it is true for any positive integral value of n.

371. The following proposition is useful in some applications of mathematics to natural philosophy : Having given that if x varies, it must be such a function of the independent variable t, that $\dfrac{dx}{dt} = ax$, where a is some quantity, not necessarily constant, which is always finite; and having given that x is zero when t is zero: then it will follow that x cannot vary, or, in other words, that x is always zero.

Denote x by $\phi(t)$. We know by Art. 101 that

$$\phi(t) - \phi(0) = t\phi'(\theta t),$$

where θ is some proper fraction.

In the present case $\phi(0) = 0$, and $\phi'(\theta t) = a\phi(\theta t)$, where a is some finite quantity. Thus we have

$$\phi(t) = ta\phi(\theta t),$$

and therefore, if $\phi(t)$ be not zero,

$$1 = \frac{ta\phi(\theta t)}{\phi(t)}.$$

But it is impossible that this result can be universally true. For since a is always finite we can take t so small that ta shall be as small as we please. And as $\phi(t)$ begins with the value zero, if it varies it must at first increase numerically with t; and therefore $\dfrac{\phi(\theta t)}{\phi(t)}$ cannot be greater than unity. Hence the result is inadmissible; and it follows that x cannot vary, or in other words, x is always zero.

372. The preceding proposition may be extended so as to involve *any number* of such supposed variables as x; we will take *three* for example: Having given that if x, y, and z vary, they must be such functions of the independent variable t, that

$$\frac{dx}{dt} = a_1 x + a_2 y + a_3 z, \quad \frac{dy}{dt} = b_1 x + b_2 y + b_3 z, \quad \frac{dz}{dt} = c_1 x + c_2 y + c_3 z,$$

where $a_1, a_2, a_3, b_1, \ldots c_3$ are quantities, not necessarily constant, which are always finite; and having given that x, y, and z are all zero when t is zero: then it will follow that x, y, and z cannot vary, or, in other words, that x, y, and z are always zero.

Denote x by $\phi(t)$, y by $\psi(t)$, and z by $\chi(t)$. Then, as in the preceding Article, we have

$$\phi(t) = t\left\{a_1 \phi(\theta t) + a_2 \psi(\theta t) + a_3 \chi(\theta t)\right\}:$$

and therefore if $\phi(t)$ be not zero we have

$$1 = t\left\{a_1 \frac{\phi(\theta t)}{\phi(t)} + a_2 \frac{\psi(\theta t)}{\phi(t)} + a_3 \frac{\chi(\theta t)}{\phi(t)}\right\};$$

and in like manner we deduce two other similar results.

But it is impossible that these results can be universally true. For suppose t indefinitely small, and let $\phi(t)$ be not less than either $\psi(t)$ or $\chi(t)$. Then the first of the three results asserts that unity is equal to an indefinitely small quantity. Hence the results are inadmissible; and it follows that x, y, and z cannot vary, or, in other words, that x, y, and z are always zero.

373. We have already given two forms for the *remainder after* $n+1$ *terms* of an expansion by Taylor's Theorem; see Arts. 93 and 110: these two forms, and others, may be deduced from one general expression which we will now investigate.

Let $\phi(x)$ and $\psi(x)$ be two functions of x which remain continuous, as also their differential coefficients between the values a and $a+h$ of the variable x; suppose also that between these values the differential coefficient $\psi'(x)$ does not vanish: then by Art. 98

$$\frac{\phi(a+h)-\phi(a)}{\psi(a+h)-\psi(a)} = \frac{\phi'(a+\theta h)}{\psi'(a+\theta h)} \quad \ldots\ldots\ldots\ldots (1),$$

where θ is some proper fraction.

Denote by $\phi(x)$ the function

$$F(a+h)-F(x)-(a+h-x)F'(x)-\ldots-\frac{(a+h-x)^n}{\lfloor n}F^n(x);$$

and denote by $\psi(x)$ the function

$$f(a+h)-f(x)-(a+h-x)f'(x)-\ldots-\frac{(a+h-x)^q}{\lfloor q}f^q(x).$$

We assume that $F(x)$ and all its differential coefficients up to $F^{n+1}(x)$ inclusive are continuous while x lies between the values a and $a+h$; as also $f(x)$ and all its differential coefficients up to $f^{q+1}(x)$ inclusive: moreover we assume that $f^{q+1}(x)$ does not vanish between these values.

Now

$$\phi'(x) = -\frac{(a+h-x)^n}{\lfloor n}F^{n+1}(x),$$

and

$$\psi'(x) = -\frac{(a+h-x)^q}{\lfloor q}f^{q+1}(x);$$

also

$$\phi(a+h)=0, \text{ and } \psi(a+h)=0:$$

thus we have from (1)

$$\frac{\phi(a)}{\psi(a)} = \frac{\lfloor q}{\lfloor n}(h-\theta h)^{n-q}\frac{F^{n+1}(a+\theta h)}{f^{q+1}(a+\theta h)}.$$

Multiply by $\psi(a)$, and put for $\phi(a)$ and $\psi(a)$ their values; then

$$F(a+h) - F(a) - hF'(a) - \ldots - \frac{h^n F^n(a)}{\lfloor n} = R,$$

where $R =$

$$\left\{ f(a+h) - f(a) - hf'(a) - \ldots - \frac{h^q f^q(a)}{\lfloor q} \right\} \frac{\lfloor q}{\lfloor n}(h-\theta h)^{n-q} \frac{F^{n+1}(a+\theta h)}{f^{q+1}(a+\theta h)}$$

$$\ldots\ldots\ldots\ldots(2).$$

This is a general expression for R, the remainder after $n+1$ terms of the expansion of $F(a+h)$ by Taylor's Theorem.

For a particular case take $f(x) = (x-a)^{p+1}$, where p is any positive number which is not less than q; then all the conditions with respect to $f(x)$ are satisfied: and we have

$$f(a) = 0, \quad f'(a) = 0, \quad \ldots f^q(a) = 0,$$

$$f(a+h) = h^{p+1},$$

and $\quad f^{q+1}(a+\theta h) = (p+1)p \ldots (p-q+1)(\theta h)^{p-q}.$

Hence

$$R = \frac{\lfloor q}{\lfloor n} \frac{(1-\theta)^{n-q}}{\theta^{p-q}} \frac{h^{n+1} F^{n+1}(a+h\theta)}{(p+1)p \ldots (p-q+1)} \ldots\ldots\ldots(3).$$

In the particular case in which $p=q$ we have from (3)

$$R = \frac{(1-\theta)^{n-p} h^{n+1} F'^{n+1}(a+\theta h)}{(p+1)\lfloor n} \ldots\ldots\ldots\ldots(4).$$

If in (4) we put $p=n$ we have Lagrange's form of the *remainder*, which is given in Art. 92; if in (4) we put $p=0$ we have Cauchy's form of the *remainder*, which is given in Art. 110.

Other particular forms may be readily obtained. Thus in (3) put $q=0$; then since $\lfloor 0$ must be replaced by unity we have

$$R = \frac{(1-\theta)^n h^{n+1} F'^{n+1}(a+\theta h)}{\theta^p (p+1) \lfloor n}.$$

Again, in the general expression (2) let $f(x) = F^n(x)$, and $q = 0$; then

$$\psi(x) = F^n(a+h) - F^n(x),$$

and assuming that $F^{n+1}(x)$ does not vanish between the values a and $a+h$, we have

$$R = \frac{h^n(1-\theta)^n}{\lfloor n} \left\{ F^n(a+h) - F^n(a) \right\}.$$

In (2) put $q = 0$; thus

$$R = \frac{h^n(1-\theta)^n}{\lfloor n} \left\{ f(a+h) - f(a) \right\} \frac{F^{n+1}(a+\theta h)}{f'(a+\theta h)}.$$

Mémoires de l'Académie... de Montpellier, Vol. 5, 1861...1863.

374.   Expand $\sqrt{(1-x^2)} \cdot \sin^{-1}x$ in powers of x.

Assume $\sqrt{(1-x^2)} \cdot \sin^{-1}x = A_0 + A_1x + A_2x^2 + A_3x^3 + \ldots$

Differentiate both sides with respect to x; thus

$$1 - \frac{x\sin^{-1}x}{\sqrt{(1-x^2)}} = A_1 + 2A_2x + \ldots + rA_rx^{r-1} + \ldots,$$

that is $1 - \dfrac{x}{1-x^2}(A_0 + A_1x + A_2x^2 + \ldots)$

$$= A_1 + 2A_2x + \ldots + rA_rx^{r-1} + \ldots;$$

therefore $1 - x^2 - x(A_0 + A_1x + A_2x^2 + \ldots)$

$$= (A_1 + 2A_2x + \ldots + rA_rx^{r-1} + \ldots)(1-x^2).$$

Equate the coefficients of x^r; thus if r be greater than 2 we have

$$-A_{r-1} = (r+1)A_{r+1} - (r-1)A_{r-1},$$

therefore $(r-2)A_{r-1} = (r+1)A_{r+1}.$

Also we can see by expanding $\sqrt{(1-x^2)}$ and $\sin^{-1}x$ and forming their product that

$$A_0 = 0, \; A_1 = 1, \quad A_2 = 0, \quad A_3 = -\frac{1}{3};$$

hence A_4, A_6, A_8, ... vanish,

and
$$A_5 = \frac{2}{5} A_3 = -\frac{2}{3 \cdot 5},$$

$$A_7 = \frac{4}{7} A_5 = -\frac{2 \cdot 4}{3 \cdot 5 \cdot 7},$$

. . . .

Put θ for $\sin^{-1} x$; thus we deduce

$$\frac{\theta}{\tan \theta} = 1 - \frac{1}{3} (\sin \theta)^2 - \frac{2}{3 \cdot 5} (\sin \theta)^4 - \frac{2 \cdot 4}{3 \cdot 5 \cdot 7} (\sin \theta)^6 - \ldots$$

See *Quarterly Journal of Mathematics*, Vol. 6, page 23.

375. Let $\phi(x)$ denote $x^n + p_1 x^{n-1} + p_2 x^{n-2} + \ldots + p_{n-1} x + p_n$, where n is a positive integer. It is required to determine the coefficients $p_1, p_2, \ldots p_n$ so that the numerically greatest value of $\phi(x)$ between the given limits $-h$ and h for x shall be as small as possible.

If we give a geometrical form to the problem, we may say that the curve $y = \phi(x)$ between the limits $-h$ and h is to deviate as little as possible from the axis of x.

The maxima and minima values of $\phi(x)$ will be determined by the equation $\phi'(x) = 0$, which is of the $(n-1)^{\text{th}}$ degree; and therefore there cannot be more than $n-1$ of such values. These values, together with the values of $\phi(x)$ when $x = -h$, and when $x = h$, will be called *extreme* values.

376. Now we admit as sufficiently obvious that there must be some definite values of the coefficients in $\phi(x)$ which solve the problem; and we shall first shew that there must be $n+1$ extreme values all numerically equal.

Suppose, for instance, that $n = 3$; then there must be 4 extreme values all numerically equal.

For if possible suppose that there are only 3 extreme values of $\phi(x)$ all numerically equal; namely, corresponding to the values x_1, x_2, and x_3 of x. Let $\psi(x)$ denote the expression

$$\mu_1 (x - x_2)(x - x_3) + \mu_2 (x - x_1)(x - x_3) + \mu_3 (x - x_1)(x - x_2),$$

and suppose μ_1, μ_2, and μ_3 to be infinitesimal constants, which are determined so that $\phi(x)$ and $\psi(x)$ may have contrary signs when $x = x_1$, when $x = x_2$, and when $x = x_3$: this can obviously be done. For instance, the sign of μ_1 must be contrary to the sign of $\dfrac{\phi(x_1)}{(x_1 - x_2)(x_1 - x_3)}$. Then $\phi(x) + \psi(x)$ differs only infinitesimally from $\phi(x)$; but when $\phi(x)$ has its extreme values $\phi(x) + \psi(x)$ is numerically less than $\phi(x)$: and so $\phi(x) + \psi(x)$ deviates less from zero than $\phi(x)$ does. Moreover the coefficient of x^3 in $\phi(x) + \psi(x)$ is unity; so that $\phi(x) + \psi(x)$ is an expression of the proper form. It follows therefore that $\phi(x)$ cannot be such as the problem requires.

The preceding argument will perhaps be more readily understood when presented in a geometrical form. The curve $y = \phi(x) + \psi(x)$ is *indefinitely close* to the curve $y = \phi(x)$; but where the latter curve deviates most from the axis of x the former curve is *nearer* to the axis of x: and thus the former curve deviates less from the axis of x than the latter curve.

In the same way we may treat the case in which $\phi(x)$ has only 2 extreme values numerically equal and numerically greater than any other value; or the case in which the numerically greatest value of $\phi(x)$ is unique.

The considerations which we have thus employed when $n = 3$ are applicable whatever may be the value of n.

Hence, as we have said, to solve the problem the coefficients in $\phi(x)$ must be determined so that $\phi(x)$ may have $n + 1$ extreme values all numerically equal.

377. Let k denote the extreme numerical value of $\phi(x)$; then we have shewn that the equation

$$\{\phi(x)\}^2 - k^2 = 0 \dots\dots\dots\dots\dots(1)$$

must have $n + 1$ values which also satisfy the equation

$$(x^2 - k^2)\,\phi'(x) = 0 \dots\dots\dots\dots\dots(2).$$

Let the $n+1$ values be denoted by $x_1, x_2, \ldots x_{n-1}$, besides $-h$ and h. We shall shew that any one of the former $n-1$ roots of (1) occurs twice in (1). For the derived equation of (1) is

$$2\phi'(x)\,\phi(x) = 0 \ldots\ldots\ldots\ldots\ldots\ldots(3);$$

and any one of the values $x_1, x_2, \ldots x_{n-1}$ is by supposition a root of the equation $\phi'(x) = 0$, and so satisfies (3).

Hence we have by the Theory of Equations

$$\{\phi(x)\}^2 - k^2 = (x^2 - h^2)(x - x_1)^2 (x - x_2)^2 \ldots (x - x_{n-1})^2.$$

But by supposition the roots of the equation $\phi'(x) = 0$ are $x_1, x_2, \ldots x_{n-1}$; hence

$$\phi'(x) = n\,(x - x_1)(x - x_2)\ldots(x - x_{n-1});$$

therefore $\qquad \{\phi(x)\}^2 - k^2 = \dfrac{\{\phi'(x)\}^2\,(x^2 - h^2)}{n^2} \ldots\ldots\ldots\ldots(4).$

Differentiate (4) with respect to x; thus we get

$$n^2\phi(x) = x\phi'(x) + (x^2 - h^2)\,\phi''(x) \ldots\ldots\ldots\ldots(5).$$

From (5) by equating the coefficients of $x^n, x^{n-1}, x^{n-2}, \ldots$ we shall be able to determine in succession $p_1, p_2, p_3, \ldots$ For thus we have

$$n^2 = n + n\,(n-1),$$

$$n^2 p_1 = (n-1)\,p_1 + (n-1)(n-2)\,p_1,$$

$$n^2 p_2 = (n-2)\,p_2 + (n-2)(n-3)\,p_2 - n\,(n-1)\,h^2,$$

$$n^2 p_3 = (n-3)\,p_3 + (n-3)(n-4)\,p_3 - (n-1)(n-2)\,h^2 p_1,$$

$$n^2 p_4 = (n-4)\,p_4 + (n-4)(n-5)\,p_4 - (n-2)(n-3)\,h^2 p_2,$$

and so on.

Thus $\quad p_1 = 0, \quad p_2 = -\dfrac{nh^2}{4}, \quad p_3 = 0, \quad p_4 = -\dfrac{(n-3)\,h^2 p_2}{8}, \ldots$

Therefore $\quad \phi(x) = x^n - nx^{n-2}\dfrac{h^2}{2^2} + \dfrac{n\,(n-3)}{\underline{|2}}\,x^{n-4}\dfrac{h^4}{2^4}$

$$- \dfrac{n\,(n-4)(n-5)}{\underline{|3}}\,x^{n-6}\dfrac{h^6}{2^6} + \ldots\ldots\ldots\ldots(6).$$

378. If in the identity at the top of page 120 we put $4t = \dfrac{h^2}{x^2}$ we shall obtain

$$\{x + \sqrt{(x^2 - h^2)}\}^n + \{x - \sqrt{(x^2 - h^2)}\}^n$$

$$= 2^n \left\{ x^n - nx^{n-2}\frac{h^2}{2^2} + \frac{n(n-3)}{\lfloor 2}\, x^{n-4}\frac{h^4}{2^4} - \dots \right\};$$

hence we infer that

$$\phi(x) = \frac{\{x + \sqrt{(x^2 - h^2)}\}^n + \{x - \sqrt{(x^2 - h^2)}\}^n}{2^n} \dots\dots\dots(7),$$

and this may be verified by shewing that this value of $\phi(x)$ satisfies equation (5).

By putting $x = h$ we find that $k = \dfrac{h^n}{2^{n-1}}$.

Assume $\dfrac{x}{h} = \cos\theta$, which is of course allowable so long as x is not numerically greater than h.

Then $\qquad \{x \pm \sqrt{(x^2 - h^2)}\}^n = h^n \{\cos\theta \pm \sqrt{-1}\sin\theta\}^n$

$$= h^n \{\cos n\theta \pm \sqrt{-1}\sin n\theta\};$$

thus $\qquad\qquad \phi(x) = \dfrac{h^n \cos n\theta}{2^{n-1}};$

that is so long as x lies between $-h$ and h we have

$$\phi(x) = \frac{h^n}{2^{n-1}}\cos n\left(\cos^{-1}\frac{x}{h}\right) = k\cos n\left(\cos^{-1}\frac{x}{h}\right).$$

379. The last result may also be obtained from (4). For put $\phi(x) = z$; then (4) gives

$$n\sqrt{(k^2 - z^2)} = \frac{dz}{dx}\sqrt{(h^2 - x^2)},$$

therefore $\qquad -\dfrac{n}{\sqrt{(h^2 - x^2)}} = -\dfrac{1}{\sqrt{(k^2 - z^2)}}\dfrac{dz}{dx}.$

Hence since $-\dfrac{n}{\sqrt{(h^2 - x^2)}}$ is the differential coefficient of

$n \cos^{-1} \dfrac{x}{h}$ with respect to x, and $-\dfrac{1}{\sqrt{(k^2 - z^2)}} \dfrac{dz}{dx}$ is the differen-

tial coefficient of $\cos^{-1} \dfrac{z}{k}$ with respect to x; it follows by Art. 102 that

$$n \cos^{-1} \frac{x}{h} = \cos^{-1} \frac{z}{k} + C,$$

where C denotes some constant quantity. Hence

$$\frac{z}{k} = \cos \left(n \cos^{-1} \frac{x}{h} - C \right).$$

But by hypothesis z must be numerically equal to k when x is equal to h; and thus C must be some multiple of π; and therefore $\cos \left(n \cos^{-1} \dfrac{x}{h} - C \right)$ is numerically equal to $\cos n \left(\cos^{-1} \dfrac{x}{h} \right)$. This gives the required result.

The problem of Arts. 375...379 is also solved in Bertrand's *Calcul Différentiel*, pages 512...519.

380. We have sometimes to determine the value of $\dfrac{dy}{dx}$ from an equation $\phi(x, y) = 0$, when x and y are such that $\dfrac{d\phi(x, y)}{dx}$ and $\dfrac{d\phi(x, y)}{dy}$ vanish; for instance, we have to do so when we are finding the directions of the tangents at a multiple point of a curve. The method of Art. 191 is liable to the objection which is there stated. In Art. 195 another method is given for the case in which $x = 0$ and $y = 0$ are the values under consideration. It is easy to make the latter method applicable for any values of x and y; by a process which is geometrically equivalent to transferring the origin of co-ordinates to the multiple point which may be supposed to be under consideration.

Suppose that $x = a$ and $y = b$ are the values to be considered. Put $a + h$ for x, and $y + k$ for y. Then the equation becomes $\phi(a+h, b+k) = 0$. Now expand $\phi(a+h, b+k)$ by Chapter XIV. Suppose that every differential coefficient $\dfrac{d^{r+s}\phi(x, y)}{dx^r\, dy^s}$ vanishes when $x = a$ and $y = b$, so long as $r + s$ is less than n. Then we may denote the expansion symbolically thus :

$$\phi(a+h, b+k) = \frac{1}{\lfloor n} \left(h\frac{d}{dx} + k\frac{d}{dy}\right)^n u + \frac{1}{\lfloor n+1} \left(h\frac{d}{dx} + k\frac{d}{dy}\right)^{n+1} v,$$

where u stands for $\phi(x, y)$ and v for $\phi(x+\theta h, y+\theta k)$, θ being some proper fraction ; and after the differentiations have been performed we are to put $x = a$ and $y = b$.

Now if we suppose h and k indefinitely small we have ultimately for determining the ratio of k to h an equation which may be expressed symbolically thus :

$$\left(h\frac{d}{dx} + k\frac{d}{dy}\right)^n u = 0,$$

or more explicitly thus :

$$h^n \frac{d^n u}{dx^n} + nh^{n-1} k \frac{d^n u}{dx^{n-1}dy} + \frac{n(n-1)}{\lfloor 2} h^{n-2}k^2 \frac{d^n u}{dx^{n-2}dy^2} + \ldots = 0,$$

where after the differentiations have been performed we are to put $x = a$ and $y = b$.

It is obvious, as in Art. 195, that when h and k are indefinitely small $\dfrac{k}{h}$ coincides in meaning with $\dfrac{dy}{dx}$ for the case in which $x = a$ and $y = b$.

381. As an example of the preceding Article suppose we have the equation $x^4 y^2 - c^2(c-x)^2(c^4+x^2) = 0$. Here when $x = c$ and $y = 0$ we have $\dfrac{du}{dx} = 0$ and $\dfrac{du}{dy} = 0$; also then $\dfrac{d^2 u}{dx^2} = -4c^4$, $\dfrac{d^2 u}{dx\, dy} = 0$, and $\dfrac{d^2 u}{dy^2} = 2c^4$. Thus we obtain

$$-h^2 4c^4 + k^2 2c^4 = 0 ;$$

therefore
$$\frac{k}{h} = \pm\sqrt{2}.$$

382. The remarks which we shall now give will illustrate an instructive mode of considering the singular points of curves. It will be seen that in effect we transfer the origin to the point to be examined, and then employ polar co-ordinates.

383. Suppose that from any point of a curve as centre a circle is described with an infinitesimal radius; then by the aid of diagrams the following statements become obvious:

If the point is an ordinary point the circle cuts the curve at two points, and the radii of the circle drawn to the two points include an angle which differs infinitesimally from two right angles.

If the point is a singular point we have other results which depend on the nature of the singularity.

If the point is a conjugate point the circle does not cut the curve.

If the point be a *point d'arrêt* the circle cuts the curve at only one point.

If the point is a cusp the circle cuts the curve at two points; but the radii of the circle drawn to the two points include an infinitesimal angle.

If the point is a *point saillant* the circle cuts the curve at two points; but the radii of the circle drawn to the two points include an angle which is neither infinitesimal nor infinitesimally different from two right angles.

If the point is a multiple point the circle cuts the curve at more than two points.

384. Now suppose that $\phi(x, y) = 0$ is the equation to the curve in a rational form. Let x and y be the co-ordinates of a point on the curve; and let $x + h$ and $y + k$ be the co-ordinates of any adjacent point.

Since $\phi(x, y) = 0$, we have, by Chapter XIV.,

$$\phi(x + h, y + k) = Ah + Bk + \frac{1}{2}(Ch^2 + 2Dhk + Ek^2) + R;$$

here A, B, C, D, E are certain differential coefficients of $\phi(x, y)$; and R may be symbolically expressed as

$$\frac{1}{\lfloor 3} \left(h \frac{d}{dx} + k \frac{d}{dy} \right)^3 v,$$

where v denotes $\phi(x + th, y + tk)$, and t is some proper fraction.

Let us suppose that A and B are not both zero; assume $A = K \sin \gamma$, and $B = K \cos \gamma$; also put $r \cos \theta$ for h and $r \sin \theta$ for k. Then the equation $\phi(x + h, y + k) = 0$ becomes

$$K \sin(\gamma + \theta) + \frac{r}{2} \left\{ C \cos^2 \theta + 2D \sin \theta \cos \theta + E \sin^2 \theta \right\}$$

$$+ \frac{R}{r} = 0 \quad \dots\dots\dots\dots\dots\dots (1).$$

It is obvious that when r is infinitesimal $\dfrac{R}{r}$ is also infinitesimal; and that the above equation is satisfied by a value of θ for which $\gamma + \theta$ is infinitesimal, and by a value of θ for which $\gamma + \theta$ is infinitesimally different from π; and by no other value of θ except such as differ from these by a multiple of 2π. Hence we have an ordinary point of the curve. Therefore for a singular point it is necessary that $A = 0$ and $B = 0$.

Suppose then that $A = 0$ and $B = 0$. The equation (1) reduces to

$$E \cos^2 \theta \left\{ \tan^2 \theta + \frac{2D}{E} \tan \theta + \frac{C}{E} \right\} + \frac{2R}{r^2} = 0 \dots\dots (2).$$

385. Suppose that D^2 is greater than CE; then we know that $\tan^2 \theta + \dfrac{2D}{E} \tan \theta + \dfrac{C}{E}$ can be resolved into real factors; and so may be expressed as $(\tan \theta - \tan \alpha)(\tan \theta - \tan \beta)$: and α and β may be supposed to lie between 0 and π. Thus the equation becomes

$$E \cos^2 \theta (\tan \theta - \tan \alpha)(\tan \theta - \tan \beta) + \frac{2R}{r^2} = 0 \dots\dots (3).$$

Now $\dfrac{R}{r^2}$ is infinitesimal when r is; therefore, denoting by η an infinitesimal angle, we see that (3) has four different solutions for θ, namely, one between $\alpha - \eta$ and $\alpha + \eta$, one between $\beta - \eta$ and $\beta + \eta$, one between $\pi + \alpha - \eta$ and $\pi + \alpha + \eta$, and one between $\pi + \beta - \eta$ and $\pi + \beta + \eta$. Thus the singular point is a *double* point, the tangents at the point being inclined at angles α and β respectively to the axis of x.

386. Next suppose that D^2 is less than CE; then we shall find that the infinitesimal circle does not cut the curve, and so the singular point is a *conjugate* point.

387. Finally, suppose that $D^2 = CE$; then equation (2) takes the form

$$E \cos^2 \theta (\tan \theta - \tan \alpha)^2 + \frac{2R}{r^2} = 0 \ldots\ldots\ldots (4):$$

the discussion of this form is rather complex, and we will only briefly indicate the results.

Suppose that $\dfrac{R}{E}$ is negative when θ is indefinitely near to α. Then denoting by η an infinitesimal angle we see that (4) has two solutions for θ, namely, one between $\alpha - \eta$ and α, and one between $\alpha + \eta$ and α. The sign of $\dfrac{R}{E}$ when θ is indefinitely near to $\pi + \alpha$ will in general be contrary to the sign when θ is indefinitely near to α, because R is in general a function of the *third* degree in $\cos \theta$ and $\sin \theta$, when r is small enough; and so there is no solution of (4) in this case besides the two already noticed. Hence the infinitesimal circle cuts the curve at two points, and only at two; and the radii of the circle drawn to the two points include an infinitesimal angle. Therefore the singular point is a *cusp;* the tangent at the cusp is inclined to the axis of x at an angle α, and the two branches are on opposite sides of the tangent.

Similarly if $\dfrac{R}{E}$ is positive when θ is indefinitely near to α we have in general a cusp of the first kind as before; the tangent at the cusp is now inclined to the axis of x at an angle $\pi + \alpha$.

But it may happen that R itself changes sign when θ is indefinitely near to α or to $\pi + \alpha$; and then our conclusion as to a cusp of the first kind does not hold. We should have in such a case to make a closer examination, and in general it would be necessary to extend our expansion of $\phi(x+h, y+k)$, and instead of R to have terms which may be expressed as

$$\frac{1}{\lfloor 3}\left(h\frac{d}{dx}+k\frac{d}{dy}\right)^{3}\phi(x, y)+\frac{1}{\lfloor 4}\left(h\frac{d}{dx}+k\frac{d}{dy}\right)^{4}\phi(x+th, y+tk),$$

where t represents a proper fraction.

388. Moreover if C, D, and E all vanish at the point (x, y), we should have to use this extended form of the expansion of $\phi(x+h, y+k)$ in order to determine the nature of the singularity.

MISCELLANEOUS EXAMPLES.

1. If a semicircle roll along a straight line, the curve to which its diameter is always a tangent is a cycloid.

2. If a cycloid roll along a straight line, the equation to the curve which its base touches is

$$\frac{x}{2a}=\left\{2+\left(\frac{y}{2a}\right)^{\frac{2}{3}}\right\}\left\{1-\left(\frac{y}{2a}\right)^{\frac{2}{3}}\right\}^{\frac{1}{2}}.$$

3. A series of circles is described having their centres on an equilateral hyperbola and passing through its centre, shew that the locus of their ultimate intersections will be a lemniscate.

4. Examine the nature of the following curves at the origin:

$$y^{4}+2ay^{2}x+x^{4}-2ax^{3}=0,$$

$$y^{4}-\frac{x^{5}}{a}+x^{4}+3x^{2}y^{2}=0,$$

$$y^{4}-4xy(ay-bx)-x^{4}=0,$$

$$y^{5}+x^{5}=2a^{2}x^{3}y.$$

5. Trace the curve $x^2y^2 + (x^2 - a^2)(x^2 - b^2) = 0$, and shew that the breadth of each closed portion is twice as great in the direction of y as in that of x. Shew also that when b approaches a as its limit, each of these portions is ultimately similar to an ellipse.

6. Trace the curve $(x^2 - a^2)^2 + (y^2 - b^2)^2 = a^4$. Shew that when $b = a$ it reduces to two ellipses.

7. If a conic section whose focus is at the pole of a given curve have with the curve a contact of the second order at the point (u, θ) the equation to the conic section will be

$$u' + \cos^2(\theta' - \theta)\frac{d}{d\theta}\left\{\frac{\dfrac{du}{d\theta}}{\cos(\theta' - \theta)}\right\} = u + \frac{d^2u}{d\theta^2}.$$

8. A given curve rolls on a straight line, explain the method of finding the locus of the centre of curvature at the point of contact of the curve and straight line.

    If the rolling curve be an equiangular spiral the required locus will be a straight line; if a cycloid a circle; and if a catenary a parabola.

9. Right-angled triangles are inscribed in a circle: if one of the sides containing the right angle pass through a fixed point, find the curve to which the other is always a tangent.

    Result. $c^2(x^2 + y^2) = (a^2 + b^2 - c^2 - ax - by)^2$,

    where a and b are the co-ordinates of the centre of the given circle and c its radius, the fixed point being the origin.

10. Determine the equation to the envelop of all the equilateral hyperbolas which have a common centre and cut at right angles the same straight line.

    Result. $x^2 + 3(axy)^{\frac{2}{3}} - y^2 + a^2 = 0$,

    where $x = a$ represents the given straight line.

T. D. C.   E E

11. Find the envelop of the axis of a parabola having a focal chord given in position and magnitude.

    Result. $x^{\frac{2}{3}} + y^{\frac{2}{3}} = c^{\frac{2}{3}}$; the origin being the middle point of the given chord, and one of the axes coinciding with that chord.

12. A system of ellipses is described such that each ellipse touches two rectangular axes, to which its axes are parallel, and that the rectangle under the axes of the ellipse is constant: shew that each ellipse is touched by two rectangular hyperbolas, the rectangle under the transverse axes of which is equal to the rectangle under the axes of any one of the ellipses.

13. A, B, are the centres of two equal circles, and AP, BQ, are two radii which are always perpendicular to each other: find the curve which is always touched by the right line PQ, and explain the result when

$$AB^2 = 2AP^2.$$

14. Trace the following curves:

$$x^3 - xy^2 + ay^2 = 0,$$

$$y^3 - 7yx^2 + 6x^3 - a^3 = 0,$$

$$y^4 + x^2y^2 - a^2x^2 = 0,$$

$$a\left(x^3 + 7x^2y + 7xy^2 + y^3\right) - x^2y^2 = 0,$$

$$xy^2 + ax^2 - a^3 = 0,$$

$$y^2\left(x - 2a\right) - x^3 + a^3 = 0,$$

$$y^5 - ax^3y - bxy^3 + x^5 = 0,$$

$$y^5 - 5ax^2y^3 + x^5 = 0,$$

$$y = \frac{x^2}{a} \pm (x - a)\frac{\sqrt{(x^2 - b^2)}}{a},$$

$$y^2\left(a + x\right) = x^2\left(a - x\right),$$

$$y = xe^{-x},$$

$$y = e^{-x}\sqrt{(x^2 - 1)},$$

$$e^{\left(\frac{y}{x}\right)^2} = \sin\frac{x}{a},$$

$$y = e^{\cos x},$$

$$r^2 \sin\theta = a^2 \cos 2\theta,$$

$$r(\theta - \pi)^2 = a\left(\theta^2 - \frac{\pi^2}{4}\right).$$

15. S and H are two fixed points, and a curve is described such that, if P be any point in it the rectangle contained by SP and HP is constant: shew that the straight lines drawn from S at right angles to SP and from H at right angles to HP meet the tangent at P at points equidistant from P.

16. If $f\left(\dfrac{x}{a},\ \dfrac{y}{b}\right)$ be a rational homogeneous function of $\dfrac{x}{a}$, $\dfrac{y}{b}$ of n dimensions, shew that the envelop of the curves represented by the equation $f\left(\dfrac{x}{a},\ \dfrac{y}{b}\right) = 1$, under the condition $ab = $ constant, consists in general of n rectangular hyperbolæ having the axes as asymptotes.

17. If any quadrilateral $ABCD$ change its form, its sides remaining constant, shew that the variations of the angles A, B, C, D are ultimately in the same ratio as the areas of the triangles BCD, CDA, DAB, ABC.

18. In Art. 274, if $p = n - 1$, we have approximately when x and y are very large

$$\frac{y}{x} = \mu_1 + \frac{b}{x}, \quad \text{where } b = -\frac{\psi(\mu_1)}{\phi'(\mu_1)} :$$

shew that if $q = n - 2$, we have by continuing the approximation

$$\frac{y}{x} = \mu_1 + \frac{b}{x} - \frac{2\chi(\mu_1) + 2b\psi'(\mu_1) + b^2\phi'(\mu_1)}{2x^2\phi'(\mu_1)} + \dots$$

Hence shew that in general the two extremities of the rectilinear asymptote are on opposite sides of the curve.

19. In Art. 275, if $p = n - 1$, we have approximately when x and y are very large

$$\frac{y}{x} = \mu_1 + \left(\frac{A}{x}\right)^{\frac{1}{2}}, \text{ where } A = -\frac{2\psi(\mu_1)}{\phi''(\mu_1)}:$$

shew that if $q = n - 2$, we have by continuing the approximation

$$\frac{y}{x} = \mu_1 + \left(\frac{A}{x}\right)^{\frac{1}{2}} + \frac{B}{x} + \frac{C}{x^{\frac{3}{2}}} + \dots$$

where $$B = -\frac{\psi'(\mu_1)}{\phi''(\mu_1)} - \frac{\phi'''(\mu_1) A}{6\phi''(\mu_1)},$$

$$C = -\frac{\chi(\mu_1) + \frac{A}{2}\psi''(\mu_1) + \left\{\psi'(\mu_1) + \frac{A}{2}\phi'''(\mu_1) + \frac{B}{2}\phi''(\mu_1)\right\} B}{A^{\frac{1}{2}}\phi''(\mu_1)}.$$

20. If (α, β) be a point of the curve $\phi(x, y) = 0$ through which pass n tangents, shew that the locus of all the tangents at that point is expressed by

$$\left\{(x - \alpha)\frac{d}{d\alpha} + (y - \beta)\frac{d}{d\beta}\right\}^n \phi(\alpha, \beta) = 0.$$

21. Shew that the theorem of Art. 91 will hold even if $\phi'(x)$ is infinite when $x = a$ or when $x = b$. Give a geometrical illustration.

22. Shew that the theorem of Art. 98 will hold even if $F'(x)$ or $f'(x)$ is infinite when $x = a$ or when $x = a + h$.

23. Shew that the formula (3) of Art. 373 will hold provided $p + 1$ is not less than q.

24. Obtain from (3) of Art. 373 the result

$$R = \frac{\lfloor q \ 2^{q+1} \theta^{\frac{1}{2}} (1 - \theta)^{n-q} h^{n+1} F^{n+1}(a + \theta h)}{1 . 3 . 5 \dots (2q + 1) \lfloor n}.$$

CAMBRIDGE: PRINTED BY C. CLAY, M.A. AND SONS, AT THE UNIVERSITY PRESS.

WORKS BY I. TODHUNTER, Sc.D., F.R.S.

NATURAL PHILOSOPHY FOR BEGINNERS. In two
parts. Part I. The Properties of Solid and Fluid Bodies. With
numerous Examples. 18mo. 3s. 6d. Part II. Sound, Light, and
Heat. 3s. 6d.

EUCLID FOR COLLEGES AND SCHOOLS. New
Edition. 18mo. 3s. 6d.

KEY TO EXERCISES IN EUCLID. Crown 8vo. 6s. 6d.

MENSURATION FOR BEGINNERS. With numerous
Examples. New Edition. 18mo. 2s. 6d.

KEY TO THE MENSURATION FOR BEGINNERS.
By Rev. Fr. LAWRENCE McCARTHY, Professor of Mathematics in St
Peter's College, Agra. Crown 8vo. 7s. 6d.

ALGEBRA FOR BEGINNERS. With numerous Examples. New Edition. 18mo. 2s. 6d.

KEY TO THE ALGEBRA FOR BEGINNERS. New
Edition. Crown 8vo. 6s. 6d.

TRIGONOMETRY FOR BEGINNERS. With numerous
Examples. New Edition. 18mo. cloth. 2s. 6d.

KEY TO THE TRIGONOMETRY FOR BEGINNERS.
Crown 8vo. 8s. 6d.

MECHANICS FOR BEGINNERS. With numerous Examples. New Edition. 18mo. cloth. 4s. 6d.

KEY TO THE MECHANICS FOR BEGINNERS. Crown
8vo. 6s. 6d.

ALGEBRA FOR THE USE OF COLLEGES AND
SCHOOLS. With numerous Examples. New Edition. Crown 8vo.
7s. 6d.

KEY TO THE ALGEBRA FOR THE USE OF COL-
LEGES AND SCHOOLS. New Edition. Crown 8vo. 10s. 6d.

A TREATISE ON THE THEORY OF EQUATIONS.
New Edition. Crown 8vo. 7s. 6d.

PLANE TRIGONOMETRY FOR COLLEGES AND
SCHOOLS. With numerous Examples. New Edition. Crown 8vo.
5s.

Dr Todhunter's Works (*continued*).

KEY TO THE PLANE TRIGONOMETRY FOR COL-
LEGES AND SCHOOLS. Crown 8vo. 10*s*. 6*d*.

A TREATISE ON SPHERICAL TRIGONOMETRY
FOR THE USE OF COLLEGES AND SCHOOLS. With numerous
Examples. New Edition. Crown 8vo. 4*s*. 6*d*.

A TREATISE ON CONIC SECTIONS. With numerous
Examples. New Edition. Crown 8vo. 7*s*. 6*d*.

KEY TO CONIC SECTIONS. By C. W. Bourne, M.A.,
Head Master of the College, Inverness. Crown 8vo. 10*s*. 6*d*.

A TREATISE ON THE DIFFERENTIAL CALCULUS.
With numerous Examples. New Edition. Crown 8vo. 10*s*. 6*d*.

KEY TO A TREATISE ON THE DIFFERENTIAL
CALCULUS. By H. St John Hunter, M.A. Crown 8vo. 10*s*. 6*d*.

A TREATISE ON THE INTEGRAL CALCULUS. With
numerous Examples. New Edition. Crown 8vo. 10*s*. 6*d*.

KEY TO A TREATISE ON THE INTEGRAL CAL-
CULUS AND ITS APPLICATIONS. By H. St John Hunter, M.A.
Crown 8vo. 10*s*. 6*d*.

EXAMPLES OF ANALYTICAL GEOMETRY OF THREE
DIMENSIONS. New Edition. Crown 8vo. 7*s*. 6*d*.

A TREATISE ON ANALYTICAL STATICS. Fifth
Edition. Edited by Professor J. D. Everett, F.R.S. Crown 8vo.
10*s*. 6*d*.

AN ELEMENTARY TREATISE ON LAPLACE'S,
LAMÉ'S, AND BESSEL'S FUNCTIONS. Crown 8vo. 10*s*. 6*d*.

A HISTORY OF THE MATHEMATICAL THEORY OF
PROBABILITY, from the time of Pascal to that of Laplace. 8vo.
18*s*.

THE CONFLICT OF STUDIES, and other Essays on
Subjects connected with Education. 8vo. 10*s*. 6*d*.

WILLIAM WHEWELL, D.D., Master of Trinity College,
Cambridge. An account of his writings with selections from his literary
and scientific correspondence. In Two Volumes. Demy 8vo. 25*s*.

MACMILLAN AND CO. LONDON.

www.ingramcontent.com/pod-product-compliance
Lightning Source LLC
Chambersburg PA
CBHW021347210326
41599CB00011B/781